S0-BPX-100

C J Leake

REPRODUCTIVE PHYSIOLOGY
FOR MEDICAL STUDENTS

C J Leake

REPRODUCTIVE PHYSIOLOGY FOR MEDICAL STUDENTS

PHILIP RHODES
M.A., M.B., F.R.C.S.(ENG.), F.R.C.O.G.
Professor of Obstetrics and Gynaecology in the University of London at St. Thomas's Hospital Medical School.

QP251
R477r
1969

1969

J. & A. CHURCHILL LTD. LONDON

254277

First published 1969

Standard Book Number
7000.1426.8

© J. & A. CHURCHILL LTD. 1969.
All rights reserved. No part of this publication may be reproduced, stored in a retrieval system, or transmitted, in any form or by any means, electronic, mechanical, photocopying, recording or otherwise, without the prior permission of the copyright owner.

PRINTED IN GREAT BRITAIN

CONTENTS

PREFACE

There are several excellent specialist works on the physiology of reproduction. On the other hand most general textbooks of physiology suitable for medical students deal with reproduction most cursorily. There have been many advances in reproductive physiology in the past few years, but they have not been easily available to the ordinary student in a short compass. This is the reason for this book. It is not designed to make a medical student into a reproductive physiologist, but to make him aware of the physiological processes underlying the phenomena that he will meet in obstetric and gynaecological practice.

The need for an introduction to the physiology of reproduction was recognized at St. Thomas's and such is the accord between the departments of Physiology and Gynaecology that a clinical gynaecologist was asked to give the lectures. It is these which have formed the basis for this book. They were designed therefore for those students who were reading for the 2nd M.B. examination. Over the years one result has been that students have been more easily able to study obstetrics and gynaecology, since their earlier course has in part prepared them for the later one. The book has inevitably expanded on the course of lectures and may now be suitable also as an introduction to reproductive physiology for those reading for the Membership of the Royal College of Obstetricians and Gynaecologists. The aim has been to give an overall view of the subject on which the reader can build deeper studies if he so wishes.

I have been more than fortunate in my two collaborators. Dr. J. R. Tighe, Senior Lecturer and Consultant in Surgical Pathology at St. Thomas's has provided all the photomicrographs and the texts to them, and he has in effect embedded a small textbook of histology of the genital tract within the main text. Dr. Maureen Young, Reader in Reproductive Physiology within the Department of Gynaecology at St. Thomas's has contributed the section on placental transfer, the chapter on Fetal Adaptations to Intra-uterine Life and that on Adaptation in the Newborn. In addition she has read the whole text and made many helpful suggestions.

My secretary, Miss J. E. Stringer, has done invaluable work in preparing the book for publication and to her and my helpful publishers I extend my thanks.

LONDON, 1969. P.R.

INTRODUCTION: The Life-cycle of Woman

Woman has a life-cycle. From conception when an X chromosome bearing spermatozoon meets and fertilizes the ovum which also contains an X chromosome the female has a different physiology from that of the male. This may seem far-fetched but the female fetus at term is smaller on average than the male, the development of the gonads is different in time and the endocrine physiology of the two sexes even *in utero* is different.

The fertilized egg passes along the Fallopian tube to the cavity of the uterus where it embeds and later the trophoblast is converted first to chorion and later still to the placenta, whilst the embryonic disc is transformed first to embryo and later to fetus. Differentiation of the embryo is a matter that occupies about the first three months of pregnancy; thereafter gestation is mainly devoted to growth of the intra-uterine contents.

When the baby is born he/she is known as a neonate for one month, and for the next eleven months is an infant. From one to two is the toddler stage, from three to five that of the pre-school child. Somewhere about the age of fourteen the school child passes through the phase of puberty and into adolescence and maturity. Maturity in this sense means sexual and reproductive maturity, and most women in our society are able to bear children from approximately the ages of 15 to 45 years. It is this phase of life that is of most immediate interest to the obstetrician and gynaecologist.

From the childbearing years the woman passes into the climacteric, colloquially known as the "change of life" when reproductive function is failing, though sexual activity may not decline until many years later. The time of cessation of the periods is called the menopause, and the post-menopausal era shades off into old age.

These phases in the career of a woman are of course individually very variable and there are few distinct landmarks in the progress, though the two outstanding ones are those of the first period (the menarche), and of the last one (the menopause). For the purposes of this book the cycle from birth to maturity through the menopause to old age will be entered at the birth of the female child, and from there her life history will be followed through her growth and maturation and pregnancy and onwards into old age.

All the times shown are very variable in individuals.

A note on the spelling of the word "fetus".

A letter to the *British Medical Journal* of February 18th, 1967, by Professors J. D. Boyd and W. J. Hamilton first drew my attention to the fact that although "foetus" had been used for more than a millennium, the more proper spelling would be "fetus", derived from the Latin "feo" = I bear. Most of us had considered that "fetus" was an American corruption of our more sanctified spelling. This may well have been so, but in fact the shorter spelling would seem to be etymologically correct. The *Shorter Oxford English Dictionary* of 1933 states "The better form with e is almost unknown in use." I have considered it to be time to revert to the better form.

Chapter I

GROWTH TO REPRODUCTIVE MATURITY

Differentiation and growth and the factors controlling them are among the fundamental processes of life. The changes involved in converting a single fertilized egg into a full-grown adult are vast and intriguing. Little is yet known about them. The changes from birth to adulthood are the present concern.

THE GROWTH CURVES

The two crude measures of growth are those of length (or height) and weight. These can be plotted on a graph against age, and the resulting curve has been known for some time. See Figs. 1.1 and 1.2.

A comparison of the growth curves of girls and boys shows that the boys in general are heavier at birth and maintain this initial "advantage" until the age of 11. From 11 to 14 the girls are heavier than the boys, but by the age of 15 the boys are ahead again and remain so. In height too the girls spurt ahead of the boys but lose their "advantage" a year earlier than they do with weight. Thus the growth spurt of girls extends from 11 to 14 as far as weight is concerned and puts them ahead of the boys, but in height they are only ahead of the boys at ages 11, 12 and 13. It must be realized, of course, that there is great biological variation in this and the above statements can only be approximately true. In general the adult male is heavier and taller throughout the mammalian class than the female, and in the Anthropoidea the male may be twice as heavy as the female. Very little is known of the growth curves of Primates, but the few data available suggest that the pattern of growth in them is like that in Man, though the menarche is earlier.

It has been suggested that growth from birth to maturity proceeds unevenly, there being spurts of growth followed by "filling-out". Those who hold this view suggest that springing-up occurs in the first year, from 5 to 7 years and from 11 to 15 years, whilst the intermediate periods are devoted to filling out when growth is relatively slower. J. M. Tanner denies that there is such unevenness and states that the growth curve is essentially even, at least until the age of 10, and will fit the mathematical

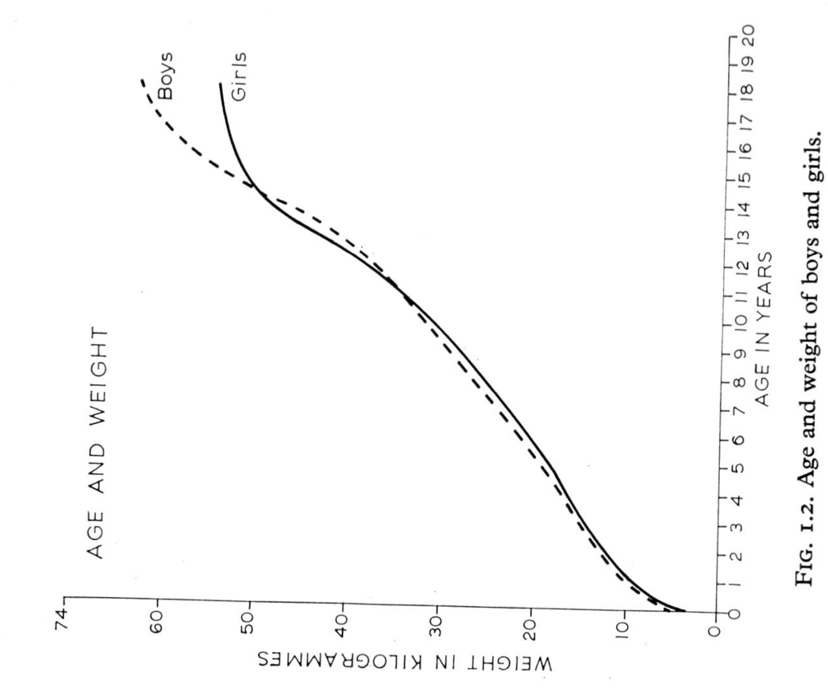

FIG. 1.2. Age and weight of boys and girls.

FIG. 1.1. Age and height of boys and girls.

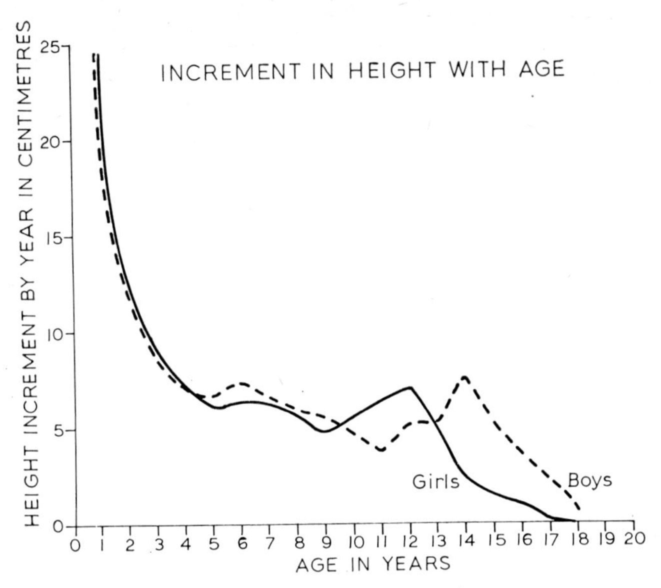

FIG. 1.3. Increment in height with age.

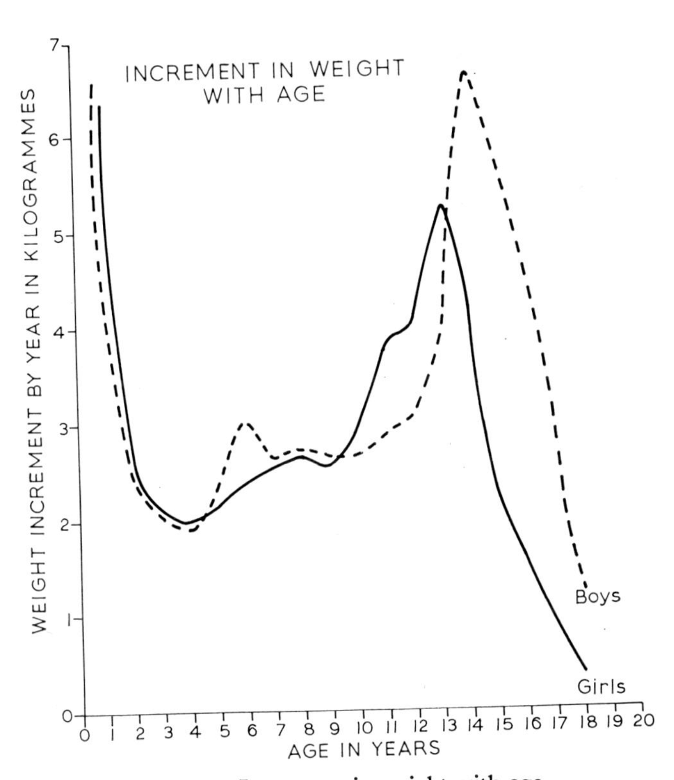

FIG. 1.4. Increment in weight with age.

Normal Body Measurements during growth. (Documenta Geigy Scientific Tables 5th edn. Data by Stuart, H. C. and Stevenson, S. S. in Mitchell-Nelson Textbook of Paediatrics, Philadelphia 1950.)

Age in years	Girls		Boys	
	Weight in Kg.	Height in cm.	Weight in Kg.	Height in cm.
0	3·36	50·2	3·4	50·6
1	9·75	74·2	10·07	75·2
2	12·29	86·6	12·56	87·5
3	14·42	95·7	14·61	96·2
4	16·42	103·2	16·51	103·4
5	18·58	109·4	18·89	110·0
6	21·09	115·9	21·91	117·5
7	23·68	122·3	24·54	124·1
8	26·35	128·0	27·26	130·0
9	28·94	132·9	29·94	135·5
10	31·89	138·6	32·61	140·3
11	35·74	144·7	35·2	144·2
12	39·74	151·9	38·28	149·6
13	44·95	157·1	42·18	155·0
14	49·17	159·6	48·81	162·7
15	51·48	161·1	54·48	167·8
16	53·07	162·2	58·83	171·6
17	54·02	162·5	61·78	173·7
18	54·39	162·5	63·05	174·5

Intermediate values are given in the original table together with percentile values. The figures above are only the mean values for purposes of discussion.

Growth increments at various ages. (Derived from table on left).

Periods of one year	Girls		Boys	
	Weight in Kg.	Height in cm.	Weight in Kg.	Height in cm.
0–1	6·36	24·0	6·67	24·6
1–2	2·54	12·4	2·49	12·3
2–3	2·13	9·1	2·05	8·7
3–4	2·0	7·5	1·90	7·2
4–5	2·16	6·2	2·38	6·6
5–6	2·41	6·5	3·02	7·5
6–7	2·59	6·4	2·63	6·6
7–8	2·67	5·7	2·72	5·9
8–9	2·59	4·9	2·68	5·5
9–10	2·95	5·7	2·67	4·8
10–11	3·85	6·1	2·59	3·9
11–12	4·0	7·2	3·08	5·4
12–13	5·21	5·2	3·90	5·4
13–14	4·22	2·5	6·63	7·7
14–15	2·31	1·5	5·67	5·1
15–16	1·59	1·1	4·35	3·8
16–17	0·95	0·3	2·95	2·1
17–18	0·37	0·0	1·27	0·8

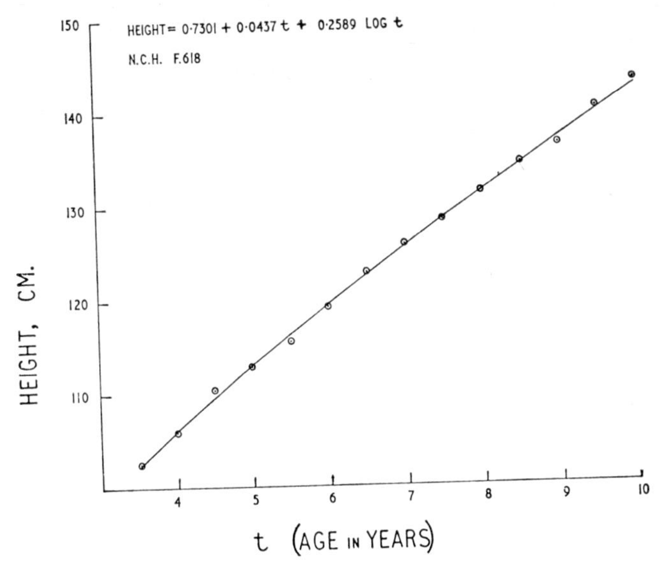

FIG. 1.5. From *Human Biology* (1964), p. 304. Section written by J. M. Tanner.

form $y = a + bt + c \log t$, where t is the age in years and a, b and c are constants. Thereafter the curve does show a definite spurt associated with puberty when the mathematical treatment is more complex.

ALLOMETRIC GROWTH

In the adult there are rough relationships between the sizes of various parts of the body, so that it is possible to say that the weight of the legs is about 37 per cent of the total body weight, or that the arm length will be say 80 per cent of the leg length. But the adult relationships do not hold throughout the period of growth. At birth, for instance, the head is relatively enormous when compared with the rest of the body whilst the legs are relatively short, but these relationships are the opposite of those in the adult. These changes in relative sizes are shown in Fig. 1.6.

Each structure and each organ has in fact its own rate of growth and so far it has been shown that there are four basic types which are skeletal, nervous, lymphoid and genital. These are shown in Fig. 1.7.

The skeletal growth curve follows that for growth as a whole. The nervous system, mainly the brain and spinal cord, grows to its maximum very quickly and reaches almost adult values by the age of 8 or 9. Lymphoid tissue, including the thymus, grows very rapidly indeed and relatively exceeds the amount in the adult by about the age of 12. After

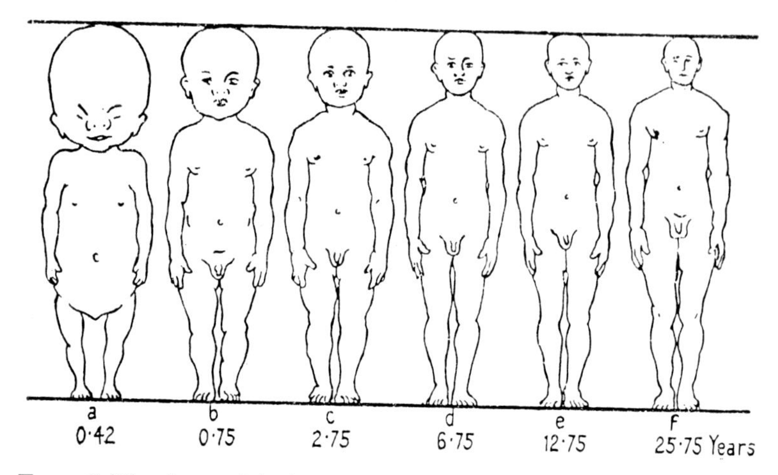

FIG. 1.6. The shape of the human being from the 5th month of foetal life to maturity. From Medawar in *Essays on Growth and Form* (1945), Clarendon Press. From *Human Growth* (1960), Pergamon Press. Ed. J. M. Tanner.

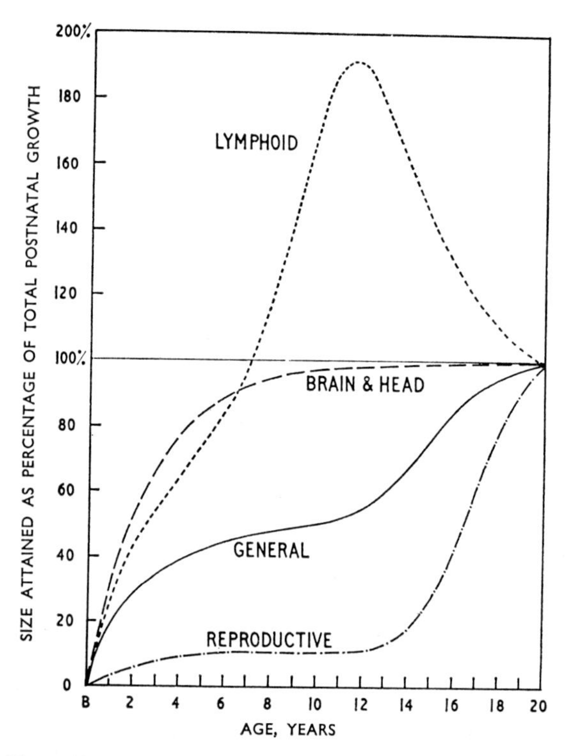

FIG. 1.7. From J. M. Tanner (1964) in *Human Biology*, Clarendon Press.

this age it involutes and comes to attain a relative adult value by the age of 20. In contrast to these forms of growth is the reproductive system, excluding the breasts. It remains relatively slow growing, almost quiescent, until the age of about 12 and then it suddenly spurts ahead. This growth is, of course, especially under the control of the oestrogen secretions from the ovary, or the androgens from the testis.

THE CONTROL OF GROWTH

Growth is controlled by genetic and environmental factors. Little is known of the genetics yet it is known that children will in general attain a stature similar to that of their parents, though if parents are very large or very small the children will approach more towards the average height and weight, that is there is a drift away from the extremes towards the norm. It would seem that the genetic inheritance puts a limit on the height which can be potentially attained, but the environment, and particularly the standard of nutrition, will determine the degree to which the potential is made actual. Comparison of heights and weights of children from different social classes and from different races shows that poorer children will in general be both shorter and lighter than those who are better endowed from birth. Also comparison of the heights and weights of schoolchildren at say ten-year intervals shows that in general with the passing years they are becoming both heavier and taller. For instance at the age of 12, in the years from 1911 to 1941 girls became about 14 lb. heavier and 2 inches taller in Britain. This has valuable implications for safe childbirth for it can be shown that reproductive results are better in taller women.

The factor in the control of growth of most interest for present purposes is oestrogen which is produced by the ovary.

GROWTH AND MATURATION AT PUBERTY

During growth till puberty the output of oestrogens from the ovary is very small. At the time of puberty this secretion is greatly increased. Its effects are widespread and are especially seen in the genitalia, but are also to be noted in breast and hair growth, fat deposition, and skeletal growth.

Height. In Turner's syndrome, also known as ovarian agenesis or dysgenesis, there is, as it were, a natural experiment in intra-uterine castration. The details of the syndrome are for the moment unimportant. Suffice it to say that the gonad entirely fails to develop, at least in the classical cases. The result is that the ovary is unable to produce oestrogen. Because of this the genitalia, that is the vulva, vagina, cervix, body of the uterus and Fallopian tubes remain infantile. Also the breasts do not develop at all, the pubic hair is scanty and the axillary hair is absent. The patient does not lay down subcutaneous fat as normal girls do at puberty.

She is also small and does not grow and so does not show the normal growth spurt that has just been considered. Moreover, the proportions of the body which are an expression of skeletal growth are boy-like and immature. This is especially shown in the width of the shoulders, the bi-acromial distance, when compared with the width of the pelvis, the bi-iliac diameter. In males the hips are usually narrower relative to the shoulder breadth than they are in females. The bony pelvis is adapted in females for the childbearing function and an expression of this is the relatively large size of the bony pelvis.

In normal puberty where there is a controlled output of oestrogens from the ovary, the girl puts in her adolescent growth spurt and this is the phase of the lanky schoolgirl. Pubic and axillary hair begin to grow, though it does not become so coarse as it does in the male. Moreover, in most women the pubic hair does not grow up towards the umbilicus, but is limited by a more or less horizontal line about 2 inches above the symphysis pubis. In about one-third of women, however, the distribution of the pubic hair is similar to that found in men without any abnormality in reproductive function.

FAT

Fat is laid down at this time in the typical feminine distribution. The fat is mainly deposited round the pelvic girdle, the buttocks, thighs and the shoulders, and to a lesser extent on the upper arms. It is of interest that the relative excess of fat seen in girls at puberty when they are compared with boys is only an accentuation of a difference that is present from about the age of 1 year. At this age girls already begin to show a slight excess of subcutaneous fat and this is maintained until puberty and beyond. It would seem that although much of the control of fat deposition must be laid at the door of oestrogen, there must be other factors involved as well. In boys the subcutaneous fat is actually decreased at puberty and they become apparently more muscular, partly for this reason. However, there is in fact a difference in muscular power as shown by strength of grip and of arm pull, and the grip in girls may be only about three-fifths of the strength of the grip of boys at the age of 17 years.

HAIR

Hair follicles are formed between the second and fifth months of fetal life. They form in almost equal numbers in all parts of the body. The differences in different parts of the body in the adult are due to differences in the hairs and their growth and not due to differences in the numbers of hairs in a given unit of area. Hair on the fetus is soft and fine and is called lanugo. Vellus is the hair which is seen in young children and it too is soft and fine though often pigmented, which lanugo is not. Later comes the terminal hair of the adult which is rather more coarse than that of the child. The change from vellus to terminal hair cannot be sharply defined, and those hairs which are neither vellus nor terminal are called intermediate. Not all follicles progress to produce terminal hair and many continue throughout life to give rise to vellus and to intermediate hair. Obvious examples of terminal hair are to be seen on the scalp and in the pubic region and axillae.

Each follicle undergoes a sequence of development throughout life. There is a phase of growth, anagen, which may last three to six years, followed by a phase of regression, catagen, which lasts about two weeks, and this is followed by a resting phase, telogen, which lasts about three to four months, before the follicle enters upon anagen once again. In many animals these phases over the whole body are together in time so that they moult when the follicles are in catagen. In Man and many other animals the phases are asynchronous in various parts of the body so that moulting does not occur. The pattern of hair in any part to some extent depends upon the numbers of follicles in the different phases. Where the hair is thick, as in the scalp, the numbers of follicles in anagen is in the region of 90 per cent. This figure falls and the follicles produce vellus when baldness occurs. In the scalp the hair grows at the rate of about 1 mm. in three days. Rates of growth are very variable in different parts of the body but may range from about 1 mm. in 10 days on the limbs to 1 mm. in 2½ days on the beard. The type of hair which a given set of follicles will grow is to a large extent determined genetically for if skin is grafted from one person to another the donor skin will continue to grow its own type of hair and will not be modified by the host.

Some hair is under the control of the sex steroidal hormones and some is not. The non-sexual hair is found on the scalp, eyebrows, eyelashes, forearms and lower legs. The ambosexual hair (implying by the prefix ambo- that it belongs to both sexes) is dependent upon oestrogenic stimulation. It determines the form of the lower triangle of pubic hair which is said to be characteristic of the female. Further ambosexual hair is found in the axillae and some limb areas. Male sexual hair is dependent upon androgenic steroids and is found in the upper pubic triangle extending towards the umbilicus, in the beard, nose and ears, and upon the body. This male type hair does not develop until the age of about 25. If women should be unfortunate enough to produce an excess of androgenic substances from their adrenal glands or ovaries they may show variations of hair pattern similar to those seen in the normal male. It is strange that even when the source of these androgenic substances is removed the hair follicles do not revert to their original state but continue to produce coarse hair.

Recession of hair in the temporal regions is characteristic of males but is also seen to some extent after the menopause in normal women. It is probably a resultant of genetic factors and endocrine influences. Men with a double dose of the gene for baldness will become bald and women with the same genotype will also become bald though at a later age than the men. Heterozygote males will become bald but heterozygote females will not, so emphasizing the importance of the endocrine factor.

There is great individual variation in hair pattern, and this is obvious between races such as the Negroes and the Mongoloids. In Europe, women from the area of the Mediterranean tend to have coarser, darker hair than those from northern regions. A survey of apparently normal young women in Wales showed that about one-third of them had a pubic hair distribution and coarse hair more characteristic of the male than of the female. Clinically these variations have constantly to be borne in mind when trying to decide if a woman is hirsute or not.

THE GENITAL SYSTEM

At puberty the outstanding changes occur in the genital system. In both sexes all the genitalia increase in size. In the female the vulva takes on an adult appearance, with the labia majora becoming rugose, instead of being smooth. The labia minora, which are a useful clinical index of the degree of oestrogenization at any age, enlarge. The vagina becomes rugose and takes on the histological appearance of maturity, whilst its cells desquamate to give a white discharge, and it becomes very acid (pH *c.* 4·5) as a result of the lactic acid within it. The acid comes from the breakdown of glycogen, present in the vaginal epithelium, by Döderlein's bacilli which invade the vagina at puberty. These are normal inhabitants of the vagina, but their source is quite unknown.

The uterus at birth is quite small and the body is rather smaller than the cervix at this time. During childhood the body of the uterus increases in size relative to the cervix so that at puberty it is about the same length as the cervix. When oestrogen has acted for a little time the body increases in size so that finally it is about twice as long as the cervix at maturity, the total length of the uterus being about 3 inches. Its muscle bulk also increases at this time. If oestrogen fails to act, as in Turner's syndrome, the body of the uterus does not grow in this way, and this is borne out by animal experiments in which castration is performed before sexual maturity. The cervical glands secrete their mucoid substance under the influence of the hormones of the ovary and this adds to the normal physiological discharge from the vagina. The lining of the uterus, the endometrium, is made to proliferate by the action of oestrogen, and made to secrete when progesterone from the ovary acts upon it after it has first been primed by oestrogen. The ripening of the endometrium followed by its regular shedding is responsible for the phenomenon of menstruation, and the first period is called the menarche.

The relationships of the main phenomena of puberty to each other are shown in Figs 1.8 and 1.9.

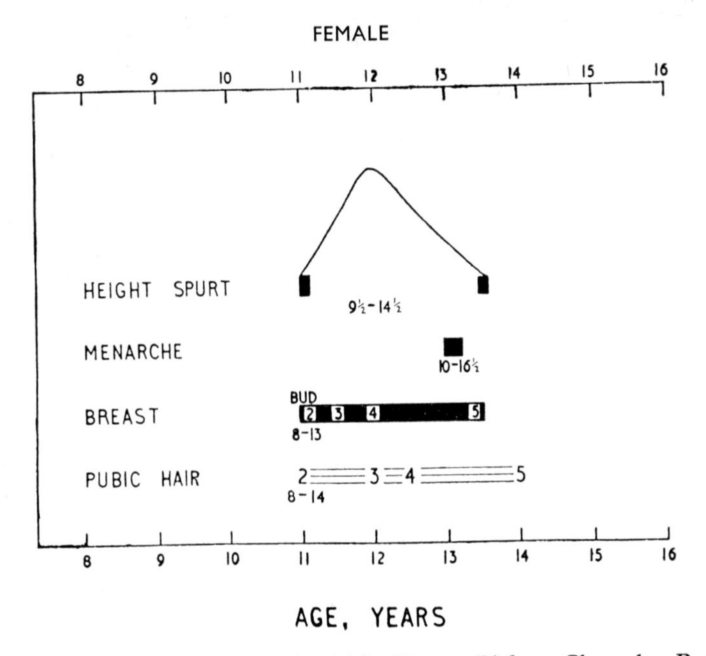

FIG. 1.8. From J. M. Tanner (1964) in *Human Biology*, Clarendon Press.

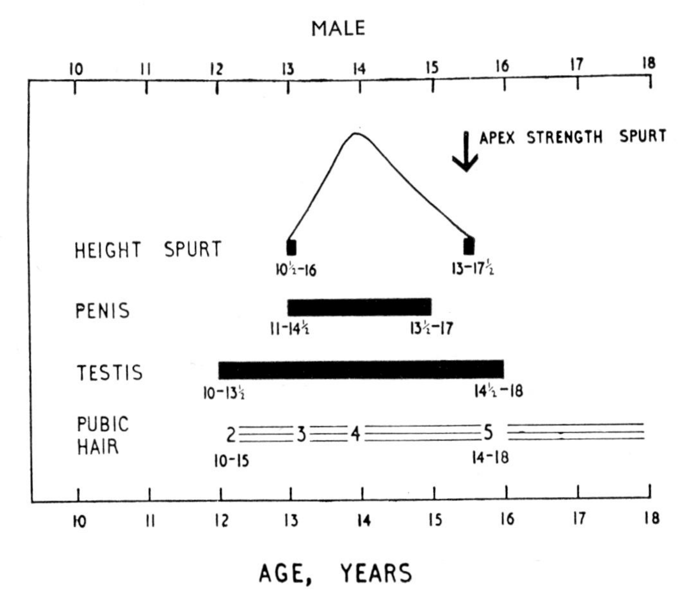

FIG. 1.9. From J. M. Tanner (1964) in *Human Biology*, Clarendon Press.

From this it will be seen that the breast buds usually develop first, to be followed, and occasionally preceded, by pubic hair growth. When these are on the way the growth spurt sets in, and the pelvic organs seem to grow in concert with the breasts, but the menarche comes towards the end of the growth spurt, by which time the uterus has reached maturity and so have the breasts. This suggests that oestrogen secretion from the ovary begins before the menarche and this is confirmed by estimations of oestrogen excretion in the urine. It should be noted that in the graph the height spurt is represented as the height gain in centimetres per year and is not a measure of the absolute height. Graphs plotted in similar fashion for boys and girls are shown, see Figs. 1.3 and 1.4, and also see Fig. 1.9, which is a graph of the main events of puberty in boys for comparison with Fig. 1.8.

SKELETON

The control of skeletal growth by oestrogens is a complex one. If there is too little hormone, growth is stunted as in Turner's syndrome, but here there are genetic defects which have a direct effect on the skeleton as in the increased carrying angle at the elbow, so that the short stature may not be a pure oestrogen effect. In other patients where there is apparently too little oestrogen circulating, but it is not altogether absent, the girls seem to grow unduly tall and long limbed. When the menarche is early, or in rare cases where there are oestrogen secreting tumours of the ovary there is first of all a rapid growth spurt, but this soon comes to an end and the final height achieved by the girl is less than it would have been if puberty had been somewhat delayed. Perhaps the simplest interpretation would be that there is an inherent tendency to growth at the epiphyses, which may be partially controlled by a host of factors among which are the oestrogens. These first appear to stimulate growth, but when present in larger amounts they induce closure of the epiphyses and bring growth in height to a stop. That is they induce calcification and ossification in the epiphyseal cartilages. The final height achieved by the individual depends largely on the length of time of the period before the growth spurt sets in. If puberty is early the girl tends to be short and if late she tends to be tall. Other factors in the control of epiphyseal growth are obviously growth hormone from the pituitary, genetic factors, the state of nutrition, especially as regards protein, calcium and phosphorus, and the state of health of the growing child. There is no doubt that illness retards growth, and when this happens calcium is laid down densely in the region of the epiphyseal cartilage and can be seen on X-ray for many years after the illness.

From the obstetric point of view the growth of the bony pelvis is of much importance since the fetus has to pass through the lower true pelvis in normal labour. Very little is known in detail of the growth of this area, though in general it would seem to follow the lines of skeletal growth as outlined, and the same factors would appear to control development here as elsewhere in the skeleton. However, oestrogen would seem to have a special effect as shown in the way the pelvis develops in females as compared with males. This is not the place to deal with them and textbooks of anatomy or obstetrics should be consulted for information about the various types of pelvis seen in women, and their dimensions. Statistically it is well known that delivery of the baby is easier the larger the bony pelvis and the better the obstetric results in terms of live and undamaged babies. It is obvious, therefore, that to obtain mothers who will fare well during reproduction it is essential to see that all girls have adequate nutrition so that their skeletal growth shall in no way be impaired, and also their general health must be a matter of concern.

PREGNANCY

During pregnancy the uterus grows enormously to accommodate its load of fetus, liquor amnii, placenta and membranes. In the patient who has not had a baby the uterus weighs about 2 oz. At the end of pregnancy it weighs about 2 lb. That is it increases in weight by sixteen times. Detailed consideration of this growth is left till later. Similar growth of the breasts during pregnancy and the storage of fat is also dealt with later.

THE CLIMACTERIC AND AFTER

After the menopause when the output of oestrogen from the ovary diminishes and almost disappears, those organs which are dependent upon the hormone tend to shrink. The ovary itself becomes smaller. The uterus atrophies so that all its muscular tissue disappears and it is just a button of tissue to be felt at the top of the vagina. At puberty the cervix is well formed and the growth of the body succeeds it under the influence of oestrogen. At the menopause the body atrophies first whilst the cervix does not shrink until later. The vaginal epithelium thins and becomes less acid whilst the Döderlein's bacilli disappear, since their normal substrate, glycogen, in the vaginal epithelium depends on an adequate amount of circulating oestrogen. The labia minora and the Fallopian tubes become smaller, but the labia majora often apparently increase in size due to the deposition of fat. The breasts become smaller in their content of gland and duct tissue, but they may become larger and pendulous due to the laying down of fat. Some weight increase after the menopause is common. It may be because of the hormone changes, but psychological factors may be operative too. There is also a tendency to increased hair growth especially in the moustache area and there may be recession of the hair from the temporal regions as in men. These effects may be due to a relative preponderance of androgens secreted by the adrenal glands.

The psychological outlook of women varies during their life cycle. The main outlines are obvious enough. The young girl is brought up to play with dolls, to do needlework and is being educated to play her role in society as a woman. She will in general become a wife and mother with a family to care for. At puberty there will come the gradual dawning of sexuality as the breasts grow and the menarche appears. In the adolescent phase heterosexual urges dominate and by the early twenties girls will usually be married. The next years are devoted to the family and the care of children whilst at the menopause and after she should be entering upon a quieter phase of life, perhaps with interests outside the home and an interest in her grandchildren. This brief outline does not, of course, touch anything like the variety of lives that are possible and satisfying for women. The hormone changes throughout life are only partly responsible for the essentially female outlook, and it is clear that the anatomical differences between men and women, such as less strength and smaller stature, must condition women's attitudes but it is certain that cultural forces are more important in determining feminine psychology than the purely chemical changes which take place at various times in the life cycle.

Chapter II
THE OVARIAN CYCLE

The ovary is the centrepiece of female physiology. It has two essential functions, to produce hormones and to produce ova. Similarly the testis must produce its androgenic hormones and produce spermatozoa. The ovary is mainly responsible for the life cycle of the woman and for other cyclic changes and therefore there is a life cycle of the ovary. It differentiates and grows in the embryo and fetus and comes to full maturity at puberty. It passes through cycles of approximately one month's duration, unless interrupted by pregnancy and lactation, till the end of the child-bearing years. After the menopause it loses its power to produce both hormones and ova and gradually becomes atrophic.

THE INDIFFERENT GONAD

The gonadal ridge is evident in the embryo when it is about three weeks old and when the length of the embryo is about 5·5 to 7·5 mm. Figure 2.1 is a composite one to show the position of the sex gland.

At this stage the gonad does not contain any gametes (ova or sperm). These are formed elsewhere at the base of the yolk sac and must migrate from there through the dorsal body wall to reach the gonadal ridge. For the gonad to be properly formed requires the interaction of the germ cells and the supporting structures of the genital ridge. Neither will develop without the other. If the primitive gametes do not reach their goal a gonad does not develop and the result is Turner's syndrome or gonadal dysgenesis. This has widespread physiological effects and the individual suffering from this disorder is sterile. But it is worthy of note here that the supporting tissues of the gonad do not differentiate and grow.

The causes of the differentiation of the ovary or testis are not fully known, but the chromosome constitution of the fertilized egg, either XX or XY is a primary determinant. Also it is probably no coincidence that throughout the whole animal kingdom of vertebrates the gonad always begins its development near the adrenal gland primordium and near the earliest kidney. Both kidney and adrenal are important in the control of electrolyte balance, and the local control of the electrolyte concentrations in the gonadal ridge may be of vital importance in the development of the gonad. Later the endocrine secretions of the gonads may be a factor in growth and they will come under the control of other endocrine glands.

At the 35th day after ovulation (i.e. probably fertilization) when the embryo is 12 mm. long the testis differentiates and can be recognized as such. For details textbooks of anatomy must be consulted. The indifferent stage of the gonad is very much more prolonged in the female and the ovary is not finally recognizable as such until the 14th week, though it is, of course, recognizable because it does not have the features of the primitive testis. At this stage the genital ridge has been invaded by blood vessels from the aorta, and the supporting tissue of the gonad, that

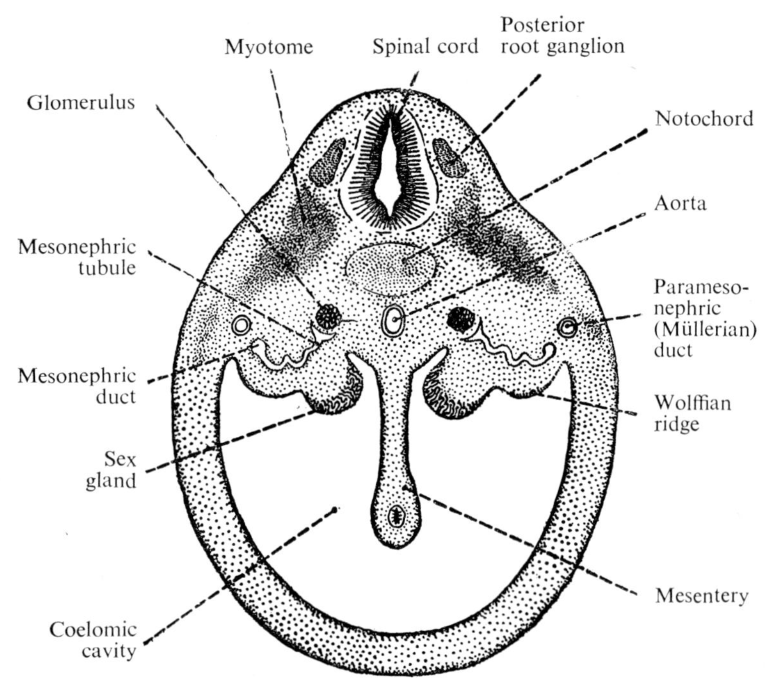

FIG. 2.1. Composite drawing from transverse sections of human embryos between 10·5 and 17·7 mm. From *Gynaecological and Obstetrical Anatomy* by C. F. V. Smout and F. Jacoby. 3rd edition, 1953. Edward Arnold, London.

is the mesenchyme, condenses round them. The whole area is stuffed full of the primitive germ cells but those near the hilum where the blood vessels enter the gonad degenerate. The stromal mesenchyme condenses in a centrifugal direction from the hilum and as it does so it comes to surround the primitive ova and the process of mesenchymal proliferation is complete at about the 30th week when it has reached the surface of the ovary. There the mesenchyme spreads out under the germinal epithelium to form the very thin tunica albuginea, which is never as well marked in the ovary as in the testis in which it is always very obvious.

By the 30th week of intra-uterine life the ovary has come to the state that it will have at birth. It is believed that there is no new formation of ova after this so that the female baby is born with a full store of ova, a stock that will be decreased by a few at every ovarian cycle. This is to be contrasted with the formation of spermatozoa which continues throughout life.

From the evidence of the intra-uterine castration of fetuses in the rabbit it seems fairly certain that the testis at least produces some secretion which influences the development of the genital tract, for a testis is required to suppress the development of Müllerian structures and turn the Wolffian system into the accessory sex organs of the male. It is not, however, certain that the secretion of the testis is an androgenic steroid. It is less certain that the ovary produces oestrogen. Removal of the ovaries from a female fetus does not greatly impair the development of the Müllerian system and the animal will be born with Fallopian tubes, uterus, vagina and vulva of essentially normal type. The finding of oestrogens in female fetuses is, of course, of no significance since these may come from the mother, and oestrogenic effects seen in the newborn such as bleeding from the endometrium and the thick vaginal mucosa are due to these hormones derived from the mother which have crossed the placenta. It is not known whether the testis or ovary are under the control of the fetal pituitary during intra-uterine life. All the trophic hormones have been found in fetal pituitaries but this is not evidence that they have any physiological effect. Nor is there evidence to suggest that the fetal gonads may be controlled by hormones derived from the mother's pituitary. The quiescence of the ovary may be due to some inherent property, but it may also be due to there being virtually no follicle stimulating hormone circulating in the mother during pregnancy. It is probable that the ovarian function only begins when the level of follicle stimulating hormone reaches a critical level. The high output of chorionic gonadotrophin, especially early in pregnancy, and the possibly high output of luteotrophic hormone may be unable to act upon the ovary unless it has first been primed with follicle stimulating hormone.

THE OVARY AT BIRTH

At birth the ovaries have descended from their primitive position near the adrenal glands and the pronephros, and have almost reached the brim of the pelvis. The downward movement is probably controlled by a gubernaculum which is a strand of tissue attached to the lowest pole of the gonad and reaching down into the perineum in the region of the definitive scrotum or labium majus. In the male the testis is pulled all the way down to the scrotum, but in the female the developing uterus prevents this so that the ovary can descend no further. In the adult the gubernaculum is represented by the round ligament of the ovary, which attaches the ovary to the fundus of the uterus, and the round ligament of the uterus which runs from the fundus through the internal abdominal ring to the labium majus. It is not certain if the gubernaculum pulls the gonad downwards or if the gonad descends because of differential growth of its tissues so that the upper pole atrophies whilst the lower pole grows into the tissues of the gubernaculum. Whatever the process it would seem probable that it is under the control of endocrine influences probably from the developing gonad itself.

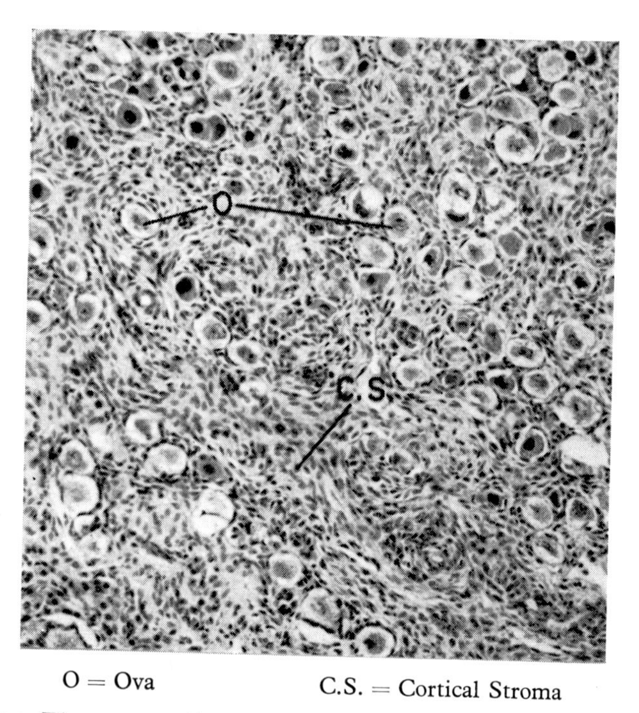

O = Ova C.S. = Cortical Stroma

FIG. 2.2. The ovary at birth. In the cortex of the ovary there are large numbers of ova scattered amongst the spindle cells. There is no suggestion of follicle formation.

The blood supply of the ovary reaches it at the hilum, and comes from the ovarian artery which is drawn out from the aorta in the region of the adrenal glands. That is, the artery traces out the descent of the gonad. Gathered round the entrance of the blood vessels is the medulla of the ovary. Surrounding this is the cortex, with the thin tunica albuginea on its surface, and covering the whole organ is the single layer of cubical cells called the germinal epithelium. This is a misnomer and derives from the days when it was thought that the ova were being continually formed throughout reproductive life from this layer. It is now known that the ova come from the base of the yolk sac and all are present at birth and are not added to throughout life.

The medulla of the ovary is small, but is of interest since it is the remnant of tissues which would have been differentiated into a testis had the fetus been a male. Within the medulla of the ovary there are still hilus cells of special type, and if these become tumorous, as very rarely they do, they produce hormones of androgenic type. Moreover, these cells have histological features similar to those of the interstitial (Leydig) cells of the testis.

The cortex is the main area of ovarian activity. At birth it consists essentially of two tissues, the oöcytes embedded in a richly cellular stroma. Both oöcytes and stroma are very obvious on histological section.

THE OVARIAN CYCLE AT MATURITY

From birth onwards the ovaries are relatively quiescent and change very little, but as puberty approaches and the pituitary gland begins to take over control of the ovaries some changes are to be seen. The changes of puberty are due to maturation of the pituitary-ovarian system. It will be realized that the changes of puberty, e.g. growth spurt, breast growth, hair growth, genital growth are spread over some years and that these changes are the result of the production of oestrogen from the ovaries. This endocrine activity is reflected in the histology of the ovary at this time.

In the year or two before the menarche (the first menstrual period) the stromal cells of the ovary condense around the oöcytes so that some of the oöcytes are surrounded by several layers of stromal cells. Some of these liquefy so that the oöcyte with a few layers of cells round it projects into a fluid filled cavity which is itself surrounded by a few layers of stromal cells. The cells round the oöcyte are called the cumulus oöphorus (little hillock of the ovary) and those round the cavity are called granulosa cells. The fluid within is the liquor folliculi.

The granulosa cells are of immense importance for they are the source of the oestrogen which controls so many of the changes of puberty. Of course to be able to have this effect they must have a rich blood supply so that each developing follicle of oöcyte and granulosa cells must have a vascular coat, and this is called the theca vasculosa or theca interna. Apart from carrying away the endocrine secretion of the ovary this blood supply is needed to bring the raw materials of hormone synthesis and to support the rapid proliferation of cells that is constantly occurring. Outside the theca vasculosa is the theca fibrosa or theca externa, which is apparently just a supporting framework for the rest of the follicle.

When a follicle has an oöcyte with its cumulus oöphorus, membrana granulosa, liquor folliculi, theca vasculosa and theca fibrosa it is said to have reached the antral stage. At puberty many follicles have come to this stage of growth and this is to be correlated with the rising output of oestrogens which can be demonstrated especially in the urine. It is of interest that the ovary has reached this stage of development before there is much evidence of pituitary activity, as shown by the excretion of gonadotrophins, and it is therefore postulated that the ovary is not fully dependent upon the pituitary during its formation of antral stage follicles. This is borne out by animal experiments, for when hypophysectomy is performed the ovaries still proceed to the formation of antra, though

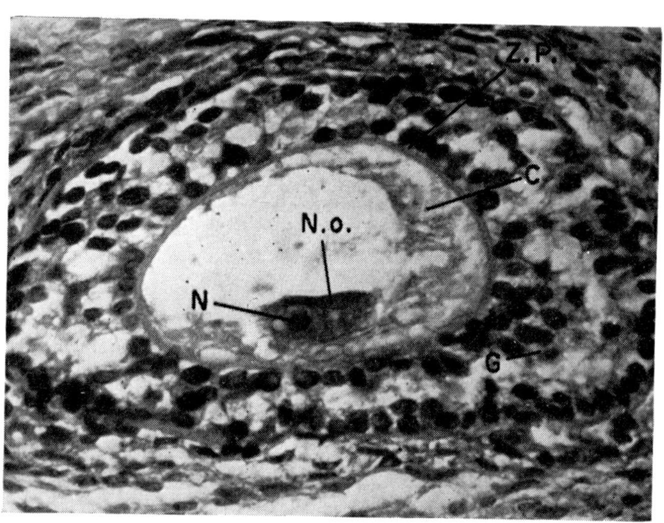

N.o. = Nucleus of ovum. N. = Nucleolus.
Z.P. = Zona Pellucida. C. = Cytoplasm of Ovum.
G. = Granulosal cells.

FIG. 2.3. Developing Graafian follicle. In the early stage of development of the Graafian follicle, the ovum can be seen with its dark staining nucleus and prominent nucleolus, surrounded by pale, vacuolated cytoplasm. The ovum is separated from the granulosal cells by a membrane, the zona pellucida. The theca interna is poorly formed at this stage of follicle development. This is the antral stage.

there is in fact very little production of oestrogen. These observations are of value in showing that the ovary has some inherent powers of differentiation but that the pituitary is essential to bring out the full potential of ovarian function.

The follicular fluid is rich in oestrogenic activity, and this suggests the importance of the granulosa cells in producing hormones. However, the presence of oestrogen in the fluid is of no importance for to have its proper actions it must be transported to other parts of the body. Possibly the oestrogen within the liquor folliculi is a store waiting to be carried away, as demand arises.

As the girl passes from puberty into adolescence the ovarian cycle passes into a regular rhythm which is disturbed only by pregnancy and lactation and some pathological states. The phases of this cycle are:

Graafian follicle ⟶ Ovulation ⟶ Corpus luteum.

Each of these must be considered in turn.

The Graafian follicle is named after Regnio de Graaf, a Dutchman who first described the bleb-like follicles on the surface of the ovary from Amsterdam in 1672. This was before the invention of the microscope and so he described only the macroscopic appearance. This is essentially the

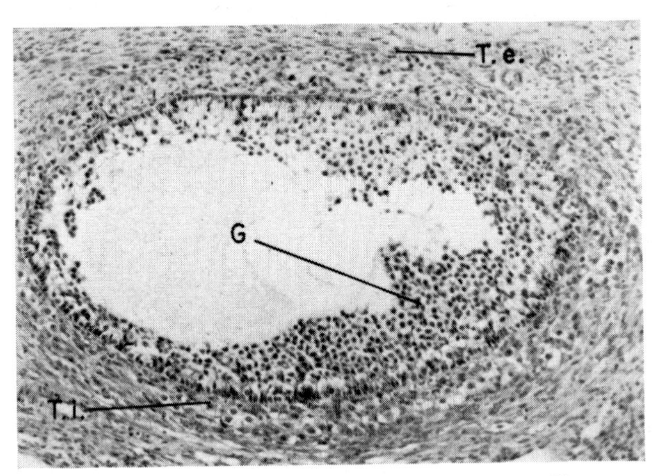

G = Granulosal cells. T.i. = Theca interna. T.e. = Theca externa.

FIG. 2.4. Developing Graafian follicle. At this stage the follicle is cystic and contains the liquor folliculi. The surrounding granulosal cells show proliferation towards one pole of the follicle and they form part of the cumulus oöphorus surrounding the ovum, which is not present in this section. The granulosal cells are surrounded by the theca interna and these in turn are surrounded by the spindle-shaped theca externa cells which merge with the surrounding ovarian stroma.

follicle already described. It consists of the oöcyte surrounded by its cumulus oöphorus projecting into a cavity filled with liquor folliculi contained by granulosa cells, and outside which are the theca interna and the theca externa. All these structures except the oöcytes are derived from the stromal cells of the ovary. The prime producers of oestrogenic hormones are probably the granulosa cells but it remains to mention some special cells which are to be found in the theca interna (theca vasculosa). They are rich in lipid material and hypertrophy at the stage of the formation of the mature Graafian follicle. There is little doubt that they have an endocrine function and they are probably of much importance in forming the secretions of the corpus luteum.

Ovulation occurs when the oöcyte surrounded by its cumulus oöphorus is expelled from the surface of the ovary into the peritoneal cavity and is then picked up by the Fallopian tube to be carried on into the cavity of the uterus. In general only one ovum is expelled from the ovaries at each ovulation. This is unexplained for the process is under hormone control, especially from the pituitary, and it is difficult to understand how one follicle only is selected from among the thousands of potential follicles

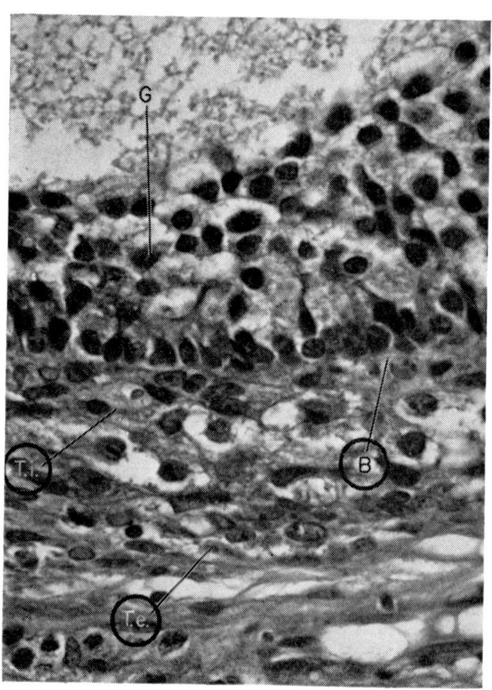

B = Basement membrane. G = Granulosal cells.
T.i. = Theca interna. T.e. = Theca externa.

FIG. 2.5. Developing Graafian follicle. The deep granulosal cells are cuboidal and they are separated from the theca interna cells by an ill-defined basement membrane. The theca externa cells merge with the ovarian stromal cells.

to be the one to provide the ovum in that particular cycle. Nevertheless, one follicle is chosen and it is marked by the proliferation of a zone of cells of the theca interna and externa near the cumulus oöphorus. It is called the theca cone. Slowly this cone makes its way to the surface of the ovary followed by the whole follicle. How this migration is accomplished is quite unknown though it is presumably due to some form of enzymatic action. Once at the surface of the ovary the follicle erodes through to the peritoneal cavity and the follicle "explodes" liberating the oöcyte with its

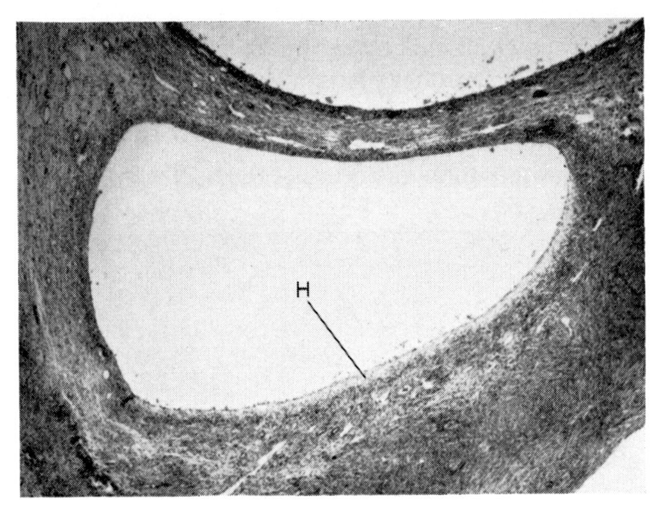

H. = Hyaline theca interna.

FIG. 2.6. Atretic follicle. In its early stages the atretic follicle is cystic, but the granulosal cells are lost and progressive fibrosis occurs in the theca interna extending from the lumen outwards.

cumulus oöphorus and the liquor folliculi. The ovum passes into the Fallopian tube. The cavity of the follicle collapses, showing that it was previously under some pressure, and the formation of the corpus luteum begins.

Despite the fact that only one follicle is selected to ovulate several follicles do begin to mature, but by some process quite unknown the

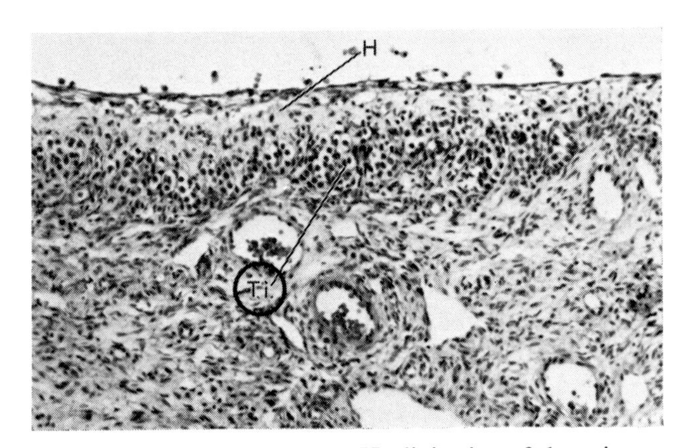

T.i. = Theca interna. H = Hyalinization of theca interna.

FIG. 2.7. Atretic follicle. The lining of the atretic follicle shows a fibrous band extending from the luminal border into the theca interna. No granulosal cells are apparent.

oöcytes within them die and such follicles without ova are called atretic. They may blow up with fluid and become small follicular cysts which are commonly seen at laparotomy. It is almost certain that they produce oestrogenic hormones and this is probably their function, for it is unlikely that the one follicle selected to ovulate could produce enough oestrogen on its own to fulfill the needs of the bodily economy.

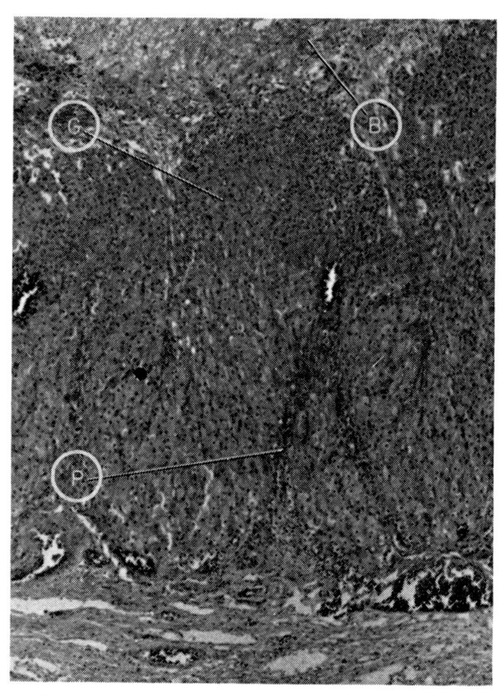

G = Granulosa lutein cells. B = Blood clot.
P = Paralutein cells.

FIG. 2.8. Corpus luteum. The mature corpus luteum is lined by large luteinised cells derived from the granulosal cells of the follicle. The theca interna also undergoes luteinisation to form the dark-staining paralutein cells. The centre of the corpus luteum consists of blood clot.

Very little is known of the intimate vasculature of the ovary, but some studies suggest that the smaller arterioles show two or three coils before reaching the follicles. If this is so it is probable that they are a device for lowering the arterial pressure before the blood reaches the developing ovarian elements. In some follicular cysts it is suggested that the arterioles are straight rather than coiled. Whether the straightness is the cause of the cyst formation or its result is not known. However, it should be obvious that the vascular supply of the ovary is constantly changing in response to the changes in the epithelial elements and whether these vascular changes are purely under the control of local influences or whether they are a direct response to endocrine changes remains to be elucidated. The total blood supply of the ovary comes through the ovarian artery, a small branch of the aorta, and possibly partly from the uterine artery, for the two arteries anastomose. The blood flow through the ovarian arteries and the blood pressure within them have not been determined. Venous blood escapes from the ovary through a pampiniform plexus. The pressure within them is very low judging from attempts to extract blood from them at laparotomy, but the flow rate through them is unknown.

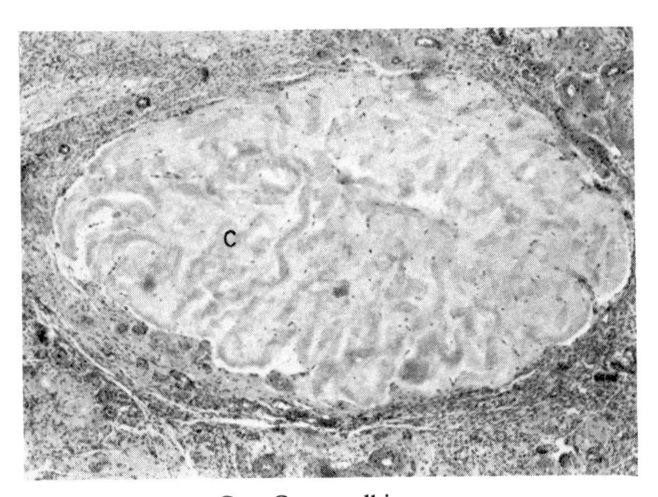

C = Corpus albicans.

FIG. 2.9. Corpus albicans. This is the scar left after organization of the corpus luteum. The convoluted outline of the corpus luteum persists in the corpus albicans.

The corpus luteum (yellow body) is formed in the place of the collapsed Graafian follicle which has now lost its ovum and surrounding satellite cells. It is yellow because of a high lipid content which is either the hormone progesterone itself or more likely its precursors. At laparotomy a corpus luteum is seen at the surface of the ovary and is red because of its great blood supply and the yellowish tinge can also be made out.

Immediately after ovulation when the tension within the follicle is diminished the granulosa cell layer and the thecae collapse into numerous folds. The corpus luteum forms within the granulosa cells and the theca interna. From having a granular structure the cells become larger, polyhedral and laden with lipid droplets, but this takes time. First the theca interna special cells take on this look and very quickly, within hours, the blood vessels grow into the granulosa along septa from the theca interna and towards the blood clot which forms as a result of slight bleeding into the cavity of the follicle at ovulation. Towards the end of the life of the corpus luteum this blood clot begins to be converted into fibrous tissue. At first the fibroblasts are delicate but later become fibrocytes. As the corpus luteum begins to regress fibrosis slowly occurs throughout the lutein layer and over the course of some weeks the corpus luteum becomes completely fibrosed being called first a corpus albicans and later a corpus fibrosum. If pregnancy occurs the corpus luteum does

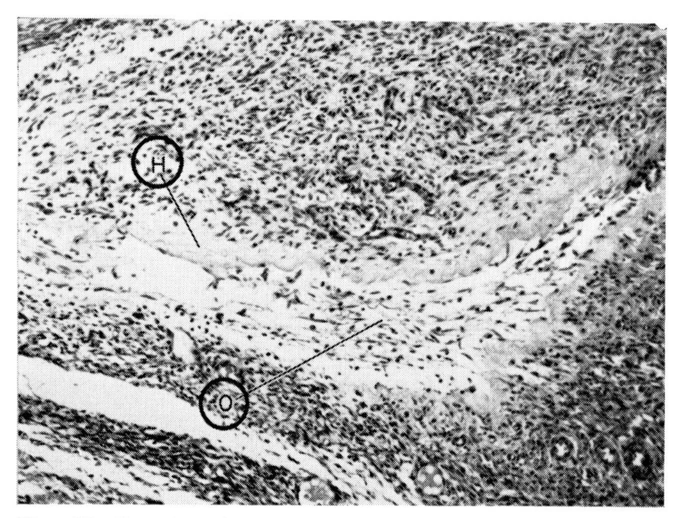

H = Hyaline theca interna. O = Organising fibrous tissue.

FIG. 2.10. Corpus fibrosum. There is complete hyalinization of the theca interna. The lumen of the follicle has been obliterated by loose connective tissue growing in from the wall.

not regress but increases in size and its cells become even more laden with lipid droplets.

The time sequence of the ovarian cycle is of importance for upon it depend all the phenomena of the menstrual cycle in the uterus and the other events associated with menstruation. Classically the ovarian cycle lasts 28 days, of which the follicular phase takes up 14 days and the luteal phase 14 days, but there is much biological variation and cycles of from 21 to 42 days or even longer are not very uncommon. In general, however, the luteal phase is relatively constant at about 14 days, so that variations in the length of the ovarian cycle are usually to be attributed to changes in length of the follicular phase.

It will be realized that a histological section of the normal ovary will show a variety of structures. There can be Graafian follicles in all phases of development and there will also be atretic follicles without ova in them. Some follicles will appear to be empty because the plane of section may pass through the antral cavity but not through the cumulus oöphorus. There will be corpora albicantia and corpora fibrosa, the remnants of formerly active corpora lutea. In the second half of the cycle a corpus luteum in various phases of development will be seen.

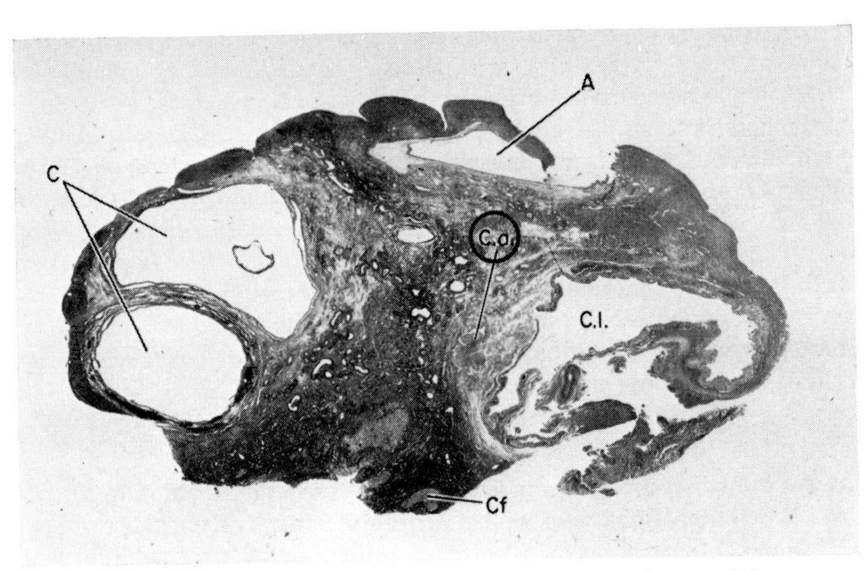

C = Cystic follicles. A = Atretic follicle. C.a. = Corpus albicans.
C.l. = Corpus luteum. Cf. = Corpus fibrosum.

FIG. 2.11. Mature ovary. This is a section of the whole ovary. It can be divided into a cellular cortex and a fibrovascular core radiating out from the hilum. A number of cysts represent developing and atretic follicles and a ruptured corpus luteum. The scars are corpora albicantes and fibrosa.

The basic cycle in the ovary is:

Graafian follicle ⟶ Ovulation ⟶ Corpus luteum

but this is modified as follows:

Graafian follicle ↗ Atretic follicle
Graafian follicle ↘ Ovulation ⟶ Corpus luteum ⟶ Corpus albicans ↓ Corpus fibrosum

Apart from these there are blood vessels, autonomic nerves, lymphatics, stromal cells and hilus cells. Also there are cell clumps known as Walthard rests whose significance is not known. On the surface of the ovary will be the germinal epithelium, a single layer of cubical cells continuous with the peritoneum at the hilum.

Because of the recurrent and great changes in the ovary which go on throughout reproductive life it is commonly the seat of various neoplasms which loom large in gynaecological practice. Moreover, the tumours do not often appear exactly like the parent cells from which they arise. Indeed, there is a bewildering variety of tumours of many kinds, in which the epithelia may look as if they come from the gut, the cervix or the Fallopian tubes, or they may be so de-differentiated that they can only be designated as adenocarcinomata. In the teratomata, which are commoner here than elsewhere, there may be pieces of skin, hair, sebaceous glands, teeth, nervous tissue, gut, muscle, cartilage, thyroid and so on. This demonstrates the wide potentiality for differentiation of some at least of the cells of the ovary. The changes that make a normally functioning ovary into one producing neoplastic tissues are not known but they might be due to abnormalities in the mitotic and meiotic processes which are a constant feature of the ovarian cycle, and these might be caused by some aberration in the complex manufacture of the sex hormones in the cells of the granulosa and the corpus luteum.

THE OVARY IN THE CLIMACTERIC

As the woman approaches the age of about 50 ovarian function begins to fail. At first ovulation ceases whilst the production of oestrogen continues. Of course if there is no ovulation the corpus luteum cannot be formed and so there can be no production of progesterone. It seems that the ovarian failure is not due primarily to failure in the pituitary hormones but is something inherent within the ovary itself. Ovulation does not cease all at once usually but there are at first only occasional failures. Later these become more frequent until finally there is no ovulation at all. This is of some clinical importance because as long as there is even occasional ovulation there is the chance that the woman may conceive and bear a child, which is not usually desired at relatively advanced ages.

Histologically the ovary reflects these changes and there will usually be no corpora lutea to be seen, though there will still be corpora albicantia and corpora fibrosa. Later there will be only corpora fibrosa. The Graafian follicles do not ovulate but mainly become atretic as their contained oöcytes die. Follicles are therefore seen in all phases of dilatation of their cavities, and some may be quite large forming follicular cysts. Slowly over the months and years even the follicles fail to form and the ovary becomes fibrotic and scarred with small pits visible to the naked eye on the surface, and the whole ovary becomes small and atrophic.

Since ovarian function does not cease all at once, the secretion of oestrogen is fitful and may come in spurts. The fluctuations in output of oestrogens, and especially when the output is low, are probably responsible for some of the symptoms that a woman may experience at the climacteric. The best known of these symptoms is hot flushes in which the neck and shoulders suddenly go red and the woman feels hot and this is followed by a phase in which she may perspire and feel cold. This demonstrates the fact that the oestrogens, besides their many other effects, also have a rôle to play in the control of vascular tone. If the output of oestrogen is high and then suddenly drops it may precipitate changes in the endometrium of the uterus and cause bleeding from the genital tract. This may take place after the periods have stopped at the

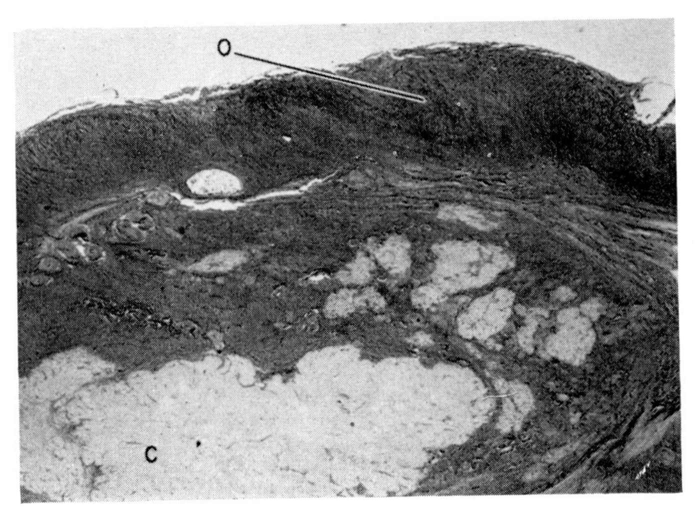

C = Corpus albicans. O = Ovarian cortex.

FIG. 2.12. Ovary after the menopause. No follicles or corpora lutea are present but corpora albicantes are apparent. The cortex is fibrotic.

menopause. It is an alarming symptom and is called post-menopausal bleeding. Because such bleeding may be due to cancer of the uterus it is invested with a very sinister significance, and it is a comfort to find that such bleeding is due only to the oestrogen spurts from the ovary.

The life cycle of a woman has been traced and the life cycle of the ovaries within her have been outlined, together with the recurrent cycles occurring monthly within the general life of the ovary. The ovary has its general effects on the woman by virtue of its endocrine secretions, the sex steroids oestrogen and progesterone. It has been seen that the effects of oestrogen are widespread and so are those of progesterone, though these are more apparent in pregnancy perhaps than in the non-pregnant woman. It is no exaggeration to say that oestrogens are responsible for the preparation of the woman for her childbearing function and that progesterone makes her capable of reproduction. Oestrogen tills the soil of the individual and progesterone is for the species as a whole. The next chapter is devoted to some aspects of these remarkable steroids.

Chapter III
THE SEX STEROIDS

The biological effects of castration in Man and other animals have been known for a very long time. In the human male especially it was noted that if the testes are removed before puberty the voice does not deepen, the man grows tall and later usually fat, the libido is impaired, the penis does not grow and the hair does not become coarse. In female animals the phenomena of recurrent sexual cycles and breeding seasons disappear when the ovaries are removed. The reason for these changes remained unknown for a long time, and it was not until 1889 that Brown-Séquard suggested that the ovary might be an organ producing an internal secretion. By 1911 Steinach had shown that castration changes in females could be prevented by transplantation of the ovaries. In 1923 Allen and Doisy showed that there were changes in the vaginal epithelium during the phases of oestrus, the sexual cycle of lower vertebrates, and using this method for bio-assay they were able to show that extracts of ovaries could produce the same changes in castrated animals. Four years later Aschheim and Zondek showed that oestrogenic substances occurred in the urine of pregnant women, and in 1929 Doisy, Veler and Thayer isolated oestrone from pregnancy urine. In the same year Corner and Allen showed that corpora lutea contained a hormone. The story of the sex hormones is largely one of the twentieth century and especially of the last forty years.

Now it is a commonplace that the ovaries produce two main varieties of hormone, the oestrogens and the progestagens, the latter being manufactured when there is a corpus luteum present. Apart from the ovaries, the placenta and the adrenal glands also produce substances with similar biological properties, and the testes at least in some animals also produce them. In almost all vertebrates investigated, oestrogens are found and progestagens are also apparently universal though this is not so certain. Even among the Mammalia where oestrogens are universal it has not been possible to demonstrate the presence of progesterone in the African elephant (R. V. Short). Oestrogens are found in some plants, e.g. willow catkins and palm kernels, but their significance here is not known. However, it is of interest that the sex steroids of the ovary (and of the testis) came on the evolutionary scene at quite an early stage and have remained up to the latest Primate. This is not to say, of course, that their functions through all that time have remained the same, but they are obviously fundamental to certain body processes, especially associated with reproduction, but they are active in other areas also.

CHEMISTRY

This is not the place to study the chemistry of the sex hormones of the ovary in detail, but some knowledge of this subject is essential to the understanding of reproduction. The oestrogens, progestagens and the androgens (male type hormones) are all to be derived from the complex steroid hydrocarbon whose name is perhydrocyclopentenophenanthrene. In this there are 19 carbon atoms and they are numbered as shown in the figure. The ring systems are lettered *A*, *B*, *C*, *D*. The phenanthrene

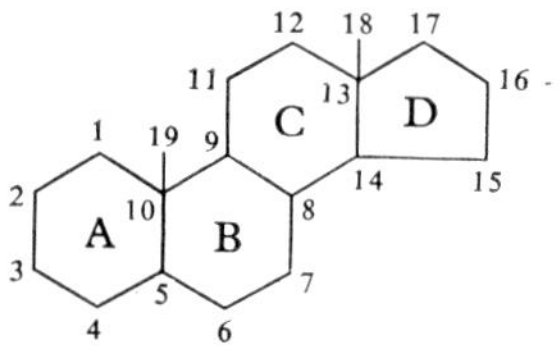

Perhydrocyclopentenophenanthrene
steroid nucleus

Fig. 3.1.

nucleus is of rings *A*, *B* and *C*, though in phenanthrene itself the rings are not saturated with hydrogen. The pentene system is ring *D* which has 5 C atoms in its structure. The "cyclo-" in the name describes the ring system and the perhydro- means that all carbon atoms are fully saturated in all four valency bonds with hydrogen. When carbon atoms are attached in a chain at position 17 (i.e. C17) the consecutive numbering continues, so that when there are 2 carbon atoms here it is a C_{21} compound, as in progesterone, and when there are 8 carbon atoms, as in cholesterol, it is a C_{27} compound. (Fig. 3.2.)

Progesterone is a C_{21} compound, androgens are C_{19} compounds and oestrogens are C_{18} compounds. In the last case there is no carbon atom at the position C19. The parent substances of these three groups of compounds are called pregnane, androstane and oestrane. The suffix -ane implies that all the carbon atoms are fully saturated. The parent compounds are hypothetical but are introduced to give order to the nomenclature of the wide variety of steroids.

From the way in which the formulae are written it would seem that the molecule is a flat one in the plane of the paper. This is not so, and some atoms attached to the main body of the molecule project above the plane of the paper and some project away from the reader. Those valency bonds which project upwards from the paper are called β and those which

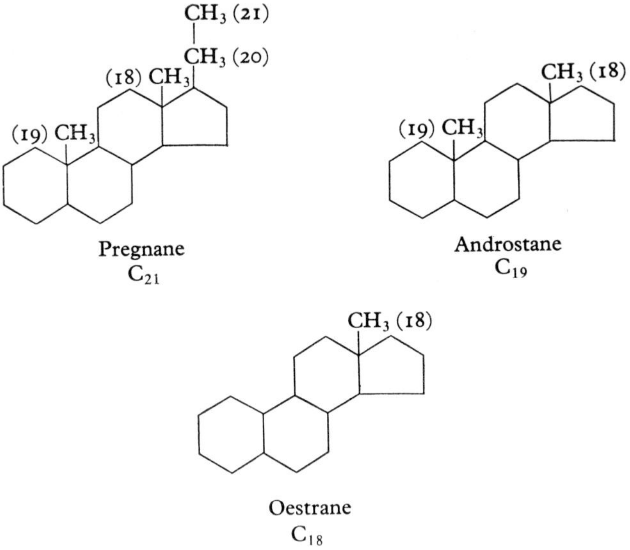

CH$_3$ (21)
CH$_3$ (20)
(18) CH$_3$
(19) CH$_3$
Pregnane
C$_{21}$
CH$_3$ (18)
(19) CH$_3$
Androstane
C$_{19}$
CH$_3$ (18)
Oestrane
C$_{18}$

Fig. 3.2.

project away from the reader are called α (α = away). This introduces the notion of spatial relationships, the subject of stereochemistry. It has great importance in all biological chemistry for spatial configurations determine many biological actions, and molecules have to fit like keys into locks in the metabolic processes of cells. Not only do certain valency

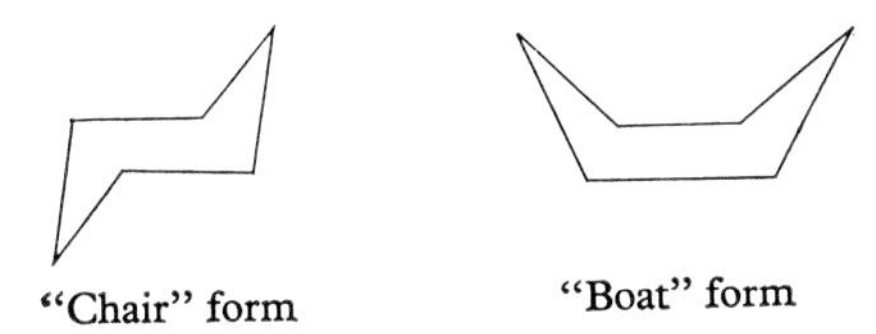

"Chair" form "Boat" form

Fig. 3.3.

^{18}CH$_3$
R
13
17
H
11
12
15
16
^{19}CH$_3$
8
14
9
H
10
2
1
6
7
H
4
5
H
3

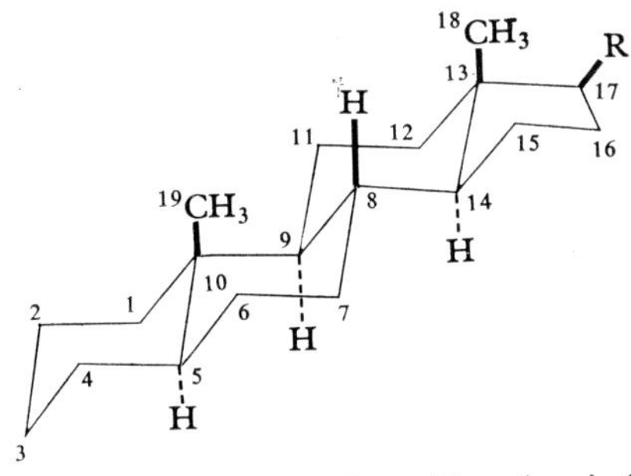

Diagrammatic representation of steroid nucleus in "chair" form.
(R = side chain)

Fig. 3.4.

bonds project away from the main body of the molecule, but the main body also has special configurations, and each ring may take up the shape of a boat or a chair as Fig. 3.3 attempts to show. The molecule, therefore, may look more like that in Fig. 3.4 than the conventional formula shows. The β bond is always shown as a continuous line, and the α bond as a dotted one. When the direction of the bond is variable it is shown as a wavy line.

OESTROGENS

Though theoretically derived from the saturated oestrane nucleus all the naturally occurring oestrogens are unsaturated in ring *A*. The substance that is believed to be the main secretion of the Graafian follicle

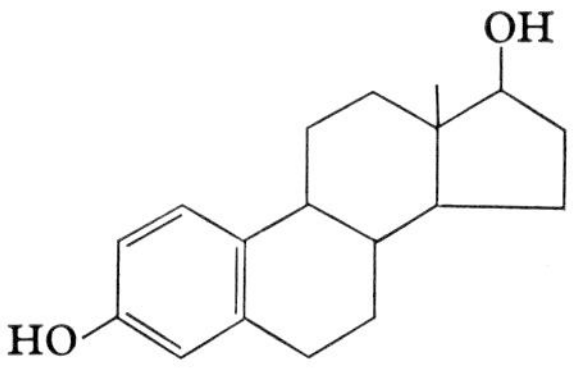

Oestradiol
(Oestradiol-17β)

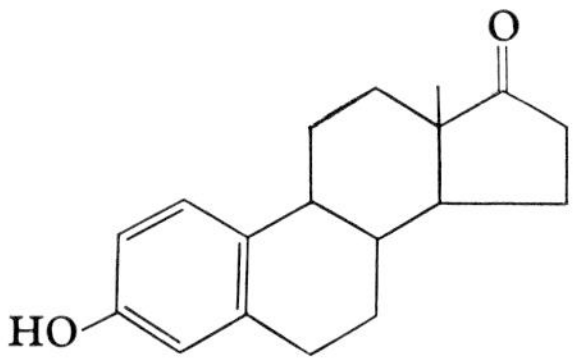

Oestrone

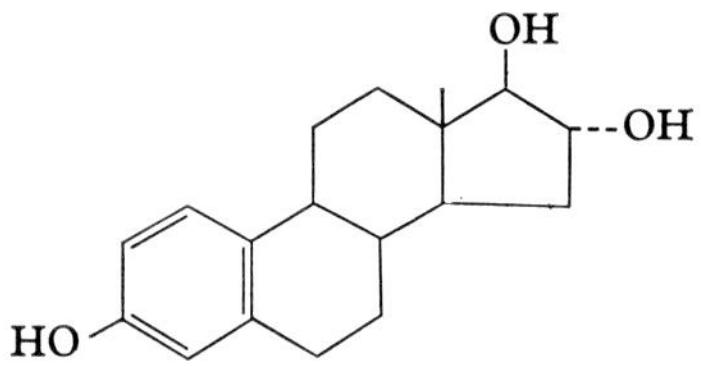

Fig. 3.5. Oestriol

is oestradiol 17-β. This seems to be interconvertible with oestrone. The name oestradiol implies that there is the basic oestrogen nucleus and the -diol means that there are two -ol groups attached to it. The suffix -ol always means that there is an alcoholic -OH group in the molecule. The 17-β means that the -OH group is attached to C17, but it is not stated where the second -OH group is. This is in fact attached at the C3 position, which seems to be very active chemically and to which there are often groups of one sort and another attached. The figure shows the formula of oestradiol 17-β, and also of oestrone. The suffix -one means that a keto-group, i.e. =O is attached, thus making it one of the keto-steroids, which are nowadays better called the oxosteroids. It should be noted that there are double bonds in ring *A* because it is unsaturated and that C18 is not specially marked but it is taken to have a methyl CH_3-

group there. Also oestradiol is the name usually used for oestradiol 17β, the suffix 17β usually being dropped. During metabolism of these primary oestrogenic hormones of the ovary they are changed to all manner of breakdown products, the chief of which is oestriol. This means that there are three hydroxy-, i.e. -OH groups, in the molecule. It will be seen that these are at positions 3, 16 and 17 and that the -OH at 16 is in the alpha position (dotted line) and the other two are in the beta position (solid lines).

These three, oestradiol, oestrone and oestriol, are the main hormones of clinical importance at the present time. This is mainly because they can be relatively easily chemically recognized especially in the urine, but also in the blood and in the liquor of the Graafian follicle. However, it must be remembered that there are at least 10 to 15 more oestrogens to be found somewhere in the human female, though at the moment they appear to be of less physiological significance than the three main ones, and certainly their biological actions on the tissues are very much less. For the time being their interest lies in their relationships to the metabolic pathways which the primary oestrogens follow. Nevertheless, it will be realized that discoveries, which come quickly in this field, may alter the view taken of their importance.

THE OUTPUT OF OESTROGENS IN THE URINE DURING THE OVARIAN CYCLE

The methods of chemical estimation of the naturally occurring oestrogens is not difficult when they are present in the pure state, but when they have to be isolated from biological fluids there is great difficulty. This is why estimations of oestrogens in blood are not often done. In the urine the technical problems are not so great so that most of the knowledge of oestrogens in the menstrual cycle is derived from estimations on the urine. Again, although it is possible to make separate determinations of oestradiol, oestrone and oestriol it is usual to estimate all three together. The chemical methods depend on the -OH groups attached at C3, and oestrone can be characterized by the =O found at C17, but the details of chemical estimation are not important here.

When all three oestrogens are determined throughout the menstrual cycle the average excretion in the urine is as shown in the graph. (Fig. 3.6a.)

This graph shows a basal excretion of oestrogens in the region of 15 μg in 24 hours at about the time of menstruation, that is days 1 to 5. As the Graafian follicles rapidly grow in the ovary the excretion of oestrogens rises to a peak at about the time of ovulation on day 12 or 13. The peak is about 60 μg in 24 hours. There is then quite a rapid fall but as the corpus luteum forms and matures there is a second peak in excretion at about day 23 to a level of 30 to 40 μg in 24 hours. Thereafter the urinary excretion falls to its original basal level of about 15 μg in 24 hours. This

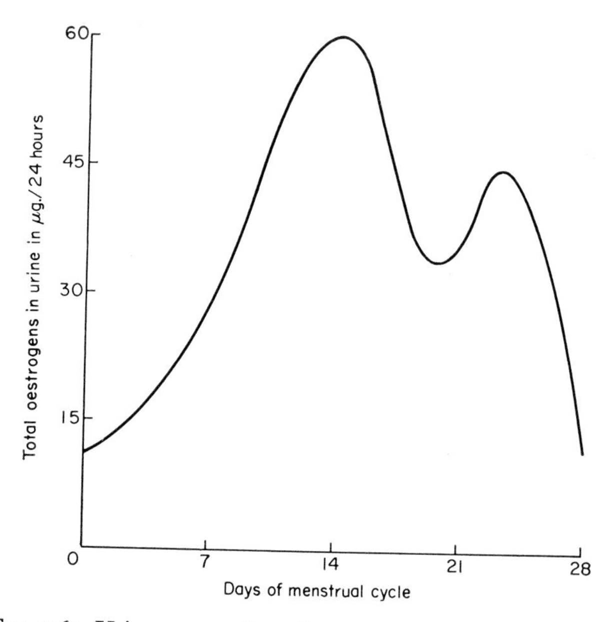

FIG. 3.6a. Urinary excretion of oestrogens in menstrual cycle.

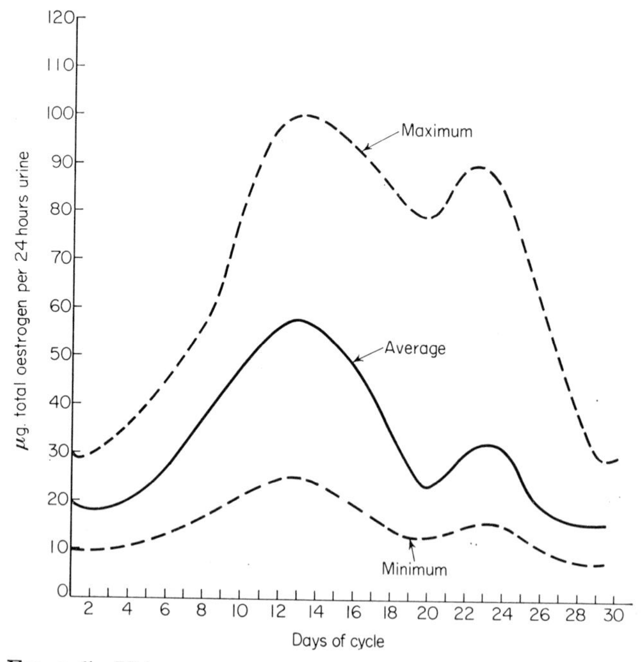

FIG. 3.6b. Urinary oestrogen excretion in the menstrual cycle.

excretion curve is evidence that the corpus luteum produces oestrogens as well as progesterone, though other elements in the ovary, and even possibly the adrenal cortex, may be adding a quota of oestrogens to the amounts found in the urine.

As with all biological measurements there is variation in the amounts of oestrogens excreted in different individuals, and basal levels may range from 2 to 28 μg, the "ovulatory peak" from 23 to 110 μg, and the "luteal peak" from 10 to 90 μg in 24 hours. It will be realized that all these values are approximate. The major contribution to the amounts of oestrogen found in the urine is made by oestriol, and the least by oestradiol, whilst oestrone is intermediate. Oestradiol and oestrone can be found in follicular fluid from the ovaries but almost no oestriol. This is evidence, therefore, that oestriol is formed by metabolic processes elsewhere in the body than in the ovary.

THE METABOLISM OF OESTROGENS

The main sites of the production of oestrogens are the gonads and the adrenal cortex. In this context it is worth recalling that the ovary, testis and adrenal all arise from the same embryological area near to the primitive kidney. It will be seen later that the placenta also is able to produce oestrogens. In all four sites, ovary, testis, adrenal and placenta, it is probable that the synthesis of oestrogens proceeds along similar lines. In outline this can be shown as:

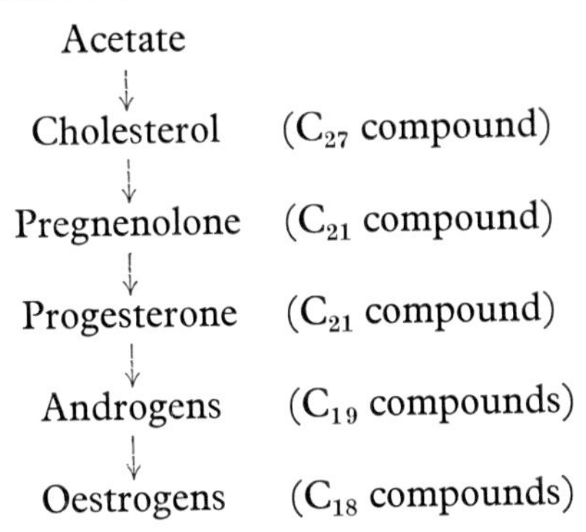

This scheme is a basic one for the understanding of steroid metabolism. Acetate is one of the products of carbohydrate, fat and protein metabolism. It is known as "active acetate" when combined with coenzyme *A* and the compound is called acetyl- coenzyme *A*. When this is available almost all tissues seem to be capable of producing cholesterol (Fig. 3.7). This is a steroid with a side chain attached at C17, and two methyl (CH_3) groups at C18 and C19. In addition the double bond should be noted between C5 and C6. By a series of enzymatic reactions the gonads, adrenal cortex and placenta can reduce the size of the side

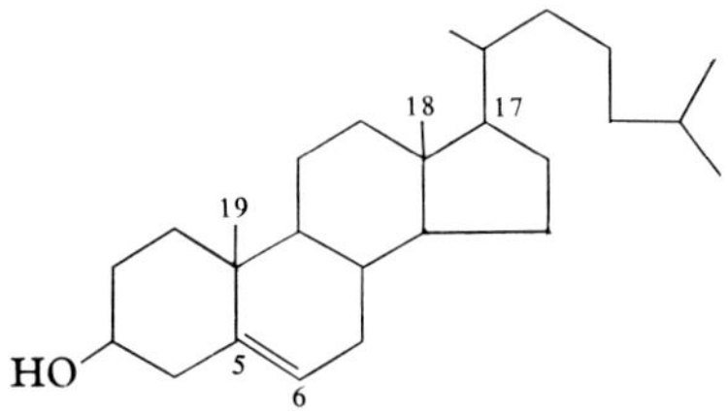

Cholesterol

Fig. 3.7.

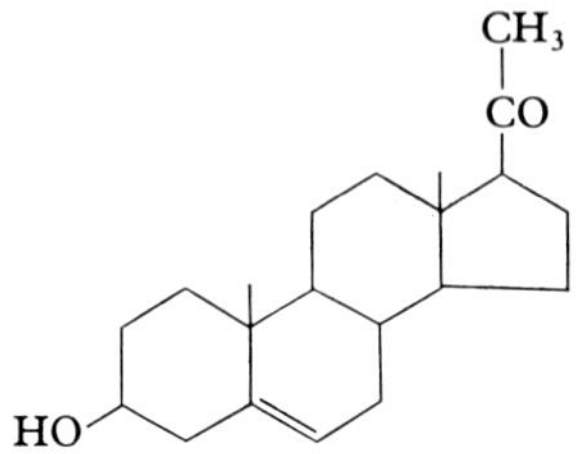

Pregnenolone

Fig. 3.8.

chain, producing pregnenolone (Fig. 3.8). This name implies that the basis of the molecule is pregnane, the saturated C21 compound, the -ene means that there is a double bond which is between C5 and C6, and the suffixes -ol and -one show that there must also be an -OH group and a =O group somewhere in the molecule. A more specific name is Δ-5-pregnene-3β-ol-20-one, showing that the double bond (delta) follows C5, that the -OH is at C3 in the beta position and that the =O group is at C20. It should be noted that this change from cholesterol to pregnenolone has only reduced the side chain attached to C17 at the apex of the five-membered Ring *D*.

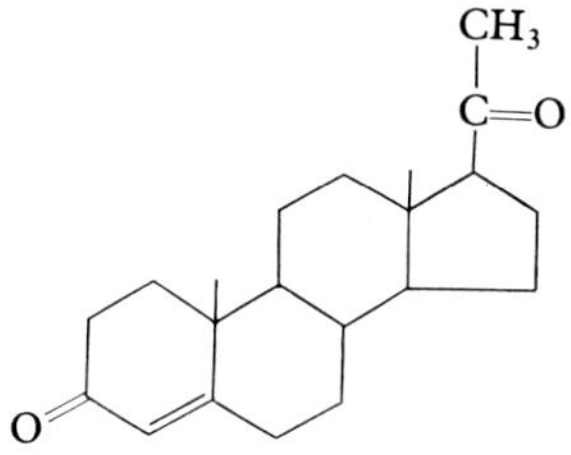

Progesterone

Fig. 3.9.

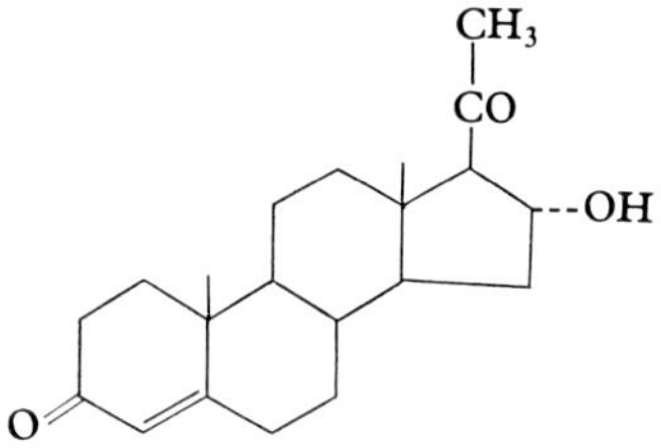

17α-Hydroxyprogesterone

Fig. 3.10.

The next major step converts the pregnenolone to progesterone. From Fig. 3.9 it will be seen that this involves reduction of the -OH group to =O at the C3 position, the shift of the double bond from C5:6 to C4:5, but no other changes. Progesterone is thus a key compound in the understanding of all the sex steroids.

Progesterone is converted to 17α-hydroxy progesterone, Fig. 3.10, which means the addition of an -OH group at the C17 position. This is, as it were, an enabling step to allow the side chain at C17 to be reduced still further in size.

When the side chain at C17 is removed it is replaced by a =O group, and this produces the very interesting compound in this context, androstenedione. It is interesting because it is an androgen, and in the testis it is the precursor of testosterone, but in the ovary it is the precursor of oestrone (Fig. 3.11).

Changing androstenedione (with its two keto- groups) to oestrone, Fig. 3.12, involves the conversion of the =O group at C3 to -OH and the unsaturation of Ring *A*, that is, two extra double bonds have to be introduced into this ring. Moreover, the double bond running from C4:5 is eliminated making room for double bonds between C1:2, C3:4 and

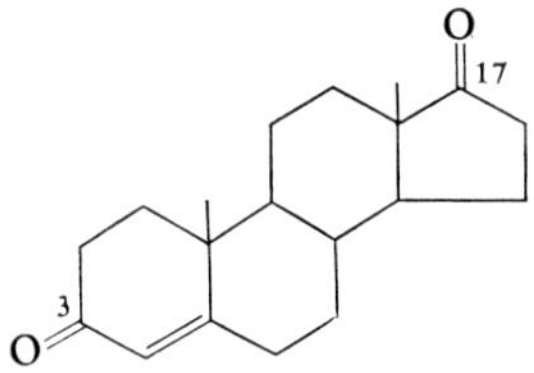

Androstenedione
Δ^4-Androstene-3,17-dione

Fig. 3.11.

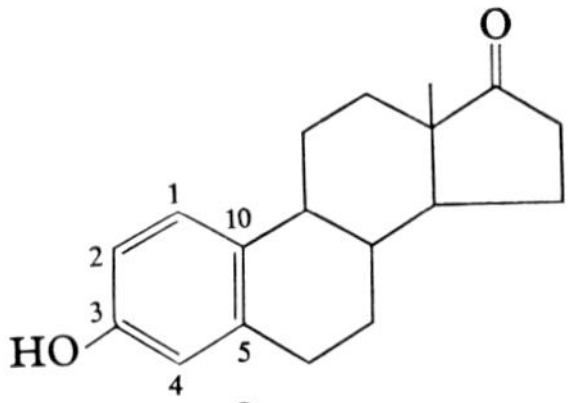

Oestrone

Fig. 3.12.

C5:10. Also the CH_3 group which was attached at C10 has been removed, converting the molecule from the C_{19} group of compounds (androgens) to the C_{18} group of compounds (oestrogens, which also have Ring *A* unsaturated with H atoms).

Oestradiol, Fig. 3.13, is manufactured in the ovary (also in the adrenal, placenta and to some extent in the testis) by changing the =O group at C17 to an -OH group. It should also be realized that oestradiol can be

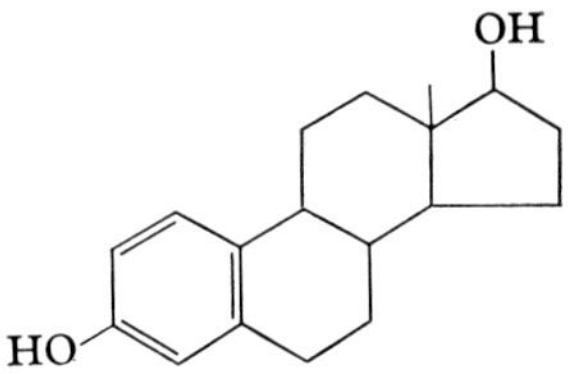

Oestradiol (17β)

Fig. 3.13.

manufactured through the formation of testosterone and this may be converted to 19-hydroxy-testosterone before reaching the goal of oestradiol.

The foregoing scheme is essentially right in outline, but the details still remain to be filled in. The general lesson is that there are no clear-cut distinctions in the metabolism of the steroids and although it may be easier to think of oestrogens, androgens and gestagens in broad biological and physiological terms, they are not in fact absolutely distinct from one another and their metabolic pathways cross at many levels. As far as the production of the sex steroids is concerned the differences between the ovary, the testis and the adrenal are quantitative only and there is no qualitative difference between them. Although the ovary in the main produces oestrogens and the testis androgens and the adrenals corticosteroids, under certain disease conditions they may produce quantities of hormones which are not normal for them, with consequent effects upon the patient.

The probable site of production of the oestrogens, oestradiol and oestrone, is the granulosa cells of the Graafian follicles (Chapter II) and follicular cyst fluid contains them. Round the granulosa cells is a rich vascular coat and the blood transports the formed oestrogens so that they may carry out their physiological actions in various sites such as the pituitary, breasts and genitalia. Their actions in the different sites may well be very different, but in general they seem to have an effect on the metabolism of protein in the cytoplasm of cells, and this they probably do by activating various enzyme systems. Certainly oestrogens are metabolized in the liver and especially this organ converts oestrone and oestradiol to oestriol, in which a third -OH group is introduced into the molecule at the C16 alpha position. It also forms many other compounds based on the oestrogenic molecule so that the blood contains a mixture of all these metabolites. In addition the liver conjugates the three main oestrogens with sulphate and with glucuronic acid. The urine therefore contains all three main oestrogens and other metabolites in the free state, but mainly in the conjugated forms.

From studies with oestrogens which have been "tagged" with radioactive hydrogen or carbon it has been estimated that only about 10 per cent of the oestrogen produced in the body is excreted in the urine. Presumably the other 90 per cent is "lost" in other metabolic pathways that are not yet elucidated. Of the oestrogens in the blood about half to three-quarters is bound to plasma proteins, mainly γ-globulin. The quantities in the blood fluctuate in rhythm with the ovarian cycle in the same way as has been outlined for urinary excretion.

EXCRETION OF OESTROGENS BEFORE AND AFTER REPRODUCTIVE LIFE

So far the metabolism and excretion of oestrogens in women in the childbearing period of life have been considered. Soon after birth the excretion of oestrogens is quite high because the oestrogen of the mother has crossed the placental barrier. Thereafter the excretion settles to a low level of about 2 μg per 24 hours till about the age of 10 when it begins to approach the adult values. After the menopause the excretion falls from its previous levels to about 3 μg in 24 hours. These figures correlate well with the histological findings in the ovary, for this organ is quiescent till the onset of puberty when Graafian follicles begin to form, though there is no ovulation, and the full cycle only begins shortly after the menarche (the first period) some years after the onset of puberty. In old age, of course, there are no follicular elements in the ovary and the oestrogen secretion declines. The question arises as to the source of the "basal" excretion of oestrogens in childhood and old age. It seems likely that it is the adrenal cortex.

PROGESTAGENS, PROGESTERONE

Progestagens are those substances which prepare the female for pregnancy. Their main effects are therefore seen on the genital system, but they also have effects elsewhere in the body just as the oestrogens do. The main progestagen in Man is progesterone, Fig. 3.14, based on the pregnane, fully saturated, nucleus. It is a C_{21} compound. (It will be re-

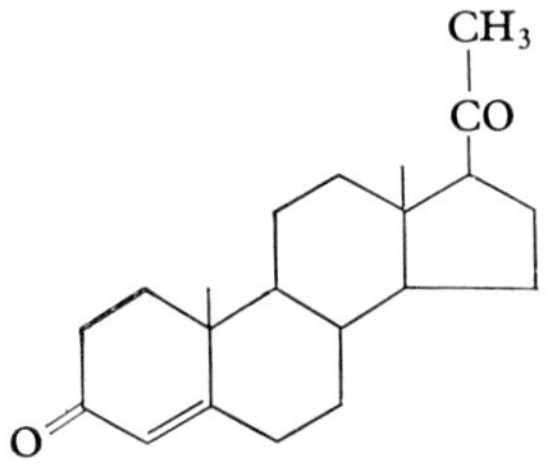

CH_3
|
CO

O

Progesterone

Fig. 3.14.

membered that androgens are C_{19} and oestrogens C_{18} compounds.) This means that it has the basic steroid nucleus of 17 carbon atoms plus two methyl groups attached at C10 and C13. There must, therefore, be another two carbon atoms attached as a side chain at C17, to bring the total complement up to 21 carbon atoms. There are two ketone =O groups attached at C3 and at C20, and a double bond between C4:5. It will be recalled that this compound, progesterone, has a very important part to play in the metabolism of all the sex steroids, even including testosterone, as well as in the production of the adrenal corticosteroids.

METABOLISM OF PROGESTERONE

Progesterone is produced by the corpus luteum of the ovary during the second half of the ovarian cycle. The evidence for this is that the hormone can be extracted from corpora lutea and that if the corpus luteum is removed the production of progesterone falls immediately to very low levels. Similar evidence shows that progesterone is produced by the adrenal glands and by the placenta during pregnancy. In all these situations the production of progesterone proceeds through the phases "acetate" → cholesterol → Δ^5-Pregnenolone → progesterone. Progesterone circulates in the blood and has its effects at distant sites. There is a definite effect on the lactic dehydrogenase (LDH) and on the diphosphopyridine nucleotide oxidase (DPNH) enzymatic systems within cells. These two enzymes, with others, are involved in the final phases of carbohydrate metabolism in the cells, and are especially concerned with the reversible lactate and pyruvate reactions of the tricarboxylic cycle. It should be realized that other sex steroids also have an effect on these systems. In general, oestrogens tend to increase the activity of the DPNH enzyme and progesterone the LDH enzyme. The exact mode of action of the sex steroids at cellular level are by no means understood and much further research is needed in this field. During metabolism progesterone is converted to pregnanediol (Fig. 3.15).

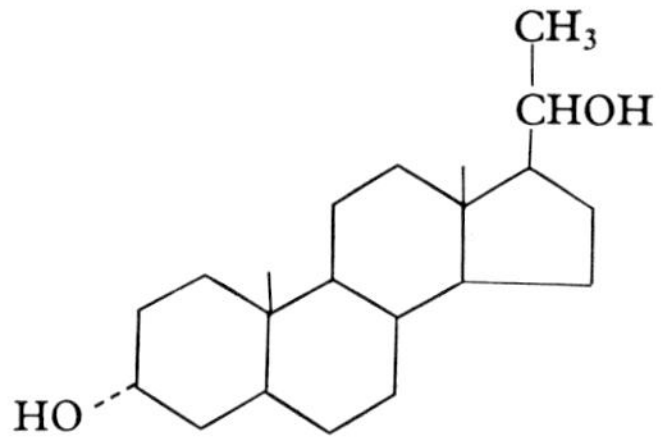

CH_3
CHOH
HO

Pregnanediol
Pregnane-3α-20α-diol
Fig. 3.15.

The proper name for pregnanediol is pregnane- 3α-20α-diol. This shows that the two -OH groups are attached at C3 and C20 in the alpha position. Moreover, the name also shows that the steroid nucleus is fully saturated and therefore has no double bonds. The change from progesterone therefore means that the two ketone groups are hydrogenated and the double bond between C4:5 has been eliminated by a hydrogen atom being incorporated. Moreover, the direction of the bond at C3 has been changed from the beta position to the alpha position. A point of nomenclature arises in connexion with the -OH group at C20. When the alcohol group is in the alpha position it is written on the right-hand side of C20. If in the beta position it is put on the left-hand side of that carbon atom. Pregnanediol is the main metabolite of progesterone that is excreted in the urine. Other steroids do contribute to this fraction, but in general it reflects the production of progesterone fairly accurately. The amounts of progesterone and pregnanediol in the blood are fairly small and difficult to estimate. If large injections of progesterone are given to a patient only about 20 per cent of the theoretical yield of pregnanediol can be found in the urine. As with oestrogens therefore the urinary excretion of pregnanediol is at best only a partial indicator of what is happening to the parent hormone, but so far because of technical difficulties it is the best index to the progesterone status of patients.

THE URINARY EXCRETION OF PREGNANEDIOL DURING THE OVARIAN CYCLE

The graph shows the varying levels of pregnanediol found in the urine during the ovarian cycle. It is measured on the total amount of urine passed in 24 hours. The amounts passed are measurable in milligrams whereas oestrogens are measurable only in micrograms in 24 hours. During the follicular phase of the ovarian cycle the excretion is of the order of 1–2 mgm. in 24 hours. A day or two after ovulation as the corpus luteum rises to a peak of activity the total daily excretion of pregnanediol rises to about 6 mgm. in 24 hours. This falls quite rapidly during the premenstrual week as the corpus luteum activity begins to wane, and the

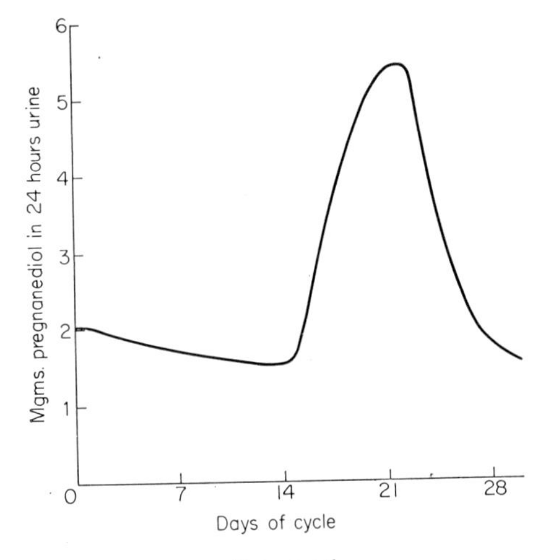

FIG. 3.16.

excretion returns to the basal level at the onset of menstruation. The range of values in the follicular phase is from about 0 to 2 mgm. in 24 hours and this presumably expresses the activity of the adrenal glands mainly. At the peak of excretion, somewhere about the 21st day of the cycle, the range of excretion of pregnanediol is from about 3 to 10 mgm. in 24 hours. Since only about 20 per cent of administered progesterone is excreted as pregnanediol it is obvious that these levels of excretion suggest that the corpus luteum at the peak of its activity may produce about 30 mgm. of progesterone per day.

The basal level of excretion of pregnanediol is found in girls before the menarche and in women after the menopause, and these observations tend to confirm the view that this is an expression of adrenal activity, since the hormone functions of the ovary are almost completely in abeyance at these stages of life.

CLINICAL APPLICATION OF URINARY HORMONE ESTIMATIONS IN THE OVARIAN CYCLE

The clinical applications of hormone estimations during the ovarian cycle are in fact comparatively few, because other simpler methods are available to determine if the ovary is going through the follicle → ovulation → corpus luteum sequence. Chief among these is diagnostic curettage of the endometrium (the lining of the uterus) which shows histological changes in sympathy with the outputs of oestrogen and progesterone from the ovary. However, knowledge of these normal outputs gives insight into the underlying abnormality when the periods are irregular, a very common complaint of patients. These will be considered further when the endometrial cycle in the uterus has been described.

Although not much used in this context, if pregnanediol excretion is relatively high it does show that the corpus luteum is functioning, and this is indirect proof that ovulation has taken place. This, of course, is fundamental for reproduction and can be of value in helping those women who complain of infertility, i.e. who have difficulty in becoming pregnant. Such women form a large part of gynaecological practice.

RELAXIN

This is a protein hormone which is probably produced by the ovary. Its physiological importance, if any, is not decided. It received its name because when injected into guinea-pigs it causes relaxation of the ligaments of the symphysis pubis so enlarging the pelvic cavity so that fetuses may more easily pass down the birth canal. It may be produced by the corpus luteum.

ANDROGENS

The name androgens implies substances which produce masculinizing effects, but the foregoing discussion should have emphasized that the relationships of the sex steroids do not allow of any sharp distinction in effects between androgens, progestagens and oestrogens. Moreover, there is great overlap in metabolism between these three groups of compounds. Nowadays the term androgens is coming to mean steroids with 19 carbon atoms in the molecule. Figure 3.17 gives a hint as to the interrelationships of the three groups of sex steroids. Down the left-hand side are shown some of the metabolic pathways that are found in the testis and the adrenal leading to the formation of testosterone. This major male hormone is metabolized to aetiocholanolone and to androsterone before being excreted in the urine. It should be realized that similar pathways are followed in the female adrenal glands and also probably in the ovary. The evidence for this in the ovary is still insecure, though there is no

METABOLIC PATHWAYS OF SEX HORMONES

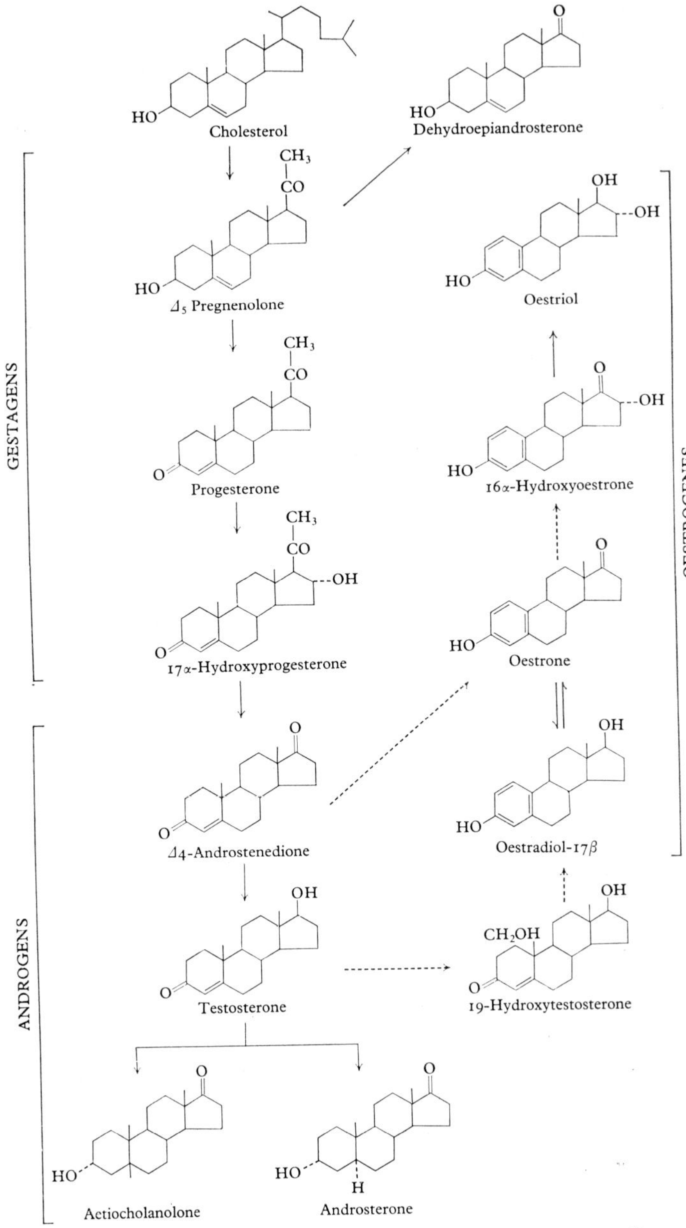

Redrawn from P. M. F. Bishop (1962) "The Chemistry of the Sex Hormones." Chas. C. Thomas. Springfield. Illinois.

Fig. 3.17.

doubt that in some disease states the ovary does produce mainly androgenic compounds rather than mainly oestrogens. On the right-hand side of the diagram are portrayed the metabolic pathways of some of the oestrogens. The dotted arrows show still unproven reactions, so that it is still not known exactly how oestrogens are derived from the androgenic series.

THE URINARY OUTPUT OF ANDROGENS

The excretion products of androgen metabolism all have a ketone group at C17, that is they are 17-ketosteroids, though the name is better 17-oxosteroids. Testosterone in particular is metabolized by the liver and probably by the kidney to its excretion products, aetiocholanolone and

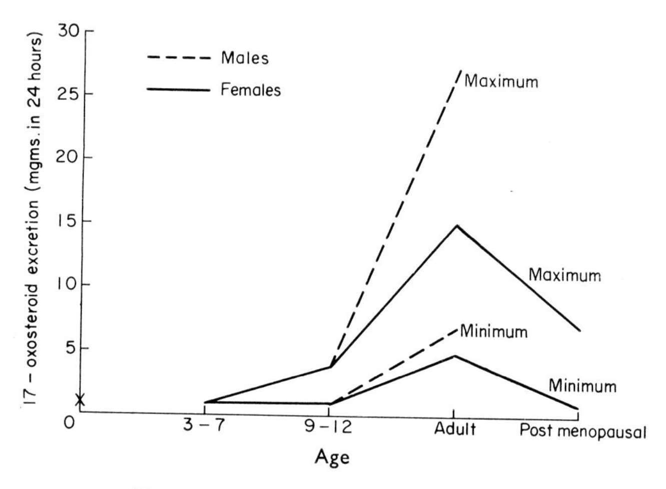

FIG. 3.18. Excretion of 17-oxosteroids.

androsterone. They can be distinguished chemically from oestrone, which also has a ketone group at C17, because the Ring *A* is not saturated in the oestrogen. Therefore the 17-oxosteroid estimation is a measure of androgen production in both males and females. It does not wax and wane in any measurable fashion during the menstrual cycle. In women of reproductive age the values in 24 hours range from about 5 to 15 mgm. In men the range is from 6 to 25 mgm. It is about 1 to 4 mgm. in male and female children before puberty and thereafter rises slowly to its maximum for each sex. In men the maximal values are found about the age of 25 years and then they slowly decline. In post-menopausal women the average values also decline and this suggests that the ovaries do contribute some 17-oxosteroids to the total output, and recent methods show that adult women produce about 1 mgm. of testosterone in 24 hours of which 0·5 mgm. comes from the ovary and 0·5 mgm. comes from the adrenal. Men on the other hand produce about 4 mgm. of testosterone per day of which 3·5 mgm. comes from the testis and the remainder from the adrenal. When testosterone is injected into a male only 35 per cent is recoverable as metabolites from the urine, the rest being "lost" in ways quite unknown. The graph shows the excretion of 17-oxosteroids throughout life in males and females and the peak of excretion during reproductive life is worthy of note for its suggests the amount of the contribution of the gonads to the androgen production during that time of life, and especially it shows something of the activity of the ovary.

In clinical practice the excretion of 17-oxosteroids is commonly done as a rough estimate of androgen status for there is a series of more or less masculinizing conditions of both the adrenal glands and the ovaries which may have to be differentiated and treated.

THE ADRENAL STEROIDS

The adrenal glands produce three main types of steroid. They are aldosterone, the corticosteroids and sex hormones. Their sites of production are probably the zona glomerulosa, the zona fasciculata and the zona reticularis respectively. Aldosterone is mainly active in the sphere of water and electrolyte metabolism, the corticosteroids in a variety of reactions associated with carbohydrate, protein and fat metabolism as well as general body reactions to stress. The rôle of the sex hormones of the adrenal are difficult to understand, but it has already been suggested that they may be important in spheres not essentially associated with reproduction and they may be of vital importance in the metabolism of many cells. It is worth noting too that there is some overlap of function between these three groups of steroids and especially the corticosteroids and the sex steroids play a secondary role to that of aldosterone in the control of water and electrolytes. These secondary effects help to explain some of the cyclic phenomena of water and electrolyte balance seen during the course of the ovarian cycle. Moreover, progesterone plays a very central part in the metabolism of all steroids and the overlap of adrenal and gonadal function can be very great. Since the steroids in all their major chemical forms are widely distributed through the animal kingdom and are so intimately involved in homeostasis and reproduction it is obvious that they must have been a major factor in evolution.

Chapter IV

THE MENSTRUAL OR ENDOMETRIAL CYCLE

THE MENSTRUAL CYCLE

The lining of the uterus is the endometrium, which is under the control of the hormones of the ovary. Since these are produced in waxing and waning cycles the endometrium waxes and wanes in time with the hormone secretions from the ovary. This is true of all mammals, but only in the Primates is the phenomenon of menstruation seen. This is the periodic shedding of the endometrium from the uterus which is seen as a blood discharge from the vaginal orifice. In all the Primates this cycle has an approximate period of about four weeks, the significance of which is unknown. Indeed, the physiological significance of menstruation is not known. It has been developed in the evolutionary sense from the sexual cycles of lower vertebrates, many of whom show the phenomenon of oestrus (oestrum = gadfly). In these lower animals oestrus occurs in regular cycles depending on the species and its environment. The vaginal bleeding which is seen in these animals at oestrus is not comparable with menstruation, for bleeding is associated with ovulation in oestrous cycles but the vaginal bleeding of menstruation occurs only after the death of the corpus luteum approximately fourteen days after ovulation. Even among the Primates menstruation is not universal and in the New World monkeys occurs only sporadically, but in the Old World monkeys and the Anthropoidea menstruation is the rule. Perhaps the significance of menstruation is to be found in the fact that it is relatively independent of the environment which determines much of the reproductive activity of lower animals; that is the breeding season of Primates is continuous whilst that of other animals is seasonal. This offsets to some extent the evolutionary disadvantage of producing only one offspring at a time. In general, lower animals have larger litters than Primates. The single offspring, at least of the higher Primates, must be seen as an adaptation to an environment which was probably arboreal. The menstrual cycle as a whole and its controlling mechanisms are evolutionary adaptations along with others such as the upright posture, binocular vision, the prehensile hand and the development of the cerebral cortex.

The ovarian cycle should be briefly recalled. At birth the ovary is stuffed with primordial follicles and there are no endocrine elements at this time nor for some time afterwards. As puberty approaches, the Graafian follicles begin to form but only reach the antral stage whilst the output of oestrogens gradually rises. At puberty ovulation takes place and the woman enters on her rhythmic cycles until the ovary ceases to function at the climacteric. During the reproductive years the ovarian cycle is of Graafian follicle → ovulation → corpus luteum and the hormones produced are oestrogens followed by oestrogens + progesterone. After the climacteric the endocrine elements of the ovary slowly disappear, there is no progesterone production from the ovary and oestrogen secretions slowly decline. The endometrium faithfully reflects these various changing hormone secretions and this is fortunate clinically for the endometrium can be obtained by curettage from the uterus and when it is examined microscopically its state can be used to infer the events of the ovarian cycle and therefore of the pituitary cycle which controls it.

THE ENDOMETRIUM BEFORE PUBERTY

The basic structure of the endometrium is of tubular glands supported by stroma. The stroma is richly supplied by vascular channels which are needed to supply the metabolic requirements of the ever-changing endometrium. Blood is needed in relatively large quantities wherever there is a hive of chemical industry such as there is in the endometrium.

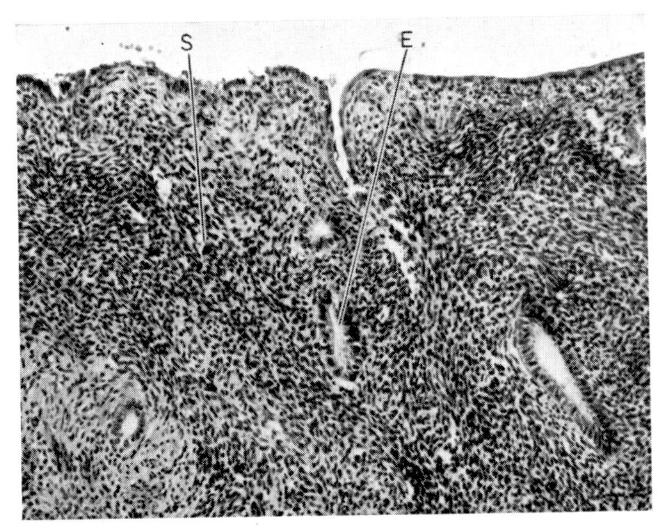

S = Stroma. E = Endometrial gland.

FIG. 4.1. Endometrium before puberty. The endometrium is inactive with small glands embedded in a stroma consisting of dark oval and spindle cells. The appearances resemble those seen in the postmenopausal endometrium.

The epithelial lining of the uterus is columnar and is one cell thick. Some of the epithelium is ciliated. The cells of the surface run down into the simple tubular glands, whose depths just dip into the muscle layer of the uterus. A feature of interest is that there is no basement membrane between the epithelial layer and the muscle as there is in other parts of the body. Resting on the muscle layer there is therefore a continuous sheet of stroma perforated only by the glands and blood vessels. Before puberty this endometrial layer is thin and not well developed. As the follicles in the ovary begin to secrete oestrogens at puberty so the endometrium responds by proliferating. The proliferation affects the glands, the stroma and the blood vessels. Just before the menarche (the first period) the three parts of the endometrium have grown so that it is about a millimetre thick. It is at this point that reproductive life really begins.

THE ENDOMETRIUM DURING THE REPRODUCTIVE YEARS. MENSTRUAL CYCLE

During the growth of the Graafian follicle the oestrogen output rises to its peak just before ovulation. The hormones cause the endometrium to proliferate so that the test-tube type glands elongate, the cells of the stroma become more obvious in their outlines and increase in number whilst the blood vessels increase *pari passu* with the growth of the cells. The whole endometrium is now about 3 mm. thick. With the formation of the corpus luteum and the production of progesterone in addition to oestrogens the rate of growth of the endometrium slows down somewhat, though the slowing down affects the stroma much more than the glands. These still grow but they have to be accommodated within a stroma which does not increase in thickness and therefore the glands come to have a wavy outline, that is in section they are serrated or saw-toothed. This gives the endometrium a characteristic pattern histologically. Moreover, these glands begin to secrete, but the nature of the secretion is not known. Before the glands have had time to grow to their serrated convoluted shape there is evidence of secretion by the formation of vacuoles beneath the nuclei, and this subnuclear vacuolation is the first sign in the endometrium of the presence of circulating progesterone. The cells of the glands also become more columnar during this phase. It will be realized that the essential change under the influence of oestrogen is proliferation and, under the influence of progesterone, secretion.

There is still much work to be done on the histochemistry and biochemistry of the endometrium and very little of value about either of them is yet known. The enzyme alkaline phosphatase is found in large quantities during the proliferative phase, and the vacuoles in the secretory phase of the cycle seem to be made up of glycogen and fat. Other isolated facts about this aspect of the endometrium are known but cannot

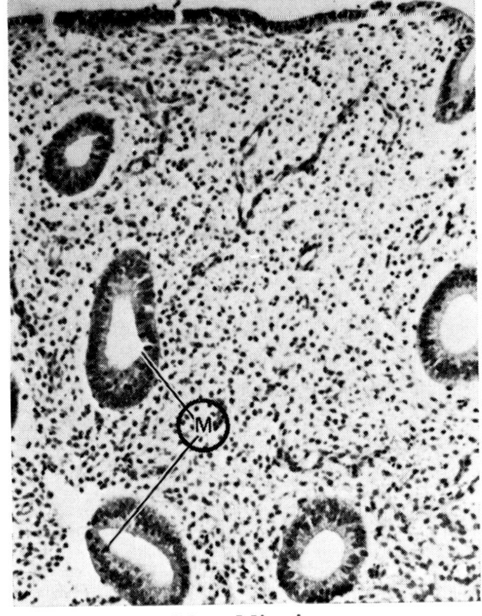

M = Mitosis.

FIG. 4.2. Proliferative phase endometrium. The glands are simple and when cut across are circular in outline. No secretion is apparent in the gland lumen. The gland epithelial cells are low columnar and show many mitoses. The stromal cells have little cytoplasm. They also show frequent mitoses.

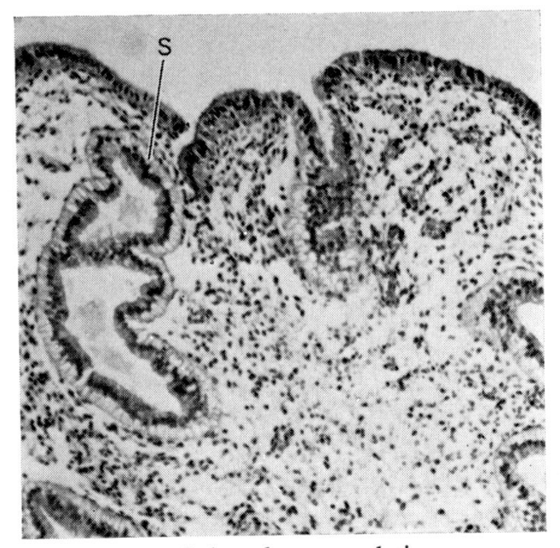

S = Subnuclear vacuolation.

FIG. 4.3. Early secretory endometrium. In the earliest phases, secretory vacuoles appear in the gland epithelial cells deep to the nuclei. The glands become more tortuous but the stroma remains of a loose delicate structure until later in the menstrual cycle.

be fitted into a coherent pattern. Just before the shedding of the endometrium in menstruation it appears to be oedematous and this is probably because of an increase in water and possibly electrolytes. At about the time of menstruation the oedema disappears and the endometrium shrinks. The cause of this is probably vascular and would seem to be due to the falling levels of oestrogen and progesterone seen at this time. The stroma just before menstruation shows well marked cells with clear outlines and they are full of glycogen.

The blood vessels of the endometrium have received much attention for the growth of the endometrium and its subsequent shedding at menstruation are presumably mediated by the vascular changes. Running

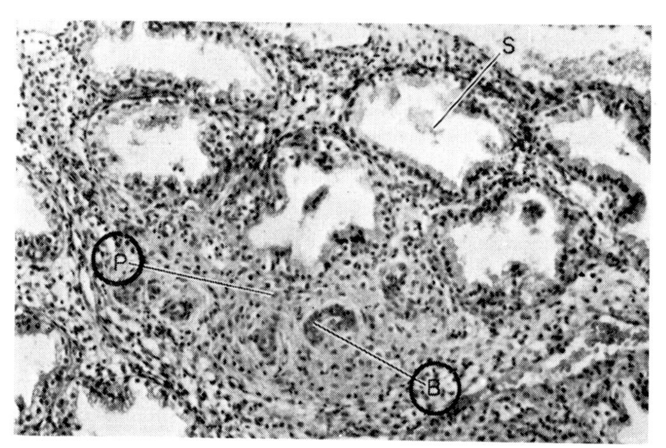

S = Secretion. P = Stromal cell; pseudo-decidual. B = Blood vessel.

FIG. 4.4. Late secretory endometrium. The glands have an irregular, "saw-tooth" outline and the lining cells are columnar. The luminal cell border is indistinct and secretion is apparent in the lumen. The stroma, particularly around blood vessels, becomes compact due to the cells acquiring abundant cytoplasm resembling decidual cells.

beneath the endometrium are the radial arteries of the muscle layer (myometrium) of the uterus. From them arise the spiral endometrial arterioles, which course up through the endometrium. Basal branches are given off from these endometrial spiral arterioles near the junction of the endometrium and the myometrium. These remain when the superficial layers of the endometrium are sloughed off at menstruation and help with the regenerative process. The spiral arterioles continue on their way to the superficial surface where they branch to form venous lakes and sometimes direct arterio-venous anastomoses. From these two terminations and from capillaries the blood is collected into venous sinuses and returned to the veins of the myometrium. The vascular pattern corresponds with the histological division of the endometrium into a basal zona spongiosa and a more superficial zona compacta. During menstruation it is the zona compacta which is shed leaving the basalis and its blood vessels behind.

The spiral arterioles are the key to the phenomenon of menstruation for they have been observed to contract and relax in the few days before the menstrual flow. During contraction the endometrium blanches and during relaxation becomes congested. This observation was made by

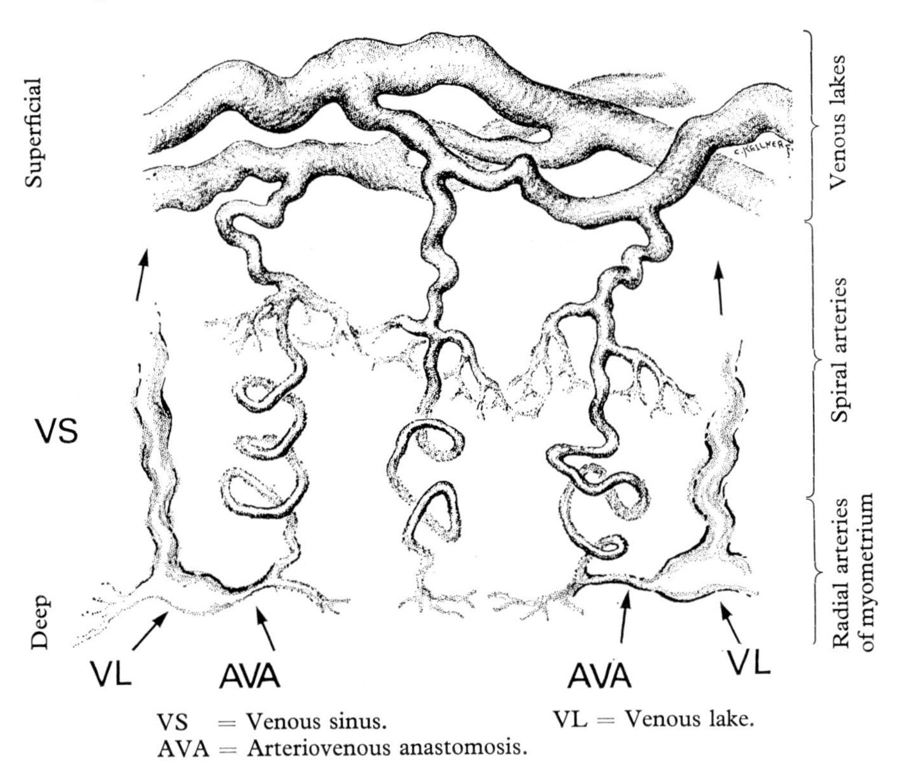

FIG. 4.5. Vascular pattern of the endometrium. From H. Okkels, *Modern Trends in Obstetrics & Gynaecology* (1950), Ed. Kenneth Bowes. Butterworth.

Markee during his elegant experiments in which endometrium was transplanted to the anterior chamber of monkeys' eyes where it could be observed. At times the constriction of the arterioles is intense and so probably leads to anoxia of such severity that the superficial layers of the endometrium are sloughed off and shed. The process of this shedding does not proceed equally in all parts of the endometrium but is rather patchy so that the endometrium comes off in shreds. Obviously the vessels break too and this is responsible for the bleeding. The reason for the constriction and relaxation of the spiral arterioles is probably hormonal for it has not been possible to demonstrate nerves in the endometrium at any time in the cycle.

The cycle seen in the endometrium is to be looked upon as the preparation for the reception of a fertilized ovum. The object of the uterus is to house the developing embryo and fetus until it is mature enough to be born and survive on its own. When the endometrium does not receive a fertilized ovum it breaks down in menstruation. Although little is known of the intimate biochemistry of the endometrium it is

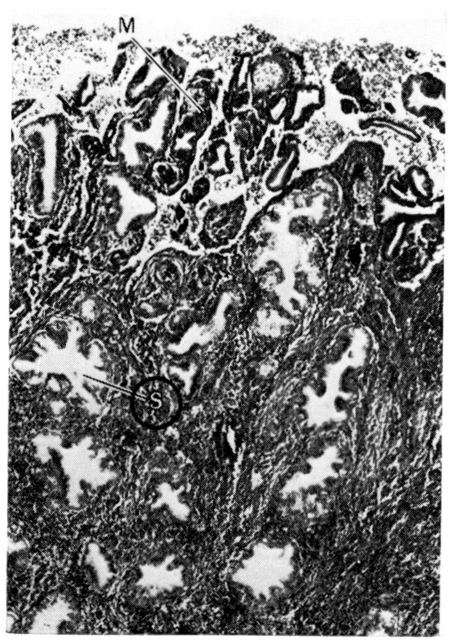

M = Breaking down menstrual endometrium. S = Secretory gland.

FIG. 4.6. Menstrual endometrium. The superficial layers of the endometrium fragment and they are discharged with the menstrual blood. Deep to this breaking-down surface, the glands have the typical late secretory pattern.

obvious that its prime function is to provide nutrition for the trophoblast on which the life of the embryo depends. In lower mammals the secretion of the glands is the sole support of the trophoblast and embryo, and even in Man it is probable that the ovum has to live off these secretions at least for a short time. Later, however, the chorion (or trophoblast) erodes through the surface layers of the endometrium to reach the maternal vascular tree from which it ultimately derives its nourishment. Probably this is the importance of the venous lakes near the surface of the endometrium and of the glycogen laden cells of the stroma. Menstruation is the failure of reproduction and it has been picturesquely said that the uterus "weeps a bloody tear" when a fertilized ovum does not materialize.

The sequence of events in the ovarian and menstrual cycles may be briefly summarized:

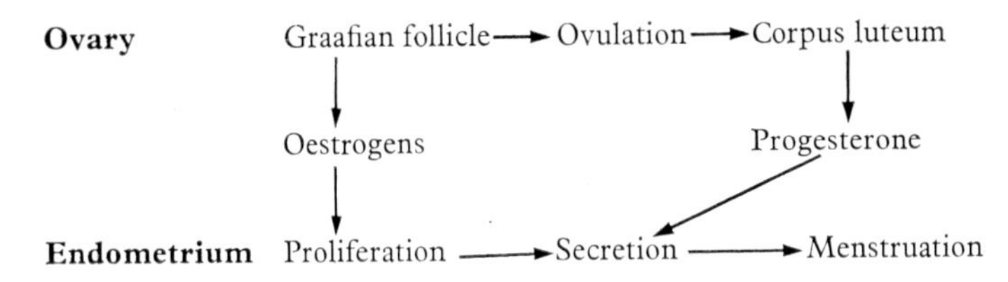

This cycle continues throughout reproductive life unless interrupted by childbirth or disease.

THE ENDOMETRIUM IN THE CLIMACTERIC AND AFTER THE MENOPAUSE

In the ovary the first failure during the climacteric is of ovulation. This means that a corpus luteum cannot be formed and so there can be no production of progesterone. In its turn this means that there can be no secretory phase in the endometrium. Thus the climacteric endometrium can only be of the proliferative type. Rarely, of course, there is sporadic ovulation and corpus luteum formation so that there will be occasional secretory type endometria during the climacteric. If the oestrogen output from the ovaries is small there will be very little proliferation so that the endometrium is thin and almost atrophic. If the output of oestrogen is large there may be massive proliferation of the endometrium and the changes then may be so great as to be pathological. Oestrogen secretion at this time is very sporadic and sometimes will be high and sometimes low and there may be no cyclic activity of the ovaries at all. Thus it is very common to find that women in the climacteric have very irregular losses of blood from the uterus. The clinical problem is that other conditions such as cancer may also cause irregular haemorrhages and they can only be distinguished from those of the normal climacteric by careful examination of the genital tract with curettage of the endometrium so that it may be examined under the microscope.

When the ovary no longer secretes appreciable amounts of oestrogen the endometrium does not respond at all and remains atrophic. There is then no bleeding from the genital tract and when the periods cease it is called the menopause. If the periods have stopped for about six months and then a loss of blood from the uterus occurs it is called post-menopausal bleeding and this too requires careful investigation since it may be a sign of cancer. Not uncommonly, however, it is due to doctors giving the patients oestrogens in too large doses so that the endometrium is made to proliferate. Alternatively there may be sudden spurts of oestrogen secretion from the ovary or adrenal gland of such magnitude that they may cause proliferation. When the oestrogen stimulus is withdrawn the endometrium breaks down in a process similar to that occurring at menstruation, with consequent bleeding *per vaginam.*

It is common to refer to any bleeding from the genital tract as a "period", but strictly speaking menstruation as a term should only be

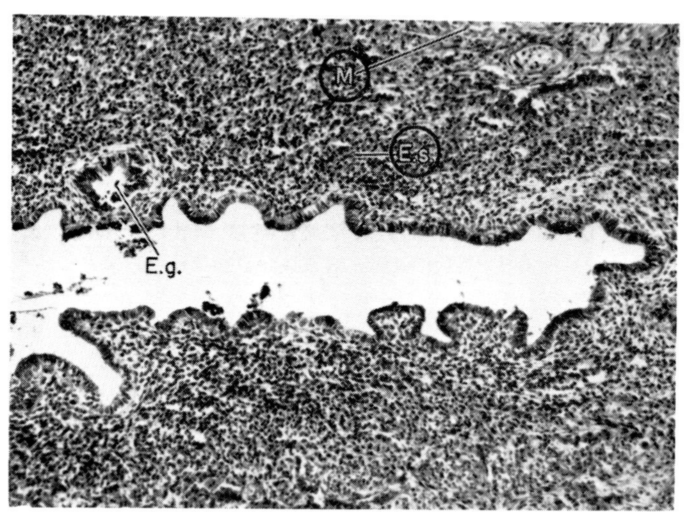

E.g. = Endometrial gland. E.s. = Endometrial stroma.
M = Myometrium.

FIG. 4.7. Endometrium after the menopause. The glands have become atrophic and mitoses are not seen in the lining cells. The stromal cells have pyknotic, oval nuclei. The appearances resemble those seen before puberty.

applied to the periodic loss of blood and endometrium which has been stimulated by both oestrogens and progesterone, that is when ovulation has occurred and the shed ovum has not been fertilized. Endometrial shedding from the withdrawal of oestrogen only is not then menstruation, but in clinical practice the distinction is somewhat pedantic, and indeed difficult, since patients themselves describe all losses of whatever kind as periods, or the menses.

CYCLICAL CHANGES IN OTHER PARTS OF THE GENITAL TRACT

The vulva. There are no obvious changes in the vulva of women during the ovarian cycle, but infra-red photography of the area which shows up the veins does demonstrate changes in vascularity. These are comparable with the changes seen in the sexual skin of many of the Primates, and especially the baboon. The vascularity and swelling due to oedema occur maximally in the first part of the cycle in the baboon under the influence of oestrogens but disappear after ovulation takes place. These changes are probably a visual signal of the receptivity of the female and so are an aid to successful fertilization.

The vagina. The vagina is a muscular tube lined by a kind of stratified squamous epithelium. This is in common with any tubes especially

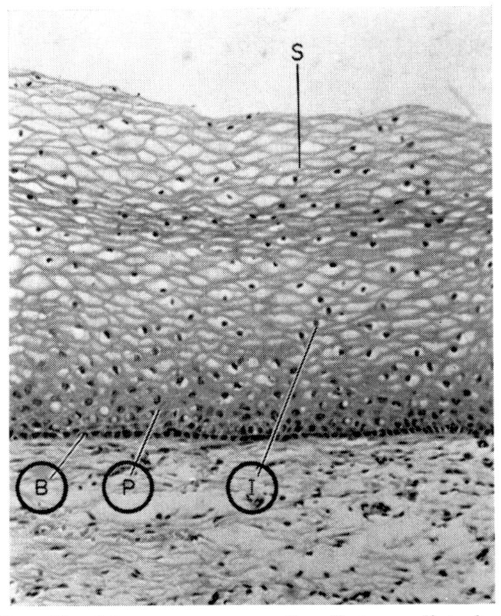

B = Basal layer. I = Intermediate layer.
P = Parabasal layer. S = Superficial layer.

FIG. 4.8. Vagina. The vagina is covered by stratified squamous epithelium resting on a vascular connective tissue stroma. Unlike the skin, the vaginal stratified squamous epithelium has nucleated superficial cells.

subject to wear and tear such as the mouth, upper oesophagus and anus. In the vagina during reproductive life there is a single layer of basal cells next to the muscle layer. The cells are small and basophilic on staining. External to this is the intermediate layer consisting of several rows of cells which are large and clear with well marked cellular boundaries. Superficial to the intermediate layer is the cornified layer similar to that seen on the skin though the keratinization is not so well marked. The more superficial cells are acidophilic and stain pink when haematoxylin and eosin are used.

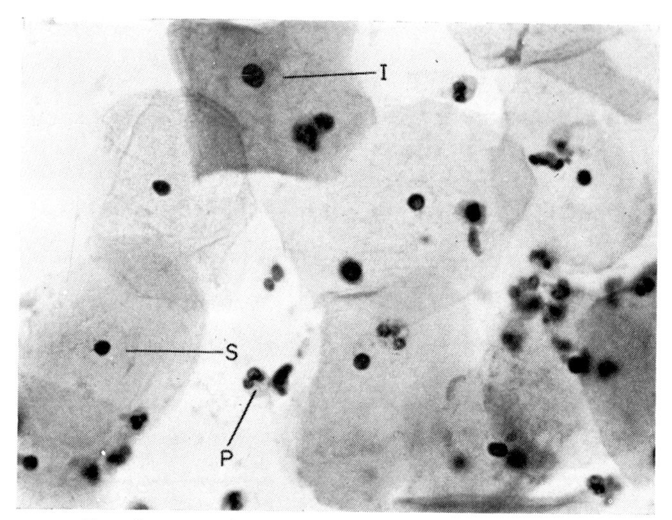

I = Intermediate cell. S = Superficial cell.
P = Polymorph.

FIG. 4.9. Vaginal smear—oestrogen phase. Oestrogen induces full maturation of the vaginal squamous epithelium. The majority of cells have small, dark, pyknotic nuclei and abundant, thin, often folded cytoplasm. These are superficial cells. The cell with similar cytoplasm but a delicate, pale nucleus, is an intermediate cell. There are several neutrophil leucocytes in the smear.

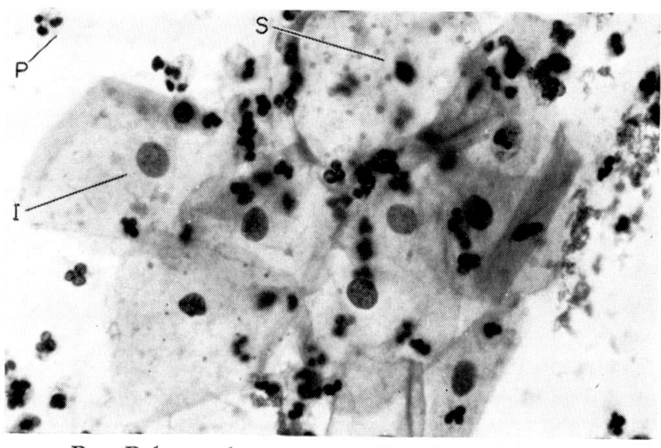

P = Polymorph. S = Superficial cell.
I = Intermediate cell.

FIG. 4.10. Vaginal smear—progesterone phase. The smear appears "dirty" with many neutrophil leucocytes. Most of the squamous cells have delicate pale nuclei with abundant, thin, folded cytoplasm—intermediate cells. An occasional superficial squamous cell, with a pyknotic nucleus, is apparent.

Cells are constantly being shed from the surface of the vaginal epithelium and they contribute to the slight white discharge which is normal in all women of reproductive age. The intermediate and superficial cells are replaced by cell division of the basal layers. It is possible to obtain specimens of the shed cells by the use of a glass tube with a suction bulb on the end of it and make smears on glass slides which can then be stained. Usually there are no basal cells to be seen but the percentages of intermediate and superficial cells vary throughout the ovarian cycle. It is usual to count 100 cells of a vaginal smear and decide how many come from the intermediate layer and how many from the cornified layer. When oestrogens predominate there is a tendency to increased

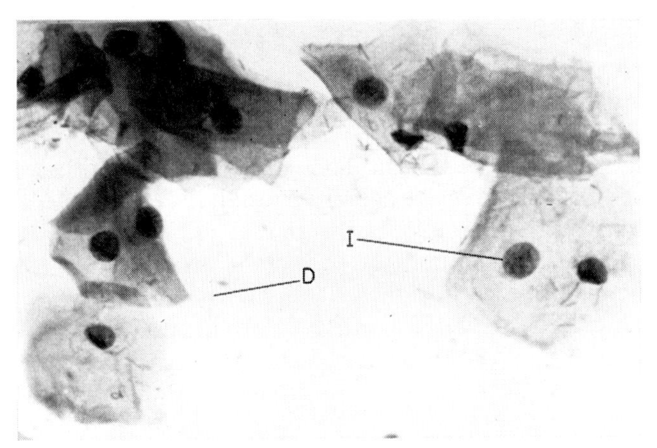

I = Intermediate cell. D = Döverlein's bacilli.

FIG. 4.11. Vaginal smear in normal pregnancy. Almost all the squamous cells are of intermediate cell pattern with delicate pale nuclei. Clumping of cells is marked. The small rods in and around the cells are Döderlein bacilli.

cornification. When progesterone predominates the numbers of cornified cells decrease. The results may be recorded in the form 0/60/40, which means no basal cells, 60 intermediate cells and 40 superficial cells. This indicates the influence of oestrogen. A reading of 0/90/10 implies the presence of progesterone since the cornified cells have decreased.

Vaginal smears have been much used for the assessment of the endocrine status of patients since it is a simple method and not so costly as chemical techniques applied to blood or urine. If smears are taken during the oestrogen phase of the cycle and also during the progesterone phase and a change is demonstrated it is evidence that ovulation has occurred. This is of importance in patients complaining of infertility and sometimes at the menopause. Since pregnancy is associated with a rising output of progesterone the smear may be used as an indirect test for pregnancy in its early phases. Some cases of early abortion may be caused by a relative deficiency of progesterone and this may be recognized by an increasing number of cornified cells in the vaginal smear.

The main stimulus to the growth of the vaginal epithelium is oestrogen secretion from the ovary. Therefore it is to be expected that the vagina of a child will be somewhat atrophic and such is the case. The basal layer is in evidence, but the intermediate and superficial layers are very thin. However, the vagina of the newborn baby is very similar to that of the adult because oestrogens have crossed the placenta from the mother. At puberty as the oestrogen secretion of the ovaries increases the vagina thickens and comes to take on the characteristics of the adult vagina, and the cells of the intermediate layer especially become laden with glycogen. This is easily recognizable when the vagina is painted with iodine for the epithelium stains a rich mahogany brown colour, which is the usual reaction of glycogen with iodine. Also during reproductive life the vagina is invaded by Döderlein's bacilli. These are small rod-shaped organisms whose source is quite unknown. They are of great importance because using glycogen as their substrate they produce lactic acid so that the vagina during reproductive life is always acid and has a pH in the region of 4·5. This serves to inhibit the growth of many pathogenic bacteria which otherwise might invade the genital tract. Before the onset of puberty the pH of the vagina is more nearly neutral at pH 7. After the menopause the vaginal epithelium becomes atrophic because the stimulus of oestrogen is withdrawn. The Döderlein's bacilli disappear and the reaction becomes neutral again. Infections of the vagina by pathogenic bacteria are much commoner after the menopause than before it. However, there are infections of the vagina which are commoner during reproductive life but it is significant that these are usually by a protozoon, *Trichomonas vaginalis*, and a fungus, *Candida albicans*. These infections are an indication of the different cultural requirements of these organisms when compared with those of the usual pathogenic bacteria.

The cervix. On its outer aspect presenting to the vagina the cervix is covered by squamous epithelium which undergoes cyclic changes just like those of the vaginal epithelium. The canal of the cervix up to the internal os is lined by tall columnar epithelium with basal nuclei. Branched glands open on to the surface of this epithelium and they secrete mucus. Their depths are buried in the stroma of the cervix which consists mainly of fibrous tissue but there is about 20 per cent of smooth muscle. There is a fairly sharp dividing line between the squamous epithelium of the outer cervix (ectocervix or portio vaginalis) and the columnar epithelium of the endocervix. This is usually at the external os. The junction between the endocervical epithelium and the endometrium above is not so clear cut.

There is no clear evidence that the cervical epithelium of the canal undergoes any cyclic changes that can be demonstrated histologically, but there is no doubt that the cervical mucus does change. It must be assumed that histological methods are not delicate enough to show changes that must undoubtedly occur. At the time of ovulation the cervical secretions are abundant and the mucus appears crystal clear and is a flowing stream. By some it has been called the "ovulation cascade". After ovulation it becomes more scanty and tacky. Experiments *in vitro* show that the mucus at the time of ovulation can be easily invaded by spermatozoa but there is greater delay when the cervical glands are under

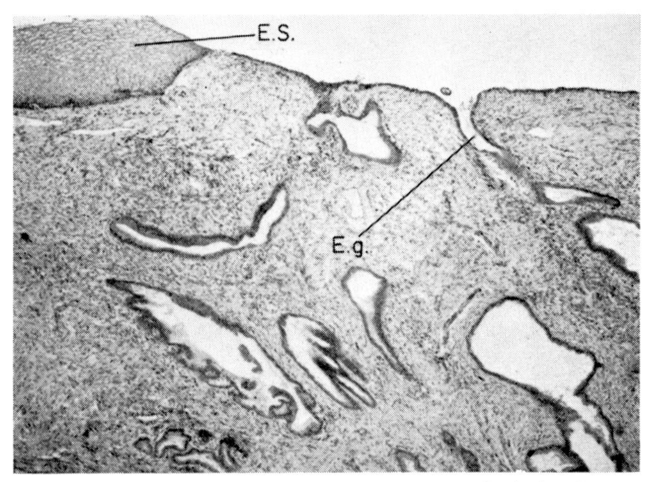

E.S. = Ectocervix. E.g. = Endocervical gland.

FIG. 4.12. Normal cervix. This is the junction of the endocervix and the ectocervix. The endocervical mucosa consists of branching glands lined by tall columnar mucus-secreting epithelium. The ectocervix is covered by stratified squamous epithelium. The stroma consists of smooth muscle and fibrous tissue.

the influence of progesterone. This is, therefore, an adaptation to procure easier fertilization. Experiments have also been done on the physical properties of the mucus and these show that the mucus may be pulled out into long threads which do not break during the follicular (oestrogen) phase. During the luteal phase the mucus tends to form small round balls which break easily when stretched in capillary tubes or between two glass slides. This tacky type of mucus fills the cervical canal during the second half of the ovarian cycle and during most of pregnancy and its function is to prevent the ascent of infection into the cavity of the uterus where the growing embryo and fetus is. It has been shown that the mucus has bactericidal properties.

The phases of the ovarian cycle can be recognized very easily by the crystallization properties of the cervical mucus. If some mucus is mixed on a glass slide with a drop of normal saline and allowed to dry a fern-leaf pattern of crystallization can easily be seen microscopically if the cervix has only been subject to the stimulation of oestrogen. When progesterone dominates the picture this fern-leaf pattern disappears. The test can be used clinically to determine if ovulation has occurred though this involves seeing the patient in the first half of the cycle and showing the fern-leaf pattern to be present and then repeating the test in the second half of the cycle to show that the pattern has now disappeared. It has also been used as a test in abortions on the same lines as

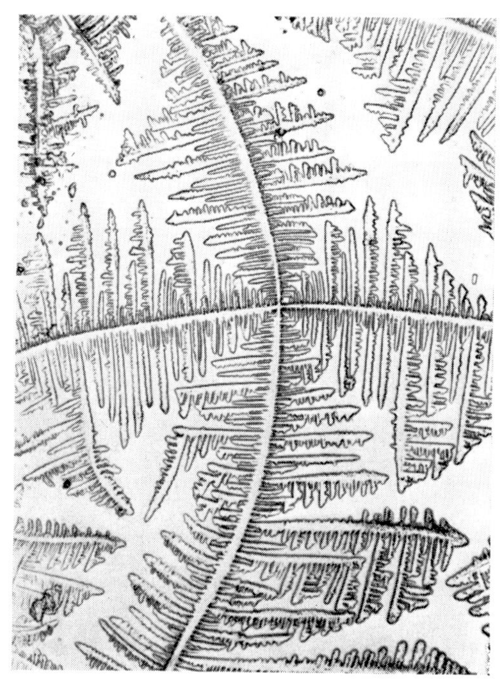

FIG. 4.13. Fern pattern of cervical mucus. During the oestrogen phase of an ovulatory cycle the cervical mucus will crystallize into a fern pattern when allowed to dry on a slide. This phenomenon is of some value in practice in assessing ovarian oestrogen production.

has been mentioned for vaginal smears. If the progesterone output is falling during pregnancy the cervical mucus will begin to show the fern-leaf pattern since the cervix is being stimulated mainly by oestrogens.

All women during reproductive life have a slight white discharge from the vagina amounting to one or two millilitres per day. This is made up of secretions from the cervix and from the desquamation of cells from the vagina. From what has been said it will be realized that the amount of this discharge may increase at the time of ovulation when cervical secretion is at its maximum.

THE MYOMETRIUM

One of the more interesting discoveries of recent years has been that the uterine musculature is in a state of constant activity and undergoing rhythmic contractions during all phases of reproductive life. The activity has been mainly studied by the use of small balloons inserted into the cavity of the uterus and connected to narrow tubes leading to pressure recording devices. During the oestrogen phase of the cycle the pressures so recorded are of the order of 0·2 mm. Hg and they come

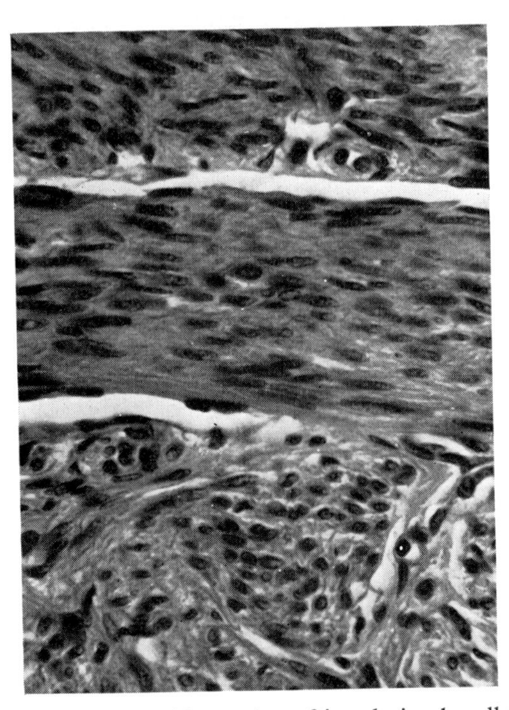

FIG. 4.14. Myometrium. This consists of interlacing bundles of smooth muscle cells. In any section some of these cells will be cut in longitudinal section and others will be cut in cross-section.

about every 30 seconds, that is at the rate of 120 per hour. Each contraction lasts about 30 seconds. These waves have been designated as of the *A* type. During the second half of the cycle so-called *B* waves begin to overlie the *A* waves. The *B* waves are of greater amplitude than the *A* waves and come rather more slowly. The pressure developed inside the uterus is of the order of 10 mm. Hg and they arise about every 2 minutes, that is at the rate of 30 per hour. This activity will be dealt with again when considering the myometrial activity of pregnancy and labour when it assumes great importance.

THE FALLOPIAN TUBES

The uterine tube is the place of fertilization of the ovum and it conducts the ovum from the ovary to the uterus. When the egg is fertilized the tubal epithelium is also probably responsible for its early nourishment

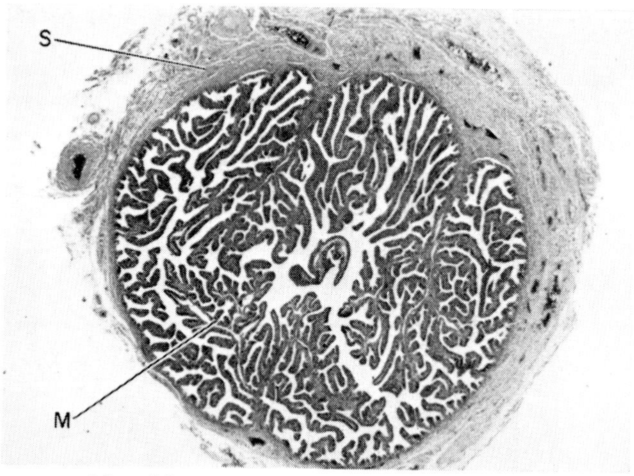

M = Mucosa. S = Sero-muscular coat.

FIG. 4.15. Fallopian tube. This is the fimbrial end of the tube where the mucosa is thrown up into complex folds. A smooth muscle coat surrounds the tube.

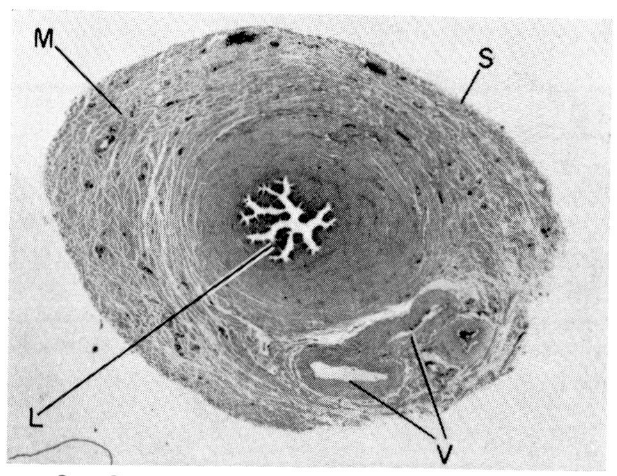

S = Serous coat. L = Lumen.
M = Muscle coat. V = Blood vessels.

FIG. 4.16. Fallopian tube. At the isthmic end, the tube has a simple structure. The lumen is narrowed by occasional mucosal folds. There is a thick surrounding smooth muscle coat.

and respiratory exchange. The tube is muscular and its rhythmic peristaltic activity increases about the time of ovulation and so correlates with it. During the luteal phase of the cycle the tubal activity is very much less and is virtually non-existent during menstruation and during pregnancy when progesterone is the dominant hormone.

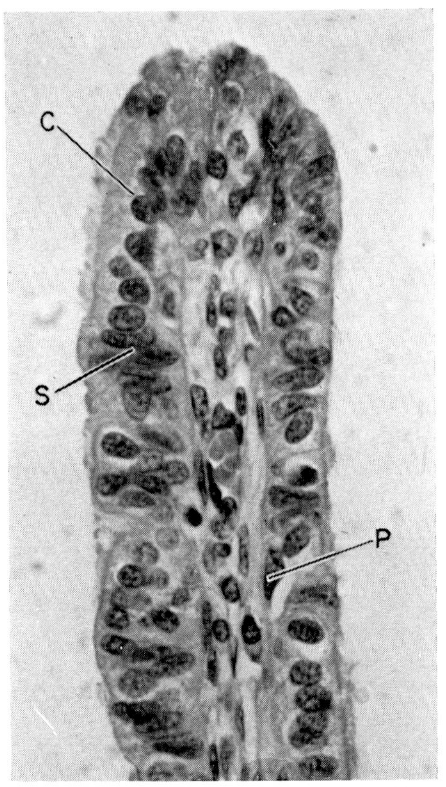

C = Ciliated cell. S = Secretory cell. P = Peg cell.

FIG. 4.17. Fallopian tube—oestrogen phase. The mucosa of the Fallopian tube has three principal cell types: (i) a pale-staining ciliated cell with an oval nucleus, (ii) a columnar, non-ciliated "secretory" cell with a nucleus elongated in the long axis of the cell and (iii) a reserve or "peg" cell resting on the basement membrane. In the oestrogen phase of the menstrual cycle, the columnar secretory cells are no taller than the ciliated cells.

Within the muscular lining is the mucosa which is thrown up into many folds especially at the outer ampullary end. The folds diminish to about four at the uterine end of the tube. The reasons for this massive plication are unknown. The epithelial cells are of three types, ciliated, secretory and "peg". The ciliated columnar cells move particles towards

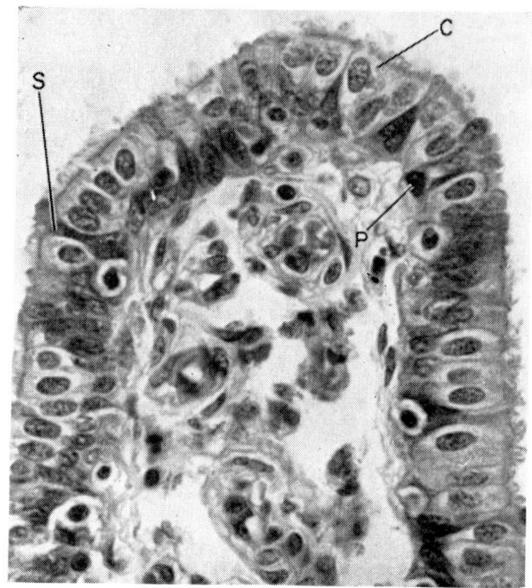

C = Ciliated cell. S = Secretory cell. P = Peg cell.

FIG. 4.18. Fallopian tube—progesterone phase. During this phase the secretory cells show a bulbous tip which protrudes above the ciliated cells into the lumen of the tube.

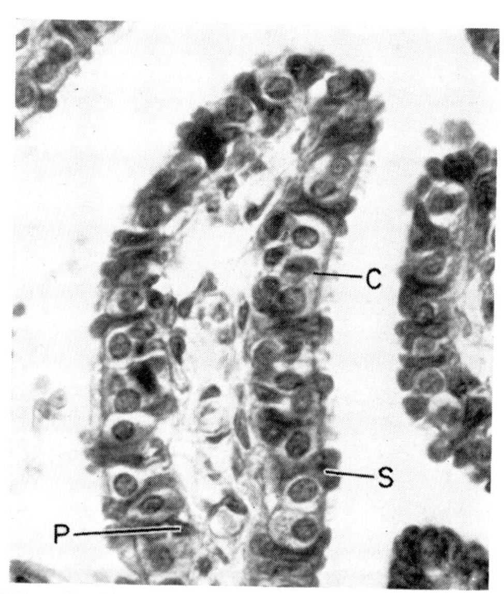

C = Ciliated cell. S = Secretory cell. P = Peg cell.

FIG. 4.19. Fallopian tube—pregnancy. This is an exaggeration of the progesterone phase of the menstrual cycle. The secretory cells stand well above the pale ciliated cells.

the uterus and are of importance together with peristalsis in transporting the ovum to the uterus. Ciliated action is at its height just before and just after ovulation as might be expected. The secretory cells are of goblet type but the composition of the secretion is not known nor is its function. It seems reasonable to suppose that the secretion is of importance for the early development of the fertilized egg and it has been suggested that the tube provides the egg with a protein coat without which it would be unable to embed in the endometrium. The secretory cells have a cycle in which they reach their maximum activity at about the time of ovulation. The peg cells have no known function but they seem to be exhausted

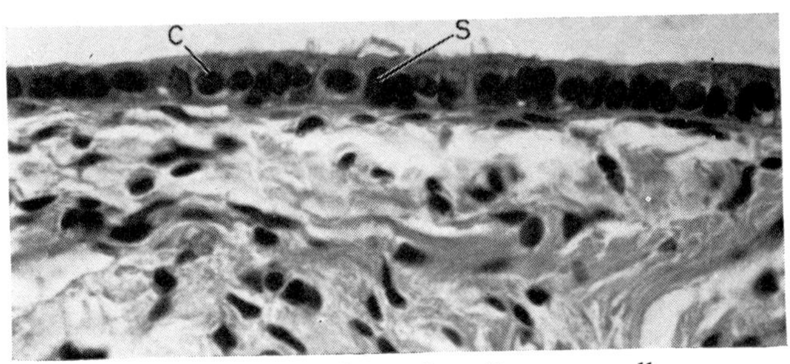

C = Ciliated cell. S = Secretory cell.

FIG. 4.20. Fallopian tube after the menopause. The mucosa is largely atrophic although the ciliated and darker secretory cells are still identified. They are now cuboidal rather than columnar.

secretory cells. Intracellular fat is at its maximum during the ovulatory period of the cycle but glycogen is most in evidence during the pre-menstrual or luteal phase.

THE MAMMARY GLANDS

The breasts have a characteristic growth curve at puberty as outlined in Chapter I. Like other tissues responsive to ovarian hormones they also have a cycle dependent upon the phases of the ovarian cycle.

The breasts consist of glandular tissue opening into a branching duct system which ultimately opens on to the nipple. These essential structures are embedded in loose fibrous tissue and fat. The fibrous tissue has to be loose to allow of expansion of the glandular tissue during activity when it may swell up to several times its non-lactating size. It should be remembered too that fat at body temperature is relatively fluid and so it too will allow of expansion of its contained epithelial elements.

The growth of the duct and glandular tissue in the pre-menarcheal period of puberty can only be caused by oestrogens. During pregnancy which is a time of progesterone relative predominance the glandular tissue is prepared for lactation, by increasing in amount. Thus has arisen

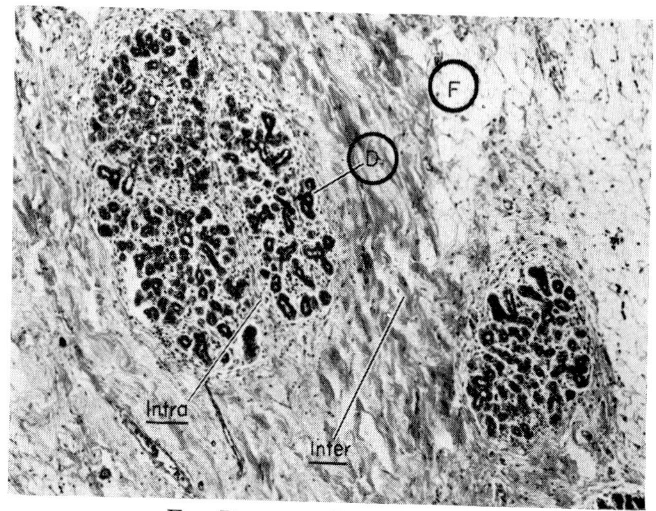

F = Fat. D = Ductules.
Intra = Intra-lobular connective tissue.
Inter = Inter-lobular connective tissue.

FIG. 4.21. Breast—adult resting. The breast consists of lobules containing many small ductules embedded in loose connective tissue. Between the lobules the connective tissue is of a more coarse pattern. Fat is abundant in the breast stroma.

the concept that oestrogen is responsible during pregnancy for the growth of the duct system and progesterone for the glandular tissue. This is an over-simplification. It is probable that during the oestrogen

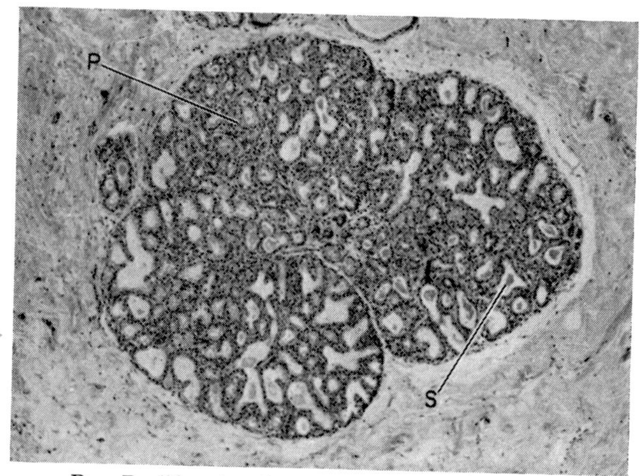

P = Proliferating ductules. S = Secretion.

FIG. 4.22. Breast—pregnancy. There is proliferation of ductules and of acini in the lobules. Some of the ductules now contain secretion.

phase of the ovarian cycle the glandular and duct tissues grow somewhat. In the progesterone phase the glands especially increase and become more lobulated and there may be some evidence of secretion. This is not accepted by all observers. Opportunities for obtaining normal breast tissue in the ovarian cycle are rare in contrast with those for obtaining tubal, endometrial and vaginal tissues. Thus comparatively little is known of the cyclical changes in the human breast.

It is worthy of passing note that only in Woman are the breasts permanent. In all other Primates the breasts only become prominent when the female is pregnant though there is probably some slight growth of the breast buds during the menstrual cycle. Of course, in domesticated

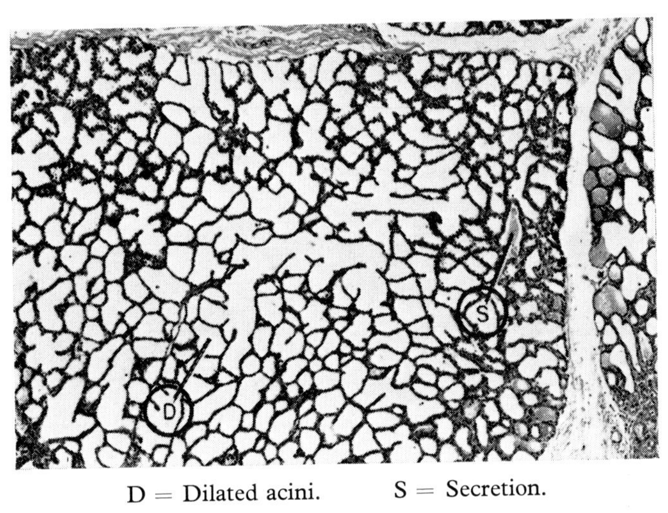

D = Dilated acini. S = Secretion.

FIG. 4.23. Breast—lactation. There is massive acinar formation and dilatation giving a cribriform pattern. When these acini are cut, the milk flows out so that most of them appear empty on section.

animals such as the cow, the udders are permanently enlarged but in them the lactating period has been prolonged by artificial selection. The permanency of the breasts in women is probably of importance in sexual selection and so may have been of importance during the course of evolution. However, it should be emphasized that the size and shape of the breasts gives very little guide to their function in lactation. Large breasts may be so because of a relative preponderance of fat whilst small breasts may be almost entirely made up of functioning tissue.

After the menopause the glandular tissue and the ducts slowly decrease so that in old age the breast may consist almost entirely of fat with very little remaining of the epithelial elements.

OTHER CYCLIC CHANGES DEPENDENT UPON THE OVARIAN CYCLE

There are many other areas of the body which feel the effects of the circulating sex steroids, but these are not so accurately detailed as those just considered. They are, therefore, best dealt with under the clinical manifestations of the menstrual cycle which form the substance of Chapter VI.

Chapter V
THE PITUITARY CYCLE

The pituitary gland through its hormones is a major controlling influence on the functions of the genitalia. The pituitary produces six hormones from its anterior lobe and two from its posterior lobe. These two lobes are better now referred to as the adenohypophysis and the neurohypophysis respectively. The constituents of these two parts are shown in the diagram. The six hormones of the adenohypophysis are thyrotrophin,

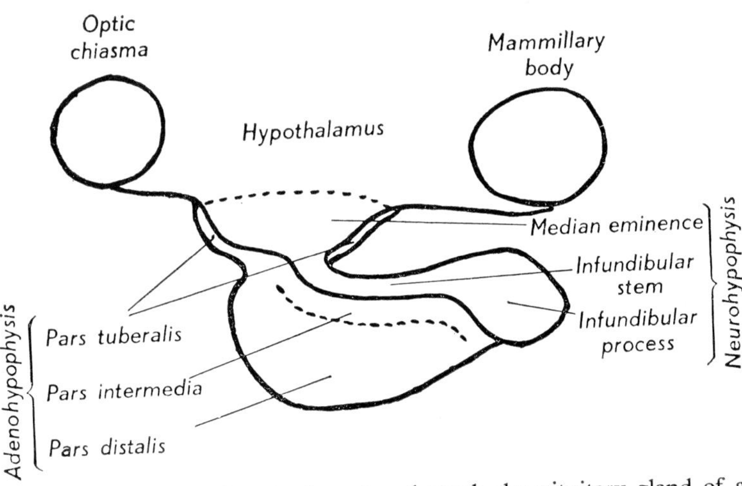

FIG. 5.1. Diagram of sagittal section through the pituitary gland of a rabbit, to illustrate the terminology of Rioch, Wislocki and O'Leary (1940). From *Neural Control of the Pituitary Gland*, by G. W. Harris. London, Edward Arnold (1955).

corticotrophin (ACTH), growth hormone, follicle stimulating hormone (FSH), interstitial cell stimulating hormone (ICSH) and prolactin. The secretions of the neurohypophysis are oxytocin and vasopressin. Other hormones are probably secreted though their function in mammals at least is not well known. Of these the melanocyte stimulating hormone (MSH) is one.

In reproductive physiology both the adenohypophysis and the neurohypophysis play a part. The hormones most directly involved are FSH,

ICSH, prolactin and oxytocin. For the present purposes of discussion of the control of the ovarian cycle only the first three are of immediate interest. FSH is obviously the hormone that controls the development of the Graafian follicle of the ovary. ICSH is the name given to the hormone that causes the interstitial cells of the testis to secrete testosterone. In the female it causes the corpus luteum to form and therefore is called luteinizing hormone or LH. Prolactin is the hormone which causes the mammary gland to secrete milk after it has been stimulated with oestrogens and progesterone throughout pregnancy and after the placenta has been delivered so that the high levels of oestrogen and progesterone have been precipitately lowered. There is little doubt about its activity in lactation, but it is also believed to play a part in the ovarian cycle. In the rat, at least, the FSH and LH acting together can cause ovulation and the formation of the corpus luteum but the corpus luteum will not secrete progesterone unless it is also stimulated by prolactin, and thus an alternative name for prolactin is luteotrophic hormone, abbreviated to LtH. The sequence of events envisaged as controlling the ovarian and menstrual cycles may therefore be depicted as:

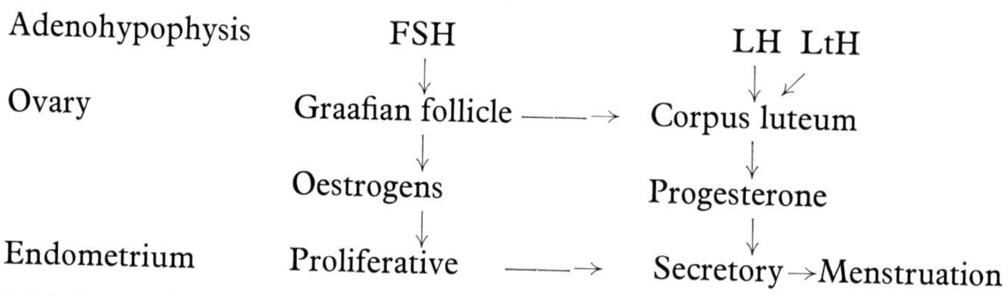

This is the first approximation to the control of the menstrual cycle, but it requires some modification and in particular some evidence to support it.

EVIDENCE FOR THE CONTROL OF OVARIAN FUNCTION BY THE ADENOHYPOPHYSIS

Most of the evidence of the control of the ovaries (and testes) by the adenohypophysis is derived from laboratory experimental animals, but experience with the human bears out most of the laboratory approach and its findings.

If the adenohypophysis is removed in an immature animal the ovaries will only come to the antral (early follicle) stage and will develop no further. Since the ovaries do not function adequately the rest of the genital tract fails to grow. If the adenohypophysis is removed in a mature animal the ovarian cycle ceases and ultimately the ovaries atrophy and the rest of the genital tract atrophies secondarily to the loss of oestrogen. Grafting the pituitary into an hypophysectomized animal restores the ovarian cycle to normal, but this only happens if the graft is put back in relation to the hypothalamus, for as will be shown later, the hypothalamus controls the adenohypophysis. Grafts into other sites such as muscle will work fitfully only, and that for only a short time. Extracts of adenohypophysis will also restore sexual cycles when injected. This is all fairly conclusive evidence of the control of the ovaries by the anterior part of the pituitary gland.

The characterization of the three hormones FSH, LH (or ICSH) and LtH has been by determining the effects of different extracts of the adenohypophysis. FSH was first discovered by Aschheim and Zondek in 1927 for they found this substance in the urine of post-menopausal women. It was shown to produce Graafian follicles in experimental animals who had been hypophysectomized, but alone, it will not cause ovulation and luteinization, nor will the follicles secrete oestrogen. In males it is believed that the FSH causes the maturation of spermatozoa but this is not yet certain. LH prepared from extracts of the pituitary is able to produce ovulation in follicles which have been primed with FSH and also LH makes the follicles secrete oestrogens. LH or ICSH is able to maintain the interstitial cells of the testis which have atrophied following hypophysectomy, and so maintain the secretion of testosterone.

The two hormones FSH and LH have been partially purified chemically. FSH is a glycoprotein containing up to about 20 per cent of glucose. In all types of FSH obtained from different species the hexose and hexosamine are firmly bound to peptides. In the sheep the molecular weight is 67,000 and in the pig 29,000. FSH is assayed biologically by its effects in increasing ovarian weight in immature hypophysectomized female rats, or alternatively by increasing the weight of the testes in hypophysectomized male rats, whilst not producing any effect on the rest of the genitalia. This last proviso is needed because it is difficult to obtain FSH free from all ICSH activity. ICSH (LH) is a mucoprotein, meaning that the molecule contains more than 4 per cent of hexosamine. It also contains the carbohydrate mannose and so is a glycoprotein. Its molecular weight in the sheep is between 31,000 and 40,000 whilst in the pig its molecular weight is 100,000. It is biologically assayed by the increase in weight of the uterus and corpus luteum formation in immature rats who have been primed with FSH, or by the repair of ovarian interstitial cells in hypophysectomized rats, or by the increase in weight of the seminal vesicles in immature male rats, or by the increase in weight of the ventral lobe of the prostate in hypophysectomized male rats. It will be appreciated that the seminal vesicles and the ventral lobe of the prostate are dependent on the supply of testosterone from the interstitial cells of the testis. The chemical variations in FSH and LH of the pituitary and their ability to offset the effects of hypophysectomy are further evidence of the activity of the adenohypophysis in normal animals.

The role of prolactin or LtH in the sexual cycles of many mammals is still a subject for speculation. If it is injected into female rats at the end of oestrus there is no doubt that it will prolong the secretion of progesterone from the corpus luteum. Because of this observation it has been widely assumed that it has the function of causing the corpus luteum to secrete progesterone in all mammals including Man. This cannot yet be accepted as proven. It has been found to be a crystalline protein and has been isolated in pure form. It also is a glycoprotein with a molecular weight of about 26,500 to 32,000. Not all authorities agree even that it contains bound carbohydrate. Assay is by observing the effect of the hormone on increasing the weight of the crop gland of the pigeon.

In the past few years gonadotrophins have been used therapeutically in women to provoke ovulation and corpus luteum formation. The results have been highly successful and many pregnancies have resulted. From this it may be concluded that the general outline of the function of the pituitary in regulating the ovarian cycle is essentially correct for women.

THE CONTROL OF THE ADENOHYPOPHYSIS BY THE OVARY

The periodicity of sexual cycles in all mammals is of immense interest because environmental factors are capable of triggering off the mechanisms whereby animals breed at the most propitious time for the rearing of offspring. Moreover, there is the additional interest of the delay in reaching sexual maturity that is the characteristic of all mammals. In lower mammals there is little doubt that external factors such as nutrition, temperature and the amount of light condition the breeding cycle, but in Man it is difficult to show that any external factors have an such effect. The periodicity seems to have been withdrawn from the environment and incorporated within the body, so freeing the reproductive capacity from full dependence upon the exterior. This may well have had great evolutionary advantage.

The periodicity of the ovarian cycle is dependent upon the pituitary cycle, but the ovary in part determines the rhythmic activity of the pituitary. When ovarian function fails after the menopause the output of FSH rises, indicating increased adenohypophyseal activity. The FSH output can be diminished by giving the woman oestrogens. It is known in animals that if the ovaries are removed there are changes in the histology of the adenohypohysis. So-called "castration cells" appear, which have been identified by histochemical techniques. The techniques show the presence of acidophil and cyanophil cells in the pars distalis. The gonadotrophin secreting cells are cyanophil and by the appearance of castration cells they seem to be grouped mainly round the periphery of the pars distalis and also in a narrow band dorsally. So distinct is this area that it has been called the sex zone. Other cyanophil cells believed to be secreting thyrotrophic hormone are situated more deeply within the substance of the gland. By more erudite studies still, it is believed by some that special cells can be found which secrete LH(ICSH), and in general these are near to those cells secreting FSH. Castration cells can be made to disappear by treatment of the experimental animal with oestrogens or by implanting ovarian tissue. The inter-relationships between the oestrogens and FSH seem to be clear. The control of LH and LtH secretion are by no means so clear but the following theory is the one most in vogue at the present time.

Graafian follicles are brought to maturity by FSH. They are probably made to secrete oestrogens by the activity of LH as well. The two hormones, FSH and LH, acting synergistically are responsible for causing ovulation and the formation of the corpus luteum. The rising output of oestrogen from the follicles is responsible for altering the quantitative outputs of the gonadotrophins from the adenohypophysis and especially by lowering the output of FSH. At the same time the rising oestrogen probably stimulates the production of LH and LtH, the latter being necessary to make the corpus luteum secrete its progesterone. As the progesterone level in the blood rises it is believed that it diminishes the activity of the pituitary so that it decreases its output of LH and LtH. Therefore, the corpus luteum loses its endocrine support and declines. When it dies menstruation ensues. Whilst it is suppressing the formation of LH and LtH the progesterone is also probably stimulating the adenohypophysis to produce FSH in readiness for the next cycle. Thus the cycle is completed by a process of positive and negative feed-back with the pituitary and the ovary each maintaining the other's cyclic activity. The endometrium reflects the cyclic output of the ovarian hormones by proliferation, secretion and finally menstruation.

The output of oestrogens and of pregnanediol in the urine have been discussed in Chapter III. In general the hormone values of gonadotrophins found in blood and urine in women support the concept outlined here, though it has not been possible to characterize FSH and LH separately, and most work on serum and urine deals mainly with FSH. The output of this gonadotrophin does rise to a peak at ovulation and then declines, so that in the urine there are about 6 mouse uterine weight units per 24 hours at the 7th day, 20 units at the 14th day, 6 units at the 21st day and almost none at the 28th day. This does not fully correlate with the theory of pituitary-ovarian relationships just given, but when it is realized that all the estimations are difficult, that the serum chemistry of the hormones is still in its infancy and that urinary outputs of hormones may only give a very indirect picture of what is happening in the blood and in the glands themselves it is scarcely surprising that the theory is still incomplete. However, it is still a reasonably useful guide in clinical practice.

THE CONTROL OF THE ADENOHYPOPHYSIS BY THE HYPOTHALAMUS

One of the questions raised earlier is that of the sexual quiescence of the immature individual, and how it is that sexual maturity is delayed until full growth has been attained. Animals without ovaries do not reach sexual maturity; nor do hypophysectomized animals. But if immature ovaries are grafted into mature animals they function as if they were adult. Therefore the delay in the appearance of puberty cannot reside in the ovary. Similar experiments with the grafting of the pituitary show that the immature gland functions perfectly well if grafted beneath the hypothalamus of a mature animal. Therefore the delay in reaching puberty does not have its origin in the adenohypophysis. The delay must reside in some further control of the pituitary-ovarian axis. This has been shown to be in the hypothalamus which controls the adenohypophysis through a portal system of vessels running from the median eminence into the anterior lobe of the pituitary. Section of these vessels together with the insertion of a plate to prevent their regeneration suppresses the sexual cycles. Destruction of the hypothalamus also does so. Stimulation of the intact hypothalamus causes the appearance of a sexual cycle and other sexual activity. Implantation of oestrogen into the hypothalamus will cause a prolonged oestrus. These various experiments show that the hypothalamus and pituitary function as a unit, and it is now believed that the delay in the appearance of puberty is, in some way unknown, a function of the nervous system. Further evidence of this is derived from the rare cases of precocious puberty induced by tumours of the hypothalamic region. It may be that at puberty the hypothalamus is stimulated from some other areas of the brain, or it could be that the hypothalamus is in some way released from inhibition which has been applied from some other part of the brain. Recent work has shown that the amygdala in the base of the lateral ventricle may have some control of the hypothalamus. When it is destroyed ovulation may ensue and it may be this area which inhibits the hypothalamus, assuming that the hypothalamus has an autonomous rhythmic function of its own. This would appear to be unlikely since there is no periodicity in male animals and there is not the fluctuating output of gonadotrophins seen in the female. This is an unexplained phenomenon because if the pituitaries of males are grafted into females under the hypothalamus the sexual cycles are maintained. Alternatively if female pituitaries are grafted into males there is no periodicity in testicular function. It would seem that the periodicity of the sexual cycles in females is a function of the nervous system and probably of the hypothalamus itself. Experiments with rats have shown that if testosterone is implanted in the hypothalamus within a day or two of birth there will be no periodicity in the sexual cycles later in life in the female. It would seem that the basic sexual activity is cyclic but that the rhythm is changed to continuous activity by male sex hormones acting on the brain. This idea gives some meaning to the observation that testes are recognizable as such in the embryo of 7 weeks but that the ovaries are not fully differentiated till the 16th week. The imprinting of the developing brain by male sex hormones probably therefore begins very early and continues throughout intra-uterine life and beyond.

The waning of the sexual cycles in the climacteric would seem to be, however, a function of the ovaries. Ovulation and corpus luteum formation cease and because of this the output of FSH is high. The pituitary is still able to respond by excessive activity to the presence of low oestrogens.

There is evidence from some animals of the intervention of the nervous system in reproductive processes. The rabbit, cat and ferret only ovulate after coitus. Ovulation in them can be prevented if the spinal cord is sectioned within a few minutes of intercourse to prevent stimuli reaching the brain. However, if the section is delayed, ovulation still takes place though there is a latent interval of several hours. The latency is believed to be due to the delay in the formation of hormones in the hypothalamus and in the adenohypophysis on which the ovulation depends. Probably the nerve impulses trigger off changes in the hypothalamic area. Further recent evidence suggests that the endometrium may produce a hormone which affects the hypothalamus. If in the sheep the uterus is removed during the luteal phase of the cycle then the life of the corpus luteum is much prolonged. (R. V. Short). It is suggested that the endometrial hormone, if such it be, is the cause of diminishing secretion of LH and LtH, perhaps through the mediation of the hypothalamus. If this inhibitor is removed by hysterectomy the LH and LtH continue to be secreted and so maintain the corpus luteum in being.

FSH and LH seem to be secreted under the influence of special hormones coming from the hypothalamus along the portal system of vessels. They have been called release factors, so there is FSH-RF and LH-RF. On the other hand prolactin or LtH seems to be secreted autonomously by the adenohypophysis and its production is normally suppressed by an inhibitory factor coming from the hypothalamus. There is clinical evidence of this in the relatively rare Chiari-Frommel syndrome, when damage to the hypothalamus releases the adenohypophysis from inhibition and the affected woman goes into permanent lactation as a result of the continued secretion of prolactin.

CLINICAL ASPECTS

An outline concept of the menstrual cycle in women has been given. It will be realized that the concept falls far short of a full explanation of all that happens and the cycle involves complex interactions of the cortex of the brain, the hypothalamus, the adenohypophysis, the ovaries and the endometrium. But for all its shortcomings it is necessary for some understanding of clinical situations. These usually present as some upset in menstrual rhythm or scantiness or heaviness of the menstrual flow. These may occur about puberty, during reproductive life or at the climacteric. Firstly in investigation it is necessary to exclude pathological changes in the genitalia themselves. When this is reasonably assured the upset must be in some disorder of the physiological control of menstruation. From what has been said this may reside in the ovaries, the adenohypophysis, the hypothalamus or some other part of the nervous system. It is unfortunate that assays of gonadatrophins are still not possible except experimentally. Moreover they are techniques of bio-assay which tend to be unreliable. Even the sex steroids which can be estimated chemically cannot be fully investigated in most hospitals because of cost. Therefore one has to be content at the moment with well-tried clinical methods of investigation together with inspired guesses based on a knowledge of the theory of the control of menstruation. One special investigation is very helpful and that is curettage of the endometrium. If this is found to be in the secretory phase then it is a fair assumption that a corpus luteum has been formed and that therefore ovulation has occurred. This would probably not have been possible had there not been an output of FSH, LH and possibly LtH. Therefore it may be assumed that the adenohypophysis and its controlling mechanisms have been properly functioning at least in the cycle being investigated. It may be that other cycles are not being properly controlled and one may have picked the one normal cycle among many abnormal ones to investigate. This is a problem of gynaecological investigation that is difficult to solve.

A further point of difficulty in the clinical situation is that endocrine disorders of the thyroid, adrenal and sometimes the pancreas may upset the function of the central menstrual axis consisting of brain, hypothalamus, adenohypophysis, ovaries, endometrium. These must all enter into the differential diagnosis of menstrual disorder. Also it is well known that emotional factors can cause menstrual upsets and this too demonstrates that the brain has an effect on the menstrual axis.

The changes in the endometrium depend upon the release of oestrogens and progesterone from the ovary. Both of these substances or their derivatives can be used therapeutically and the endometrium forced to go through a sort of menstrual cycle. This may be valuable, but when these steroids are given they suppress the functions of the ovary and of the pituitary, so that ovulation does not occur and natural hormones are not produced or at least are produced in diminished amount. However, sometimes it is believed that the cyclic use of oestrogens and progesterone may exercise some effect upon the pituitary and so force it to produce gonadotrophins. One oestrogen-like compound called clomiphene seems to have been shown to do this. Only recently has it been possible to use the gonadotrophins therapeutically. They are difficult to use and very expensive so that they are still very much in an experimental stage. They have been reserved for women who want to have children and yet have some degree of ovarian failure in that they do not ovulate, though often they do produce some oestrogen. The treatment when carefully worked out has been very successful and women have been made to ovulate and produce enough progesterone to maintain a pregnancy. There can be no better criterion of the normal functioning of the genitalia than a successful pregnancy, which is what is desired by most women.

Chapter VI

CLINICAL FEATURES OF THE MENSTRUAL CYCLE

There are many concomitants of the menstrual cycle which occur in normal women. Some of them are very helpful in understanding the clinical situation, and may for instance be used to predict whether or not ovulation is occurring.

THE MENSTRUAL FLOW

The normal amount of loss from the uterus at menstruation is very variable but usually lies between 50 and 150 ml. It consists of blood, endometrium and mucus. In normal menstruation the blood does not clot. It probably does clot within the uterine cavity but is then lysed by the enzyme fibrinolysin which is present in the endometrium. If the blood loss is unduly heavy then the shed blood does clot. The reason probably is that there has not been enough fibrinolysin produced to liquefy the clot, or the enzyme has not had sufficient time in which to act.

In bleeding disorders such as thrombocytopenic purpura or von Willebrand's disease it might be expected that women would lose large amounts of blood at each period, but in fact this does not occur. This observation suggests that there are special mechanisms for the control of loss of blood in the uterus which are not operative in other sites. Little is known of them. On the other hand it is suspected that some women may lose excessively at menstruation because of local defects in the clotting mechanism. The evidence is as yet tenuous, and difficult to investigate since there are at least twelve factors concerned in clotting. But therapeutically the use of epsilon-amino-caproic acid (EACA) may help to diminish excessive losses of blood in some women. This substance is known to stimulate some aspects of the complex clotting mechanism in other sites.

If a woman loses about 50 ml. of blood at menstruation this is equivalent to a loss of about 25 mgm. of iron. If she loses 150 ml. of blood this means a loss of about 75 mgm. of iron and she may then be very likely to suffer from iron deficiency anaemia if the iron in her food is below 7 mgm. per day and such relative dietary deficiency is not too uncommon.

The loss of electrolytes at menstruation is very small, amounting to about 5 mEq. of Na^+ and 5 mEq. Cl^- and 2·5 mEq. K^+ in a total loss of 50 ml.

THE LENGTH OF THE CYCLE

The length of the menstrual cycle is taken as the number of days from the first day of one period to the first day of the next. This is because only the first day can be fixed with any accuracy. The last day of the period is often one on which there is diminishing flow of blackish altered blood of small amount. Such bood may have been shed from the uterus some time before and now be escaping from the crevices and folds of the vagina. The average length of the cycle is about 28 days, but there is great variation in this. Even women who believe that they are menstruating every 28 days often are found to have a cycle length varying between about 25 and 35 days when they keep careful menstrual calendars. Individual women may also have perfectly normal cycles of 5 or 6 weeks or even up to 3 months, and at the other extreme many have a cycle length of 21 days. Each woman's standard is herself and she must not be compared with some theoretical norm. The essential function of the genitalia is reproduction, and cycle length is not necessarily a guide to fertility, though of course infrequent periods suggest infrequent ovulation and so less opportunity to become pregnant.

THE DURATION OF THE PERIOD

The average length of a period is about 5 days. But here again there is great variation. The range of normality may be from 1 to 8 days and as with the length of the cycle each woman is her own standard. Variations from her own norm may be of importance. The duration of the flow is not necessarily a guide to fertility.

THE MENARCHE

The average age of the menarche is about 13 to 14 years. But the range of normality is from 9 to 17 years. The factors involved in the onset of the periods have been investigated for some time. It seems fairly well established that the age of the menarche has decreased over the last 150 years from an average of about 17 years to its present level, and moreover it would appear that the age at which the first period occurs is going down by about 4 months in every 10 years in this country at the present time. Roughly the same is true in all countries where studies have been made. There is no foundation for the belief that girls in hot climates menstruate at an earlier age than they do in the more temperate climates, and the best guess at the reason for the falling age of the menarche is improved general nutrition and perhaps other social factors. Some work in Eastern Europe has suggested that if a girl normally lives in a city she will tend to begin menstruation in the winter, whereas if she lives in a country district the menarche will tend to be in the summer months. There also seems to be a correlation with the height above sea level at which she lives. The higher the altitude the more will the menarche be delayed and it may be delayed by three months for every 100 metres of altitude. It is said that girls blind from birth menstruate on average earlier than the normally sighted. In lower animals it is well known that light has an effect on the sexual cycles via the retina and brain. There is also some genetic control of the time of the menarche and this is partly shown by the fact that identical twins will menstruate for the first time within three or four months of each other, whilst unlike twins may show a variation of up to eighteen months or more.

The first few cycles after the menarche are often associated with failure to ovulate and so there is an era of relative infertility should intercourse take place at this time, as it may in some primitive human groups and even in our more sophisticated society.

THE MENOPAUSE

The age of the menopause varies between about 45 and 55 years. Until recently it was thought that the average age was about 48 years, but evidence is accumulating that the average is rising slightly and may now be about 50. With the falling time of the menarche this means that the potential reproductive life of women is increasing. Also in the past it was believed that if a girl started to menstruate early then her menopause would be relatively late. There is no evidence to support this view and the menarche seems to bear no demonstrable relationship with the menopause.

The way in which the periods come to an end is very variable. Rarely they just stop. More often there is some irregularity. The essential failure in the genital system is that of ovulation which ceases before the endocrine function of the ovaries. The endometrium may therefore be subject to fluctuations in the ovarian output of oestrogens with consequent effects on the endometrium. Whether or not any given pattern of menstruation is abnormal must be a matter of individual determination and requires the help of a gynaecologist.

In the male there is no proven climacteric, though libido and fertility slowly decline over the years, but not uncommonly potency and ability to produce children continues into old age.

OVULATION PAIN

Almost all women at some time in their lives experience pain in one or other iliac fossa at the time of ovulation. It is often called Mittelschmerz which is the German for "middle pain". This is its characteristic for in a 28-day cycle with ovulation occurring on the 14th day the pain is felt in the middle of the cycle. If the cycle is longer the pain is usually felt about 14 days before the next period, since the life of the corpus luteum is about 14 days and variations in the length of the cycle are nearly always due to changes in the length of the follicular phase and not the luteal. It is thought that ovulation usually occurs from each ovary alternately and a few women do feel Mittelschmerz in alternate iliac fossae, but some always feel it in one side with each cycle and some feel the pain in alternate months in one side only.

The cause of the pain is not known. Theoretically it could be due to the rupture of the follicle on the surface of the ovary or to the slight amount of blood that inevitably is shed into the peritoneum at ovulation. Alternatively it might be due to an increase in peristaltic activity of the Fallopian tube on the same side as the ovulating ovary. None of these explanations is entirely satisfactory, especially since a woman may suffer from the pain for several months on end and then it disappears and whether it recurs or not depends on no discoverable causes. The pain is not usually severe and lasts for a few hours and sometimes for a whole day.

OVULATION BLEEDING

In all women red blood cells may be found in the cervical mucus by microscopic examination at the time of ovulation. In some women, and indeed in nearly all of them at some time in their lives, the bleeding may be sufficient to become obvious to the naked eye, and necessitate wearing a sanitary pad. The bleeding is usually slight in amount and only a drop or so. Sometimes it may be more and last for a day or two, but in general it lasts for only half a day.

It is almost certain that the blood found at this time derives from the endometrium. It will be remembered from Chapter III that oestrogen production from the ovary rises to a peak at ovulation and then falls quite rapidly until it rises again as the corpus luteum comes to maturity. There is no doubt that bleeding from the endometrium can be caused by a rapidly falling oestrogen level and this is probably what is taking place in a woman suffering from ovulation bleeding.

The phenomenon is of interest evolutionarily because in most mammals, except for the Primates, bleeding at about the time of ovulation is the characteristic feature of oestrus, and it is this bleeding that tells the stockbreeder when to mate his animals or keep them away from males if breeding from a particular animal is not at that time required.

OVULATION MUCUS

The secretion of the cervical glands changes throughout the menstrual cycle under the influence of the ovarian hormones (Chapter IV). Especially the mucus increases in amount and becomes crystal clear and "runny" at the time of ovulation. Many young women notice this increase in vaginal discharge, which is also added to by an increase in cellular desquamation from the vaginal walls. The crystal clear mucus can often be seen through a speculum cascading from the external os of the cervix at this time in the cycle.

Before ovulation the cervical mucus will crystallize in the presence of normal saline to a fern-leaf pattern which is easily recognizable under the microscope (see Fig. 4.13). After ovulation when the cervical glands are under the influence of progesterone the crystallization pattern disappears and only amorphous debris is visible.

DYSMENORRHOEA

Dysmenorrhoea means difficult menstruation. In practice it comes to mean pain with the periods. Such pain is felt by virtually all women at some time in their lives, and especially from the ages of about 16 to 25, but mainly in the age group 17 to 20. It is not usually in evidence for the first year or so after the menarche and it is believed that the reason for this is that ovulation may not occur at this time. It seems certain that dysmenorrhoea only occurs when the uterus is under the influence of progesterone, and this only happens when there is a corpus luteum which can only form if ovulation has occurred. The pain is a cramp-like one and the height of the pain probably corresponds with the uterine contractions. The pain is felt just above the symphysis pubis and in the lumbo-sacral region. Sometimes in severe cases pain is referred down the inner side of the thighs towards the knees. It comes on during the first day of the period and lasts for about 8 to 12 hours though there is much individual variation.

The cause of the pains felt with the periods is still not elucidated. There is no doubt that progesterone increases the force of uterine contractions when these are compared with those of the oestrogen phase, but the pressures developed within the uterine cavity are only of the order of 15 to 20 mm. Hg. However, the intramuscular pressure might be a great deal more and sufficient to cause ischaemia and so pain. Probably ischaemic pain is the best explanation of the clinical phenomena, though there are many other theories such as partial blockage of the cervix, and allergies especially to some hypothetical toxin developed during the breakdown of the endometrium. Abnormalities of the uterus have been proposed as causes but in by far the majority of women suffering from the disorder none can be demonstrated. The theory of cervical obstruction is likewise untenable since a sound can be passed through the cervix under anaesthesia and in any case the menstrual fluid is sufficiently liquid to flow through quite a narrow canal. When the menses are associated with clots, which are said to be hard to pass through the cervix, it is almost certain that they form in the vagina on the distal side of the hypothetical block, since the clots may have quite a large volume and the cavity of the uterus has a volume of only about 7 ml.

All pain depends on central nervous system connexions, and particularly those with the thalamus and cortex. There is no doubt that the degree of suffering which a given painful stimulus causes is dependent on the previous upbringing of the sufferer. This is especially true in dysmenorrhoea, and psychological factors loom large in this condition. This does not deny the reality of the pain, which is real whether it is generated or added to by the psychological state of the patient. The ways in which the mind interferes with physiological function are still largely unexplored though the relationships of body and mind are of great importance in the practice of medicine.

Daughters of dysmenorrhoeic women often suffer from painful periods, and girls who work together often influence one another's attitudes so that the numbers of women suffering from dysmenorrhoea and needing time off work may steadily rise in an office or factory. Those patients who marry, often lose their dysmenorrhoea shortly after marriage, and it is very uncommon for dysmenorrhoea to persist after the birth of a baby. Of course it is possible that both marriage and childbirth bring about physiological changes in the genitalia, but it is certain that they bring about psychological changes.

PREMENSTRUAL TENSION

It is not uncommon for women in their teens and twenties to suffer from a series of symptoms during the premenstrual week or fortnight. This periodicity is evidence that the symptoms are due to the action of progesterone mainly and it is well established that they do not occur when there is no ovulation.

The main symptoms of premenstrual tension are referable to the higher levels of the nervous system and include headache, irritability, depression, lack of concentration and poor memory. Disturbances of behaviour may also be seen for it has been shown that suicide is commonest in the luteal phase of the cycle, and in girls' schools punishment is more often meted out in the second half of the cycle than during the first half. Various crimes such as shop-lifting and prostitution are more often found in women in the luteal phase. This may be because they are more easily detected at this time rather than because of an increased propensity to commit the crimes.

The breasts may often be tingling and painful during the premenstruum and this may be associated with the central nervous system changes. The cause of the discomfort is not known but may be due to increased vascularity or enlargement of the glandular tissue itself. The condition is often called premenstrual mastalgia. There is no evidence that these breast pains or the symptoms of tension are any commoner in women known to be neurotic, though as with all syndromes involving a whole person the psychology of the individual has a part to play.

Some women feel "bloated" in the abdomen during the premenstruum, and some develop slight oedema of the feet. The bloating may be due to an increase in the gas in the alimentary tract, but some women seem to increase the amount of tissue in the lower half of the abdomen at this time. This may be due to retention of fluid. Studies of water and electrolyte balance during the cycle are still not conclusive. Some believe that there is an unusual gain in weight in the week before menstruation and this may amount to 1 to 3 lb. This weight is lost by an increased formation of urine during the period. Experimentally in animals it can be shown that the sex steroids do have an effect on the renal handling of water and of sodium. In general they cause a retention of sodium in the tubules, but this is not invariable and under some circumstances progesterone may cause an increase in the excretion of sodium. If sodium is retained, and some studies with radioactive sodium in women tend to suggest that it is, then water will be retained to maintain isotonicity of the body fluids. Other investigations have shown that weight gain in women is not always to be found in the premenstrual time but is spread evenly throughout the cycle and so the matter is still *sub judice*. It would seem that there is no necessary relationship between premenstrual tension and any water and salt retention that there may be.

Attempts have been made to correlate sexual feeling or libido with the events of the menstrual cycle. On *a priori* grounds it might be expected that libido would be greatest at the time of ovulation so that successful fertilization might take place. This would accord with the oestrus of animals. In fact there seems to be no general pattern of libido with the cycle. Some women undoubtedly do have an increase in sexual appetite at ovulation but just as many feel increased libido at other times, such as just after or just before a period, and some experience no set cyclic changes in libido at all. Sexual responses are more conditioned by psychological factors than by hormones.

The body temperature is elevated by about 0·6 to 0·8° Fahrenheit during the second half of the menstrual cycle. This is due to some action of progesterone, the exact nature of which is not known, though it is likely that it must be at some point in the hypothalamus. At menstruation the temperature falls to basal levels again. The rise in temperature can be generated by injections of progesterone.

About 10 per cent of all married women will complain about an inability to conceive at some time in their lives. The first essential for conception is the production of an ovum so that this must be proved as far as possible. There are no definite proofs of ovulation except pregnancy, but if it can be shown that progesterone is being produced then this is evidence of ovulation without which a corpus luteum does not form.

Rarely Mittelschmerz may give a clue as to ovulation, but all other tests depend on the presence of progesterone. This can be inferred from dysmenorrhoea, premenstrual tension and mastalgia, changes in cervical mucus, the rise in temperature at mid-cycle, the demonstration of significant amounts of pregnanediol in the urine or the histological demonstration of secretory changes in the endometrium. A further sign may be that of the regular recurrence of acne on the face and back especially.

Clinically the menopause may be associated with mental changes and also with the so-called "hot flushes". These are felt as a hot sensation beginning in the upper chest region and rising into the face. The skin may become quite red due to the dilatation of skin vessels. Very soon the sensation of heat passes off to be replaced by one of coldness and there may be beads of sweat on the forehead, face and chest. It is probable that these effects are caused by a low level of circulating oestrogen, though some have attributed the symptoms to the high level of FSH found at this time. However, "hot flushes" do not occur in Turner's syndrome with ovarian agenesis when the level of oestrogens is low and the FSH high. It would seem that the response can only be obtained in a patient whose vascular system has been conditioned to the presence of oestrogen in fairly large quantities. It seems most probable that the "hot flushes" are mediated through the vasomotor centre rather than through a direct effect on the blood vessels. This is still a further piece of evidence of the profound effect of the sex steroids on central nervous system function.

Chapter VII

THE PRODUCTION OF GAMETES

Samuel Butler is credited with the saying that "a hen is an egg's way of producing another egg". The species, and the whole of evolution, is dependent upon the constant reshuffling of the genetic material of which it is the storehouse. Without the formation of gametes and their subsequent fusion at fertilization life grinds to a halt, and if the genetic material is not rearranged evolutionary progress stops and the species is liable to extinction if there should be comparatively slight changes in the environment.

Early in development of the embryo special cells are laid aside in the region of the root of the yolk-sac. In Man this can be recognized at about 24 days after fertilization. During the course of the next week they migrate through the dorsal body wall and ultimately come to rest in the gonadal ridge in the region of the pronephros and the developing adrenal gland close to the dorsal mesentery of the gut and near to the aorta. The cells can be observed by time-lapse photography in the mouse to move in amoeboid fashion. When they have reached the gonadal ridge sex cords of epithelium from the surface of the gonad grow in to reach them and incorporate them in their epithelium. At this stage the gonad is indifferent and it cannot be said whether it will develop into a testis or an ovary. It is probable that the direction of development of the primitive gametes is partly determined by the environment which they meet in the gonadal ridge. Throughout the vertebrate kingdom the gonad is always to be found next to the coelomic cavity and close to the excretory ducts and the adrenals. These elements are greatly concerned with the internal environment particularly as regards water and electrolytes.

The embryonic testis is recognizable at 37 days after fertilization because the sex cords condense to form seminiferous tubules and become separated by mesenchyme, and also because of the growth of the thick tunica albuginea which forms under the coelomic epithelium. By about 98 days from fertilization the sex cells are recognizable as spermatogonia.

The embryonic ovary is not recognizable as such till later than the testis, though by 37 days from fertilization the sex cords are not so distinct as they are in the testis and there is no development of the tunica albuginea. It is not till 16 weeks (112 days) from fertilization that primitive ova can be recognized and these ova are called oögonia. They are surrounded by primitive granulosa cells. Later still, at about 20 weeks from fertilization (140 days) the stroma of the ovary develops greatly. The ovaries can soon be seen to be stuffed with ova (oögonia) in a dense stroma.

At birth the testis has an appearance similar to that of the adult, though often it has not yet fully descended into the scrotum. Spermatozoa are not produced until puberty, but the seminiferous tubules are thereafter able to produce spermatozoa for the rest of the life of the individual.

The ovary at birth is as just described and it does not begin to produce many Graafian follicles until just before puberty, and ovulation tends not to occur until a year or so after the menarche. The ovary contains at birth all the sex cells it will ever have and they are not added to throughout life. The number of oögonia at birth has been estimated as about 800,000. Allowing for a reproductive life of say 40 years with ovulation occurring every month only 480 ova will be expelled from the ovary. The rest become atretic. This is one expression of the prodigality of nature with sex cells. In all species they are produced in their thousands, and even millions with spermatozoa, but very few are able to come to fruition at fertilization.

SPERMATOGENESIS

The basic anatomy of the testis is shown in the figure. The testis is divided into compartments by septa. Spermatozoa are produced by the epithelium in the convoluted parts of the seminiferous tubules. From here they pass into the lumina of the tubules, traverse the rete testis, the efferent ducts and epididymis before they reach the vas deferens. From here they go past the seminal vesicles and are ejaculated through the prostatic and more distal urethra.

A cross section of a seminiferous tubule shows that there is a basement membrane lined by spermatogonia. In successive irregular layers within this are primary spermatocytes, secondary spermatocytes, spermatids and spermatozoa. The essential point in the process of the formation of spermatozoa is between the primary and secondary spermatocytes, for it is here that meiosis, reduction division, occurs. The spermatogonia and primary spermatocytes therefore contain the diploid number of chromosomes, 46, whereas all stages after this have only the haploid number, 23. Moreoever, half the spermatozoa contain an X chromosome and half a Y chromosome, so that the haploid number is really 22 + X or 22 + Y. The spermatozoon is the determinant of the sex of the individual at fertilization in the human.

Embedded within the layers of the seminiferous tubule are the Sertoli or nurse cells. As the spermatozoa come to maturity they attach themselves to these cells before being cast into the lumen. They are especially interesting cells for there is some evidence that they may be a source of oestrogens, certainly in some cases of intersex. The main endocrine secretion of the testis, testosterone, comes from interstitial cells embedded between the seminiferous tubules. They become especially prominent if the tubules die, or are atrophic. The interstitial cells are recognizable by their slightly yellow colour, probably due to their content of steroid. Also they may show Reinke crystals, which under some circumstances are visible in the ovary too.

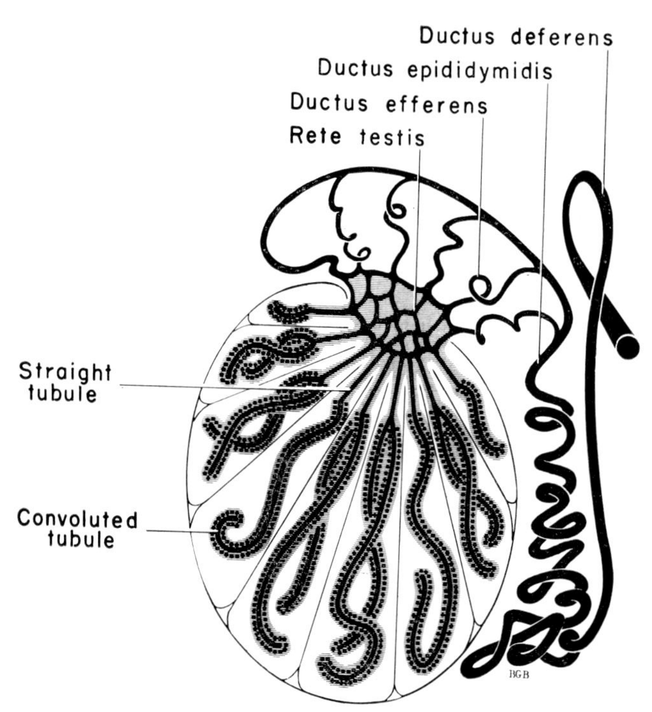

FIG. 7.1. From *Greenhill's Obstetrics*. W. B. Saunders Co. Pa. and London (1965).

Each primary spermatocyte gives rise to two secondary spermatocytes and each of these to two spermatozoa, that is each spermatocyte of the primary kind gives rise to four cells. This is comparable with the production of four cells from the primary oöcyte, but whilst the cytoplasm is divided evenly between spermatozoa this is not the case with the ova.

OÖGENESIS

For the sake of simplicity the ovary at birth has been described as being full of primitive germ cells packed in stroma. This is not exactly true and in fact there are the beginnings of primordial follicles progressing

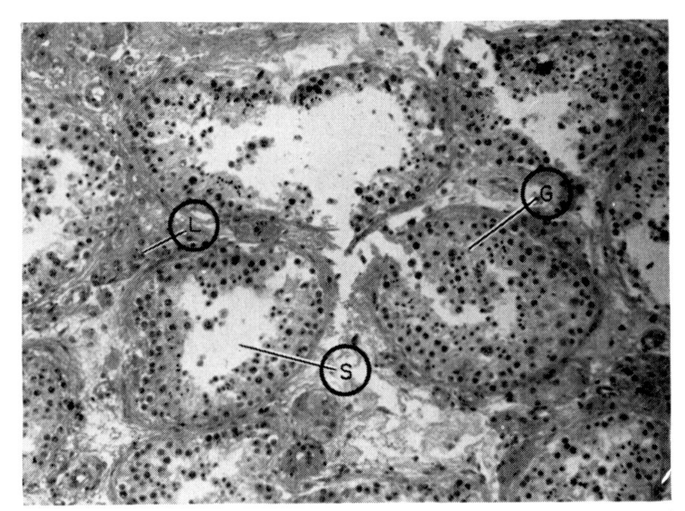

L = Leydig cells. G = Germinal epithelium. S = Seminiferous tubule.

FIG. 7.2a. Testis. The testis consists of seminiferous tubules set in loose connective tissue. The seminiferous tubules have a thin surrounding layer of fibrous tissue. Within the tubules is the germinal epithelium maturing into spermatozoa. The interstitial connective tissue contains Leydig cells, the source of androgenic steroids.

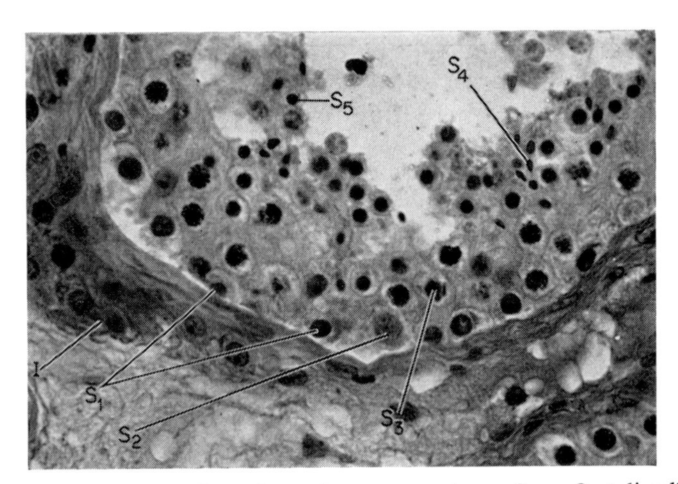

I = Interstitial cell. S_1 = Spermatogonia. S_2 = Sertoli cell.
S_3 = Spermatocyte. S_4 = Spermatozoa. S_5 = Spermatid.

FIG. 7.2b. Spermatogenesis. The basal layer of cells in seminiferous tubule consists of spermatogonia and Sertoli cells. The spermatogonia proliferate to give primary spermatocytes. These undergo a reduction (meiotic) division to form secondary spermatocytes which in turn divide to form the spermatids. The spermatids mature into spermatozoa.

towards the antral phase. This proceeds throughout childhood and the only essential difference between a pre-pubertal and a post-pubertal ovary is that the post-pubertal one shows the phenomenon of ovulation, that is each month one ovum is shed from the surface of the ovary. It will be realized that the change from the immature ovary to the mature is a gradual one with changing emphasis on the numbers of follicles and the amount of granulosa cells and stroma. The reduction division of meiosis takes place before the formation of primordial follicles so that all the oöcytes have the haploid number of chromosomes through childhood and even before birth, or more exactly the oöcytes are in a resting phase of meiosis, that is pachytene, in which the chromosomes are coiled round one another.

During reproductive life the oöcytes in their follicles have to undergo maturation and ripening before they are in a state preparatory to ovulation and ultimately to fertilization. The first obvious changes are in the cytoplasm which increases in size. The ovum must apparently carry its own store of nutrient materials before it becomes nourished by the maternal circulation. This is in accord with the ova of all vertebrates. In reptiles and birds in which development of the embryo is external to the body of the mother the egg must carry with it the whole of the nutritional requirements of the developing offspring. This is only needed for the very earliest stages of development in the mammals, for it is only at this very early time that the embryo is dependent mainly upon its own resources. Very quickly usually its needs are supplied by the mother's uterus either from the secretory glands of the endometrium or more directly from her blood stream, depending upon the species.

The oögonia, the first primitive sex cells, divide to become primary oöcytes, and both these groups of cells contain the diploid number of chromosomes. In the change from primary to secondary oöcytes comes the reduction division of meiosis to the haploid number. In most other cells as the nucleus divides so the cytoplasm is shared equally between the daughter cells, but this is not so in the development of the ovum. In the change from primary to secondary oöcyte the cytoplasm is unevenly divided so that most of it goes to one cell. This is the one which will become the ovum. The smaller cell is the first polar body. Before fertilization is possible the ovum must divide again to extrude the second polar body. This may not occur until the egg is penetrated by a spermatozoon, and so takes place in the Fallopian tube. At the time of ovulation the ovum therefore still has one division to undertake. It carries its first polar body within its zona pellucida and after its division in the tube the second polar body is extruded into the same space. Moreover, the first polar body may also divide so that just external to the cell membrane of the ovum there may be three polar bodies. The ovum in effect has the cytoplasm of four cells.

MITOSIS AND MEIOSIS

Mitosis is the process of cell division in all but the sex cells. It is divided into the four stages of prophase, metaphase, anaphase and telophase. During prophase the nucleolus disappears and the chromosomes become visible as coiled threads. There are 46 chromosomes in each somatic cell of Man and 23 come from the father and 23 from the mother. The

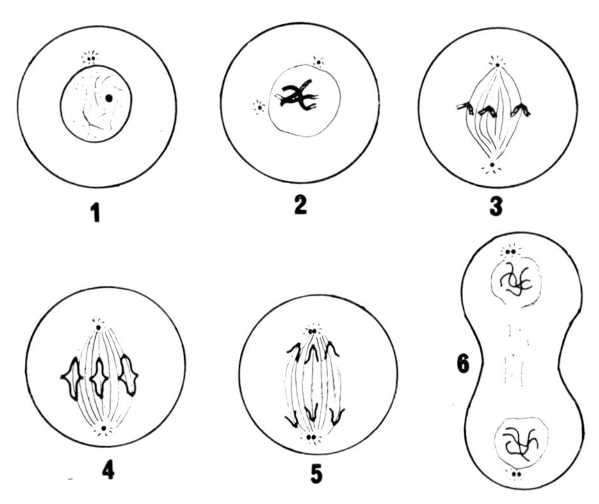

FIG. 7.3. Diagram of mitosis.

1. Resting cell. The centriole near the nucleus is paired. The nuclear membrane is distinct. There is a prominent nucleolus in the nucleus.
2. Prophase. The centrioles separate. The nuclear membrane fades. The nucleolus disappears and chromosomes become distinct as coiled filaments.
3. Metaphase. A spindle forms between the centrioles. The paired chromatids are attached to the equator of the spindle by their centromeres.
4. Metaphase-anaphase. The centromere divides and the chromotids separate towards the pole of the spindle.
5. Anaphase. The chromosomes have reached the pole of the spindle. The centriole divides.
6. Telophase. A nuclear membrane begins to form around the chromosomes and a waist forms in the cell cytoplasm eventually dividing the cell into two daughter cells.

centriole just external to the nucleus divides into two and each half moves to opposite poles of the nucleus and the two halves are joined by threads to form the spindle. At the fattest part of the spindle, which is like a rugby ball, is the equator.

In metaphase a clear zone in each chromosome can be seen. This is the centromere. All the chromosomes now become attached to the equator by their centromeres.

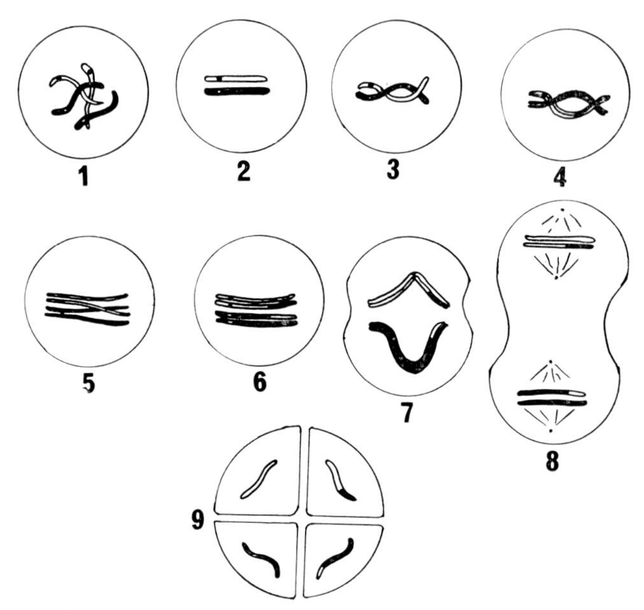

FIG. 7.4. Diagram of meiosis. Meiosis consists of two divisions. 1–8 is the first reduction division. 9 represents the second meiotic division.

1–5. Prophase.

1. Leptotene. The chromosomes become distinct and filamentous.
2. Zygotene. Homologous pairs of chromosomes come to lie along-side one another.

3 & 4. Pachytene. The chromosomes become twisted on one another and each chromosome splits into two chromatids but the centromeres do not divide.

5. Diplotene. The homologous chromosomes repel each other and cross-over occurs where the chromatids are coiled.
6. Metaphase of the first division.
7. Anaphase of the first division. The centromeres do not divide so that the pair of chromatids of each chromosome move to each pole of the cell, resulting in a reduction by half in the number of chromosomes.
8. Telophase of the first division.
9. The second meiotic division proceeds without a pause and resembles a mitotic division in its stages.

In anaphase the centromeres split so that each chromosome is in two longitudinal halves, each of which in some way repels the other so that the two halves of each move to opposite poles of the nucleus. Thus the genetic material is equally divided between the two daugher cells. Each half chromosome replicates itself so that a new chromosome is formed consisting of its usual two halves.

In telophase the chromatids (the half chromosomes) can no longer be seen by the usual microscopic methods. The nuclear membrane is re-formed and the cytoplasm constricts so that the two daughter cells are now complete.

Meiosis is rather more complex. During it the number of chromosomes has to be reduced from the diploid number of 46 to the haploid of 23 so that at fertilization the diploid number is restored.

The same four stages of prophase, metaphase, anaphase and telophase are seen as in mitosis, the major differences between the two types of cell division being in prophase and anaphase.

Prophase is divided into four parts. In leptotene the chromosomes become visible. In zygotene the homologous pairs of chromosomes come to lie alongside one another. One of each pair comes from the mother and one from the father. The centromeres remain separate. In pachytene the two halves of each chromosome (chromatid) entwine round one another and this is a most important event for it is at this time that pieces of each chromatid break off and re-arrange themselves so that there is an exchange of material between the two previously homologous chromosomes. The resulting chromosomes are now not the original maternal and paternal ones but each in some measure consists partly of maternal and partly of paternal genetic material. In diplotene the two reconstituted chromosomes repel each other and broken pieces of chromatid join on to an appropriate chromosome.

Metaphase is exactly comparable in the two processes. Now in meiosis the newly constituted chromosomes becomes attached by their centromeres to the spindle.

In anaphase the centromeres do not split as they do in mitosis. The pairs of newly constituted chromosomes move to opposite poles of the nucleus. Thus each half now contains 23 chromosomes, whereas in mitosis each half contained 46 chromatids which would soon replicate to form chromosomes. Moreover, in mitosis, the genetic material was half from the father and half from the mother. In each chromosome of the daughter cells of meiosis the genetic material is mixed and some comes from the mother and some comes from the father and the extent of each depends on the crossing-over that has occurred when the maternal and paternal chromosomes have come together during zygotene and pachytene. The degree of interchange of material in zygotene is the determining factor of the variability of the offspring of a fertile mating. Each fertilized egg carries genetic material derived from its maternal grandmother and grandfather and from its paternal grandfather and grandmother. The amount of material derived from each is apparently in the lap of the gods or more specifically in the events of pachytene and diplotene.

The events of meiosis occur only between primary and secondary spermatocytes and oöcytes. Before and after this cells divide only mitotically.

THE DETERMINATION OF SEX

In special preparations designed to show chromosomes each chromosome can be matched with another for size and for the position of the centromere. This is the basis of the Denver classification of chromosomes according to size and shape. The figure shows this. However, in male cells, one pair cannot be matched and these two are the basis of sex. The larger of the two is the X chromosome which has been derived from the mother, and the smaller one is the Y chromosome. In females on the other hand there are two X chromosomes, one from the mother and one from the father. Since these two chromosomes must be present in the oögonia and primary oöcytes it is obvious that the cells of the ovum must also all have an X chromosome. Since the spermatogonia and primary spermatocytes all contain an X and a Y chromosome it is obvious that the spermatozoa can contain either a Y chromosome or an X chromosome if normal meiosis has occurred. These two types of spermatozoa must be produced in equal numbers. At fertilization the X bearing ovum may be invaded by either an X bearing or a Y bearing spermatozoon. If the X bearing sperm fertilizes the egg a female results with two X chromosomes per cell, and if

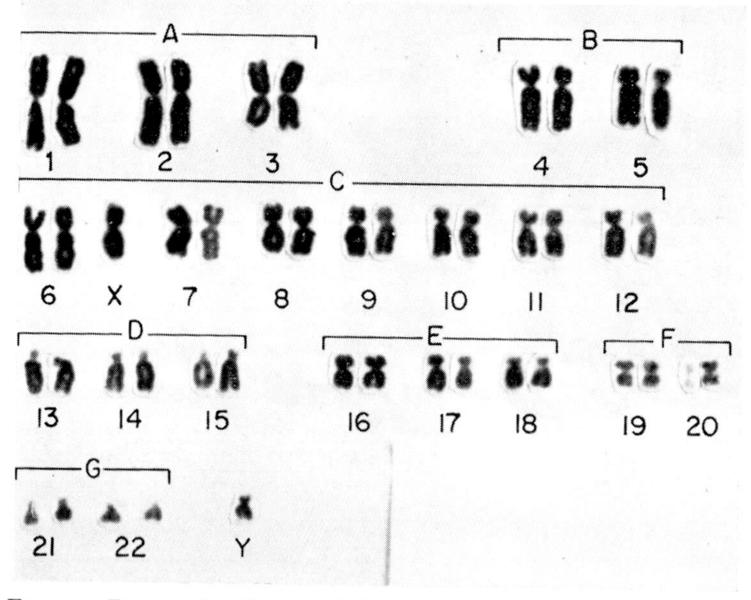

FIG. 7.5a. Denver classification of chromosomes. Normal male karyotype. Each chromosome is split into two chromatids joined by a centromere. According to the size of the chromosomes and the position of the centromere, the chromosomes are placed in one of seven groups for the autosomes and one pair of sex chromosomes. The X chromosome is placed in group C and the Y chromosome is placed in group G.

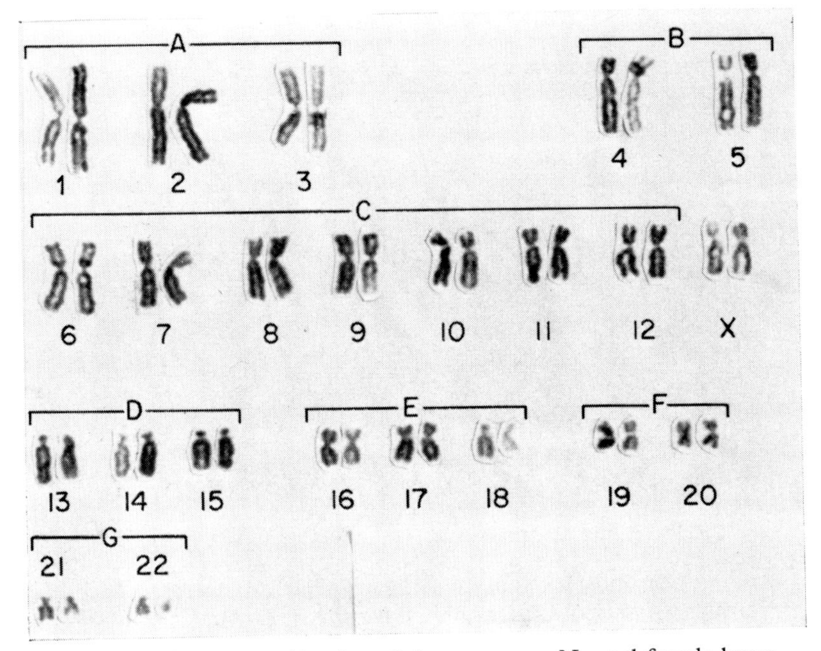

FIG. 7.5b. Denver classification of chromosomes. Normal female karyotype. The two X chromosomes are placed in group C.

a Y bearing sperm invades the egg then a male results containing an X and a Y chromosome in each of his cells. The chromosomes other than the X and Y are called autosomes. In Man the male is called the heterogametic sex and the female the homogametic. It is of interest that in birds the position is reversed and it is the female who is heterogametic (XY) and the male who is homogametic (XX).

SEMINAL FLUID

The fluid which is expelled from the male urethra into the genital tract of the female at sexual congress contains many other constituents besides spermatozoa. Tracing the path of the spermatozoa it is theoretically possible that these other materials may have been contributed by the rete testis, the efferent ducts, the epididymis, the vas deferens, the seminal vesicles, the prostate or the penile urethra. The bulbo-urethral or Cowper's glands open into the proximal cavernous part of the urethra. It is probable that the rete testis and the efferent ducts are conduits only.

The epididymis is a tightly coiled tube which when dissected out may be as much as 6 metres long. It is lined with a secretory columnar epithelium and seems to be the main storehouse for sperm. In animal experiments it can be shown that the sperm obtained from the distal part of the epididymis have a greater fertilizing power than those from the

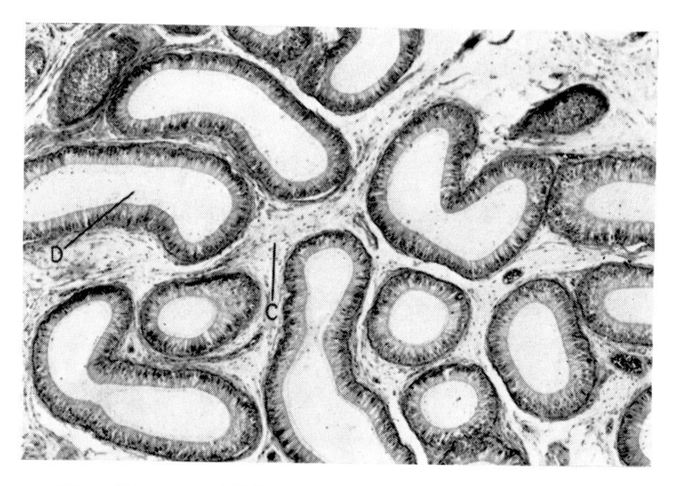

D = Ductus epididymis. C = Connective tissue.

FIG. 7.6. Ductus epididymis. The ductus is highly coiled and lined by tall, pseudostratified ciliated columnar epithelium. Spermatozoa may be found in such tubules.

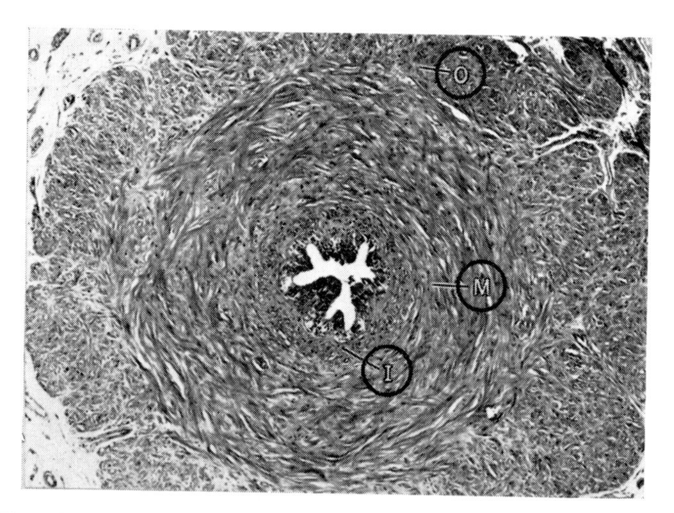

O = Outer longitudinal muscle. M = Middle circular muscle.
I = Inner longitudinal muscle.

FIG. 7.7. Vas deferens. This is a muscular tube, the lumen being lined by pseudostratified ciliated columnar epithelium. The muscle coats can be separated into inner longitudinal, middle circular and outer longitudinal layers.

proximal end. In some way the secretions of the epididymis bring the sperm to maturity and enhance their physiological role. The vas deferens is lined by columnar epithelium and has a thick muscular coat. The epithelium is not ciliated. It would seem likely that it does not add significantly to the contents of the seminal fluid. Its muscle wall is mainly of significance during the reflex contractions in it during coitus.

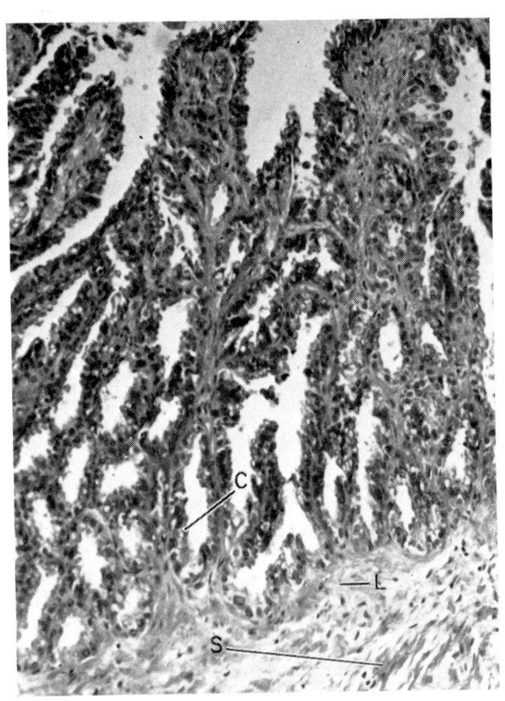

C = Columnar epithelium. S = Smooth muscle.
L = Lamina propria.

FIG. 7.8. Seminal vesicle. The lining epithelium is thrown up into folds which fuse to form irregular spaces. The lining epithelium has two layers, a superficial columnar cell and a cuboidal basal cell. Many cells contain brown fat pigment. Surrounding the mucosa is a thick smooth muscle coat.

The seminal vesicles are lined by columnar epithelium which secretes a yellowish fluid which is alkaline and viscid and contributes much of the volume of the ejaculate. The prostate gland has many small acini opening by ducts into the urethra and the glandular tissue is surrounded by a fair amount of smooth muscle. Its secretion is thin and has a characteristic smell which gives seminal fluid its peculiar odour. The prostatic secretion is acid (pH 6·4) and contains calcium, citrate and acid phosphatase in quantities of 30 and 150 mEq. per litre and 100–1200 units per 100 ml.

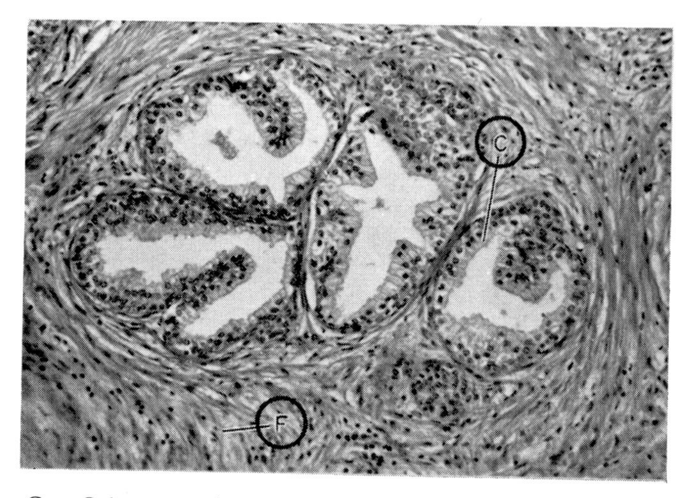

C = Columnar epithelium. F = Fibro-muscular stroma.

FIG. 7.9. Prostate. This is a tubulo-alveolar gland embedded in a fibro-muscular stroma. The glands are lined by inner, tall columnar cells and an outer layer of cuboidal or flattened cells.

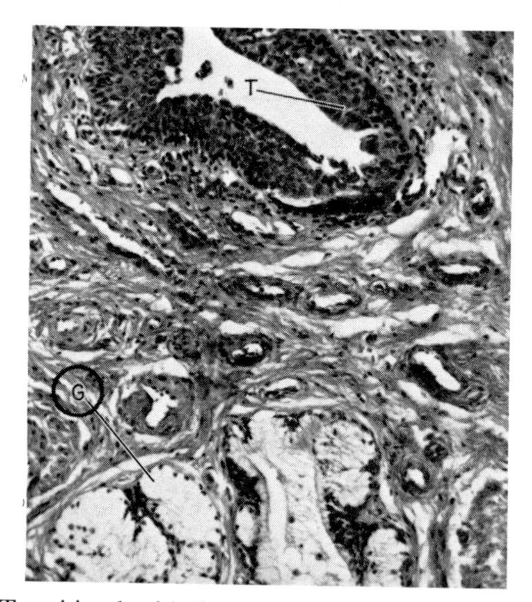

T = Transitional epithelium. G = Glands of Littré.

FIG. 7.10. Male urethra. This is largely lined by transitional epithelium or stratified columnar epithelium. Outfolding of the mucosa forms the lacunae of Morgagni into which open the glands of Littré. The lamina propria contains large numbers of blood vessels.

respectively. In sexual excitement the amount of acid phosphatase may be as much as 4,000 units per 100 ml. The bulbo-urethral glands secrete a mucoid fluid under conditions of sexual excitement.

Seminal fluid for clinical examination is usually obtained by masturbation. There are certain generally accepted standards of normality in various characteristics of the fluid by which an estimate of male fertility may be made. There are many factors in the ability of a couple to produce

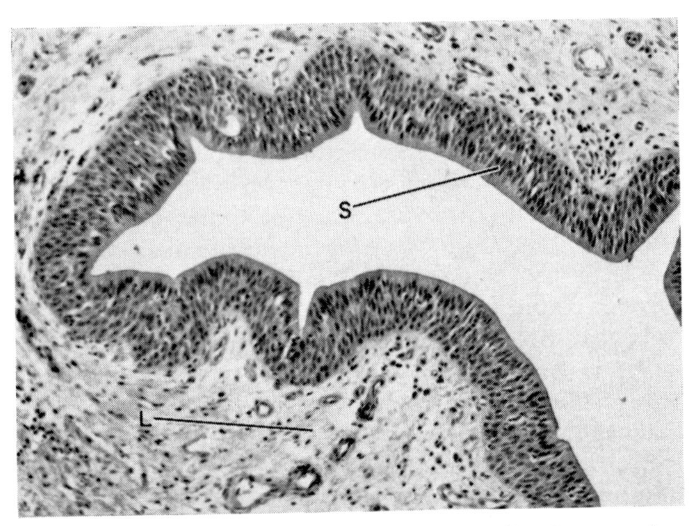

S = Stratified columnar epithelium. L = Lamina propria.

FIG. 7.11. (For comparison with Fig. 7.10.) Female urethra. The urethra is lined by stratified squamous epithelium or by stratified columnar epithelium, as here. The lamina propria is richly supplied with blood vessels. Outpockets of mucous glands, corresponding to the glands of Littré, may be found in the female urethra.

a zygote (fertilized egg) and the quality of the sperm is only one amongst them. The following are the rough standards:

Volume of the ejaculate	2–5 ml.
Density of sperm	40–100 million per ml.
Morphology of sperm	60–80 per cent should be normal.
Motility of sperm	50 per cent should be motile after incubation for 1 hour at 37°C.

The volume of the seminal fluid varies between about 1 ml. and 6 ml. Much depends on the previous period of continence. If there has been no ejaculation for the past 4 days or so the volume will approach the maximum. If ejaculation has taken place fairly frequently the volume will fall. Not all of the volume is made up by spermatozoa and indeed much of it comes from other parts of the genital tract, especially the seminal vesicles. Shortly after the fluid is passed it clots but within a short time at body temperature it liquefies again. The exact mechanisms by which these changes occur is not known but the seminal plasma (the liquid portion) contains fibrinogen and thromboplastin but not prothrombin or thrombin. Fibrinolysin is responsible for the liquefaction of the clot. Also there is calcium contributed by the prostate but the large amount of citrate probably prevents it from taking a part in clotting reactions. The seminal plasma also contains hyaluronidase in large amounts (100 units per 100 ml.). This is an enzyme capable of acting on the hyaluronic acid found in mucus so liquefying it. It is believed that this action may allow the sperm more easy progress through the cervix and uterus and tube to the ovum. Moreover, the zona pellucida surrounding the ovum may be mucinous, when the hyaluronidase might be of value in facilitating entry of a spermatozoon into the ovum. Fructose is present in high concentration in semen and this is probably for the nutrition of sperm. They break it down to lactic acid, using the same metabolic pathways involving ATP and ADP and creatine phosphate that other cells do.

Recently discovered substances in the seminal fluid are prostaglandins. They are lipids which were thought to originate in the prostate. However, further research has shown that they are to be found in many other tissues as well, notably the brain and lung. They probably have widespread physiological importance and they seem to be manufactured from the essential fatty acids, that is the unsaturated ones linoleic, linolenic and arachidonic. Injection of some prostaglandins, of which about six have been isolated, into the brain of a cat may cause it to pass into a catatonic state and there is hope here that some mental abnormalities might in the future be explained in terms of disorder of the metabolism of prostaglandins. Their importance in the lungs and in the genital tract may be related to their power of affecting smooth muscle activity. Some of them increase contraction of smooth muscle and some cause it to relax. There is no doubt that they have the ability to affect both uterine and Fallopian tube activity *in vitro*. It is suggested that they may do this physiologically by being absorbed from the vagina at coitus. If this is shown to be so it may be a demonstration of a "hormone" being produced by one person having an effect on the organs of another. It seems likely that these prostaglandins may not provoke antibody formation since they are of lipid nature when hormones based on protein would do so. Research in this field is urgently needed but will be difficult since the female genital tract also produces prostaglandins whose physiological rôle is not known.

The density of spermatozoa in the seminal fluid is presumably an expression of the activity of the seminiferous tubules together with the storage activity of the epididymis and also the mechanisms which transport them to the external urethral meatus. With repeated ejaculations over a short space of time the sperm count diminishes, but increases again after abstinence. The seminiferous tubules probably require a minimal level of testosterone to keep them active. Production of spermatozoa will be diminished in hypogonadal states, but can be improved by injections of the male hormone. There is usually no sperm production in Klinefelter's syndrome where the sex chromosome content of the cells is XXY. This indicates that both genetic and endocrine factors play their part in sperm production. Warmth applied to the scrotum also diminishes sperm counts, though the effect may not be seen for some weeks after repeated bathing of the scrotum with warm water. Certain forms of suspensory belts for men and the so-called Y-front underwear may hold the testes so close to the body that the temperature of them is raised and so diminish the sperm count. Varicocoele too, an enlargement of the pampiniform plexus of veins supplying the testes, causes diminution of sperm production probably by raising the local temperature. It has been known for many years that the cryptorchid (hidden testis), that is the patient whose testes are undescended, does not produce spermatozoa. This also may be due to their being exposed to the general body temperature within the abdomen. If one testis descends to the scrotum it will produce sperm. Careful experiments have shown that the temperature of the testes is in general up to 5°C lower than the general body temperature. Experiments with certain drugs such as diamines and also repeated injections of testosterone have shown that sperm production can only be suppressed over the course of several (about eight) weeks. When the testes are allowed to recover from the effects of these drugs the sperm count may not return to normal for up to ten weeks. This is a demonstration of the long cycle required for the production of sperm. The experiments with repeated injections of testosterone seemed to show that sperm production was inhibited by suppression of the pituitary and thus prevented the endogenous production of testicular hormones, and also they suggest that testosterone alone is not enough to maintain spermatogenesis.

Recently male sterilization by removal of a short segment of both vasa deferentia has become more common, for the purpose of family limitation. Such males may remain fertile for some weeks after the operation since spermatozoa may stay viable within the genital tract. Contraceptive methods may therefore be needed for about 8–12 weeks after the surgery is performed.

The normal spermatozoon has a head, neck, body and a long tail. In surface view the head is roughly oval, but in profile is flattened at the tip. It is nearly all nucleus and the only cytoplasm is a thin layer known as the head cap overlying its front end. The neck is the slight constriction behind the head and the anterior centriole is found there. Arising from the posterior centriole are several filaments which are grouped together

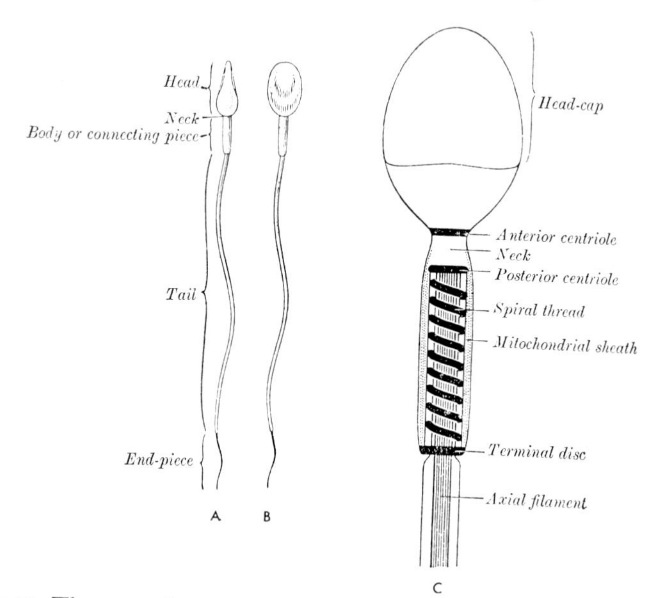

FIG. 7.12. The normal spermatozoon. From *Grays Anatomy*, 29th edition.

to form the axial filament which is prolonged into the tail. In the body of the spermatozoon the axial filament is surrounded by a spiral thread of mitochondria. Both the body and the tail are enclosed by a thin layer of cytoplasm, but at the very end of the tail this disappears so that the end

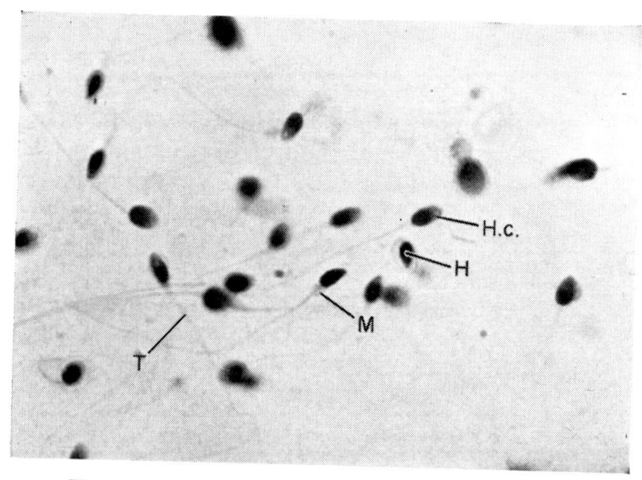

T = Tail. M = Middle piece.
H = Head. H.c. = Head cap.

FIG. 7.13. Seminal fluid. There are many spermatozoa in this fluid. These spermatozoa have a dark head with a pale head-cap. A pale middle-piece joins the filamentous tail to the head.

of the tail is naked axial filament only. The average length of the spermatozoon is 52–62 μ, of which the tail forms 40–50 μ, the head 4–5 μ and the body 6 μ.

Not all the spermatozoa even in a fertile specimen of seminal fluid have a normal morphology and it is very common to see sperm with abnormally shaped heads and deformed tails or even with two tails. It seems unlikely that these abnormal forms would be capable of fertilization. There seems to be no relationship between the percentage of abnormal forms and the production of abnormal children or abortions which might be thought to be an expression of genetic defects. It will be seen from the table of standards for seminal fluid that not more than about 40 per cent of the spermatozoa should be abnormal if fertilization is expected to be successful.

Spermatozoa will move through a not too viscous medium by the action of their tails. That is, they swim. They move at the rate of a few millimetres per hour. If they are put into a slowly moving stream they orientate themselves so that they swim against the current. However, there is no certain evidence that there is a current within the female genital tract, though the cilia of the Fallopian tube sweep material towards the uterus and it would seem probable that any current would be towards the exterior in the body of the uterus and in the cervix. The reasons for the orientation of the sperm in a stream are not fully worked out. It is partly a matter of the hydrodynamic shape of spermatozoa. For swimming to occur there must be expenditure of energy which is supplied by glycolysis of fructose and during activity there is a measurable though small uptake of oxygen. Motility is lost in an acid medium and this explains why seminal fluid that is kept for long shows diminishing activity of sperm, for the metabolism of fructose produces lactic acid. Fructose is present in seminal fluid to the extent of about 300–700 mgm. per 100 ml. The pH of freshly expelled semen is 7·2–7·4. The sperm are probably protected from the acid vagina (pH 4·5) by the large volume of the ejaculate and by being passed very quickly into the slightly alkaline cervical mucus. Motility of spermatozoa is enhanced after they have been in contact with the secretions of the epididymis.

Chapter VIII

SEXUAL RESPONSE AND INTERCOURSE, SPERM TRANSPORT

The meeting of a spermatozoon with an ovum is brought about by sexual intercourse, in which the erect penis of the male is introduced into the vagina. During the height of the sexual excitement of the male, seminal fluid is ejected into the fornices of the vagina and around the cervix. From there spermatozoa are carried up through the cervical canal and the uterine body until they reach the Fallopian tube and it is there that fertilization takes place.

Sexual response is determined by psychology, the nervous system connexions with the genitalia, the endocrine background of the individual and the local anatomy of the genitalia.

Conditioned reflex responses to a variety of stimuli are built up from the moment of birth up through the phases of childhood and into maturity and beyond. The growth of sexuality is a matter of psychology and will not be dealt with here. Suffice it to say that phases of infantile, pubertal, adolescent and mature sexuality can be recognized and though their nature is mainly psychological they are conditioned in part by physiological processes. The psychological and physiological stimuli to sexual response overlap greatly and during courtship and marriage, the usually accepted social pattern, the response may be triggered by such factors as proximity, bed, sounds, clothing, nudity, and scents among a host of others. Animal experiments have shown that removal of the cerebral cortex does not greatly impair the sexual responses of the female, but in the male the responses are much damaged by this operation. It is suggested that it is the testosterone of the male that is responsible for this effect, and there is no doubt that the sex steroids can and do cause different reaction patterns in the nervous system. For instance, the acyclic activity of the hypothalamus, pituitary and testis in the male are due to the "imprinting" effect of androgens in the early weeks of extra-uterine life. When these organs are not so acted upon but only exposed to the action of oestrogens the sexual rhythms are truly cyclic. If a male animal, e.g. cat, is castrated and then treated with high dosage of oestrogens then its responses become more female in character, that is instead of displaying mounting activity as a normal male does, it tends to take up a position of crouching, much as the normal female does when ready to be mated. It is doubtful whether such experiments are directly applicable to Man where psychological responses may completely override the physiological ones. This is seen in some cases of intersex, when it is possible that the genetic, gonadal and endocrine background is male, but because the individual has been brought up as a female, the sexual responses are female. However, there is no doubt that the administration of sex steroids for therapeutic purposes to males and females may alter sexual behaviour. Oestrogens given to males may reduce libido, whilst testosterone given to either males or females may heighten it. On the other hand hormones would seem not always to be essential to human sexual responses for they can be seen in apparent normality in eunuchs and in Turner's syndrome or ovarian agenesis where there is no ovarian tissue.

It is well known that there are various areas of the body which when stimulated are more likely than others to cause sexual arousal. Such are the lips in kissing, the breasts and the genitalia. Also the back in the region just below the scapulae is said to be such a zone. It is noteworthy that the breasts and the genitalia grow under the influence of sex steroids.

It is evident that lower levels of the nervous system are involved in sexual responses. In the cat removal of the ovaries does not impair sexuality provided that a pellet of oestrogen is implanted into the hypothalamus. Rabbits and ferrets have been shown to ovulate only after coitus and the hypothalamus has been shown to be a part of this process. Rarely in women hypersexuality develops when there is a tumour in the temporal lobe of the brain, and it will be remembered that there is evidence that this area has some control over ovulation, the control being exercised through the hypothalamus. In animals, damage of the temporal lobes may impair sexual responsiveness. In monkeys stimulation of the caudal thalamus may cause erection of the penis. In Man lesions of the spinal cord causing paraplegia may be attended by uncontrollable erections of the penis and this suggests the presence of a lower sexual centre in the sacral part of the cord. There can be little doubt of the prime importance of the central nervous system in sexual activity, even though the details of its activity are still far from clear.

SEXUAL INTERCOURSE

The phases of sexual intercourse have been divided into excitement, plateau, orgasm and resolution.

Excitement may last from a few minutes to some hours. During this time it may wax and wane. For the male the essential feature is the erection of the penis, together with psychic tension. Erection is caused by the erectile tissue of the penis becoming engorged with blood. The outflow of blood from the organ is prevented by contraction of the ischiocavernosus muscles which are closely applied to the crura of the penis where they lie along the ischio-pubic rami. Their nerve supply is derived from the internal pudendal nerve, whose roots are from S 2, 3 and 4. The nervi erigentes are parasympathetic also from S2, 3 and 4 and they reach the penile blood vessels through the pelvic plexuses. They probably decrease vasomotor tone and allow the erectile tissue to fill rapidly with blood. These two sets of nerves belong to the efferent side of the reflex arc with its centre in the sacral segments of the spinal cord. Local afferents run from the penis and adjacent areas along somatic nerves to the sexual

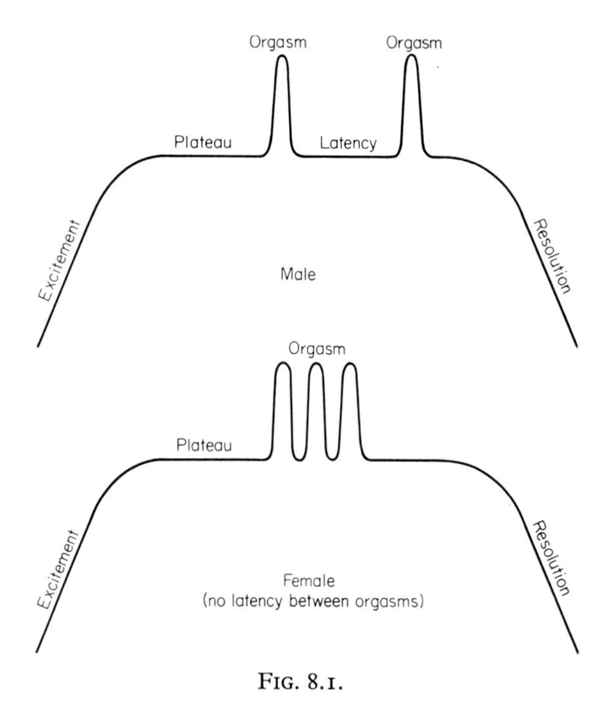

FIG. 8.1.

centre, but the most powerful influence on this area comes from the higher centres in the brain through conditioned reflexes which have been built up during the previous life of the person.

In the female the phase of excitement may also be characterized by erection of the clitoris and of the labia minora, which are largely made up of erectile tissue. The physiological mechanisms are essentially the same as those outlined for the male. However, there may be few physiological changes in the female during excitement. In general women are slower to arousal than men and erection and other changes may not be seen until the plateau phase. It will be realized that the division of sexual intercourse into phases has no complete reality since each shades off imperceptibly into the other. Perhaps mainly during the plateau phase the breasts enlarge by as much as one-fifth in volume and the nipples become erect. A little later the areola surrounding the nipple also becomes engorged so that the apparent size of the nipple then diminishes, as the areola encroaches upon its base. As the excitement phase gathers strength there is often a flushing of the skin which begins in the region of the xiphisternum and then spreads upwards over the breasts and the neck and up to the face.

C.c. = Corpus cavernosum. C.s. = Corpus spongiosum.
A = Artery. T = Tunica albuginea.
U = Urethra.

FIG. 8.2. Cross-section of penis. The erectile corpus cavernosum is surrounded by the dark-staining tunica albuginea. The urethra and surrounding corpus spongiosum has a much less well developed tunica. Above the corpus cavernosum are the dorsal artery and vein. There is abundant subcutaneous connective tissue without fat. The skin covering the penis is thin.

The first phase of excitement, comparable to erection of the penis in the male, is the transudation of fluid through the vaginal wall so that this canal and the introitus become moist. How this transudate forms is not known since there are no glands in the vagina. It used to be thought that the secretions at the introitus came from Bartholin's glands but more recent work suggests that these glands secrete mainly during the orgasmic phase.

During excitement there is a general heightening of muscular tone throughout the body.

These changes in the genitalia, breasts, skin and muscle are maintained into the plateau phase. In the region of the introitus the orgasmic platform becomes more obvious. This is a region of tumescence just within the introitus though the stiffly erect labia minora make the platform seem as though it is well within the vagina. The platform is made up

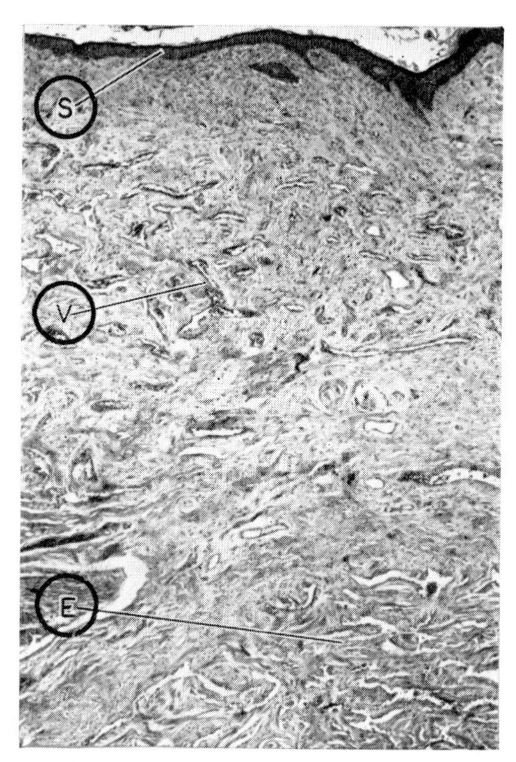

S = Stratified squamous epithelium.
V = Vascular connective tissue.
E = Erectile cavernous body.

FIG. 8.3. Clitoris. The surface stratified squamous epithelium is the same as that covering the labia minora. In the underlying connective tissue there are many thin walled blood vessels. Vascular erectile tissue comparable with that seen in the penis forms the cavernous body.

of the bulbs of the vestibule, which are erectile tissue surrounding the introitus covered by the bulbo-cavernosus muscles. Immediately beneath these bulbs is the inner edge of the levator ani, or more strictly the pubo-coccygeus in this part.

The female is now physiologically prepared for the orgasmic phase. In the male with penile erection at its peak the corona of the glans is suffused and bluish red and the testes become pressed in close to the perineum by the action of the cremaster muscle pulling up the testis and by the contraction of the dartos muscle underlying the skin of the scrotum. He also shows the skin flush and the myotonia.

The penis in the vagina is moved in and out and this causes sensations of pleasure in the male and female. As the thrustings become more

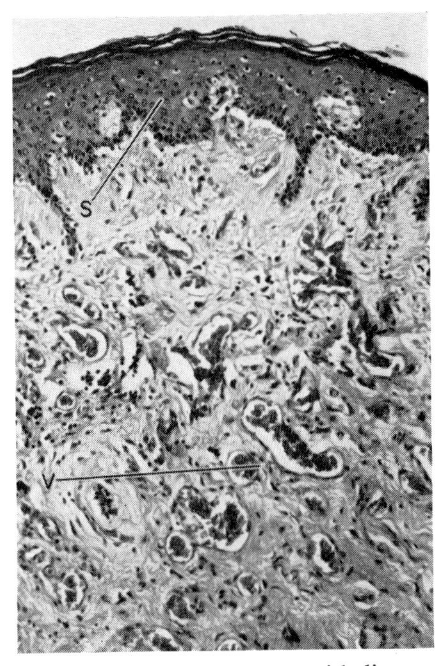

S = Stratified squamous epithelium.
V = Vascular connective tissue.

FIG. 8.4. Labium minus. There is a surface covering of stratified squamous epithelium with prolongation of epithelium into the underlying connective tissue. The latter contains many thin-walled blood vessels. Occasional vestigial sebaceous glands are found but hairs are not present.

frequent each partner experiences orgasm. The male orgasm is easily recognizable by the ejaculation of seminal fluid. Ejaculation is brought about by the rhythmic contractions of the vasa deferentia, the seminal vesicles but above all by the bulbo-cavernosus which compresses the dilated urethra at the proximal end. This muscle may contract vigorously for 5 to 12 times and cause the ejection of the seminal fluid under some pressure. At the same time as the ejaculation is proceeding the levatores ani also contract and also the external anal sphincter. After the ejaculation the myotonia gradually recedes and there is a feeling of physical and mental lassitude. It is worthy of note that the male cannot immediately return to another orgasm but requires a latent period of some minutes to start again. This contrasts with the female who requires no such latent period. Immediately after orgasm the male may show a fine perspiration all over the body.

The female has a similar orgasmic pattern and when she reaches her climax the orgasmic platform undergoes 5 to 12 rhythmic contractions with a periodicity of 0·8 sec. These may or may not be in exact time relationship with those of the male. It has been thought that some excitation came from the shaft of the penis rubbing against the clitoris, but the work of Masters and Johnson (1966) shows that the clitoris is

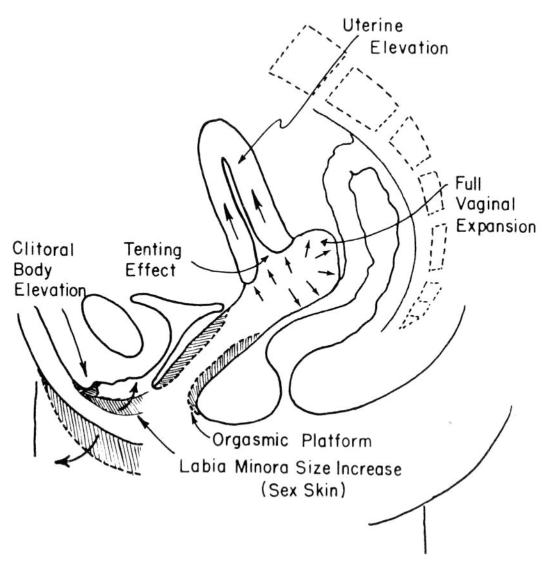

FIG. 8.5. From W. H. Masters and V. E. Johnson *Human Sexual Response.* J. & A. Churchill, London (1966).

withdrawn towards the symphysis pubis during sexual intercourse and is away from the penis. The contractions of the orgasmic platform are due to the underlying levatores ani, and the external anal sphincter also contracts. The woman too shows the gradual decrease of muscle tone after intercourse as well as the fine perspiration and gradually over the next few minutes there is a reversal of all that has happened before, so that the tumescence of the labia minora and of the clitoris and vestibular bulbs disappears and the breast changes are also reversed.

During the plateau and orgasmic phases the upper vagina distends and the uterus becomes engorged and enlarges whilst the cervix moves up in the abdomen perhaps because of some change in intra-abdominal pressure. After orgasm when there is a pool of seminal fluid in the upper part of the vagina the cervix moves down again into the pool and the vaginal walls close down once more. There is some doubt as to whether the uterus contracts and by its contractions aspirates seminal fluid into its cavity. Some experiments using radio-opaque dye during sexual excitement seem to show that seminal fluid is not in fact aspirated, but animal experiments show the presence of spermatozoa within the Fallopian tubes within minutes of sexual congress, and it is certain that the sperm could not have swum that far in the time since their normal rate of progress is only a few millimetres per hour.

It should be realized that orgasm is not achieved at every intercourse by every woman and there are many reasons for this, but especially they are cultural. However, it is certain that orgasm is not an essential prerequisite of fertilization and many women have large families without ever having an orgasm.

Sexual intercourse is a time of intense physical and psychological tension which is quickly relieved by the orgasm. There are general physiological changes attending the tensions, and it has been shown that the respiratory rate may rise to 40 per minute, that the pulse rate may reach 100 to 170 per minute, and that the blood pressure may rise 30 to 80 mm. Hg systolic and 20 to 40 mm. Hg diastolic. Knowledge of the magnitude of these changes may be of importance in advising sufferers from respiratory and cardiac disease, though the duration of these changes is relatively short.

SPERM TRANSPORT

Spermatozoa removed directly from the testis are non-motile. Only after contact with the secretions of the epididymis are they motile. They probably reach this structure by the pressure of production of new spermatozoa from behind. Thereafter, they are able to swim, but in fact they are mainly propelled along the vasa deferentia by the rhythmic contractions of these tubes. From the proximal urethra they are propelled along the rest of the urethra by the bulbo-cavernosus muscles.

Spermatozoa cannot survive for long in the acid vagina (pH 4·5), but this acidity is locally reduced by the large volume of the seminal fluid to which is added some cervical secretion which is almost neutral or slightly alkaline. Animal evidence shows that spermatozoa may be found in the Fallopian tubes within about 10 minutes of copulation, so that they must be actively transported through the uterus and cervix presumably by uterine contractions. It is imagined that during orgasm the seminal fluid is sucked out of the vagina and into the uterus. It is, however, true that orgasm is not essential for fertilization to occur but it may be that uterine contractions take place independently of orgasm, or it may be that spermatozoa can swim through the uterus without the aid of myometrial activity.

An interesting recent finding has been that of prostaglandins in the seminal fluid. These are lipids containing essential fatty acids. There are several of them and they have an effect on smooth muscle and particularly on the uterus. Some cause it to relax and others make it contract. It has been suggested that some prostaglandins might be absorbed from the vagina and make the uterine muscle contract so aiding the transport of spermatozoa through the uterus, but this is an unproven assumption, but if proved it will be of great interest since it would be an example of a hormone being produced in one person and having an effect in another. However, this is a well-known phenomenon in pregnancy.

Further aid in sperm transport may be given by the hyaluronidase which is present in large amounts in the seminal fluid. It acts upon mucoid substances making them more permeable and so may ease the path of the swimming spermatozoa through the uterine cavity. The fructose content of seminal fluid is probably for the nourishment of spermatozoa which use this carbohydrate in their metabolism.

CLINICAL FEATURES OF SEXUAL INTERCOURSE

Impotence in men is the inability to have an erection of the penis. The cause is most often psychological. Ejaculatio praecox is premature ejaculation of the seminal fluid, sometimes even before intromission. This too usually has psychic origins. Sexual frigidity and sexual anaesthaesia are common in women and again are usually due to some psychological abnormality. Such conditions do not preclude women from having sexual intercourse since they are not such active participants as men and engorgement of the genitalia in them is not an essential of the act.

Infertility may be due to there being no spermatozoa in the seminal fluid. This can be due to a failure in production by the testis as in some cases of Klinefelter's syndrome (XXY) or it can be due to inflammatory obstruction of the vasa deferentia. Testicular biopsy will demonstrate which is the factor involved, and injection of radio-opaque dyes into the vasa in the groin will demonstrate any blockages. Very little is known of infertility in men due to failures in the production of substances which would normally be found in seminal fluid.

Infertility may be due to cervical factors and spermatozoa may be unable to gain access to the uterus and so to the Fallopian tube. This can be tested either by mixing cervical mucus with the husband's spermatozoa and seeing whether they invade the mucus as they should. Alternatively a post-coital test may be done. In this the woman presents herself in the clinic shortly after intercourse and specimens of discharge are taken from the vagina and the cervical canal for examination under the microscope. Normally spermatozoa should be present in the cervical canal. The mechanism of "cervical hostility" is not yet fully known, but it is possible that there may be a local antibody reaction to the husband's spermatozoa.

If this hostility can be shown, then the husband's sperm can be introduced into the uterine cavity above the cervix; that is artificial insemination by husband (A.I.H.).

A matter of interest is whether intercourse and especially orgasm can cause ovulation to occur. It is known in rabbits and ferrets and many other animals that ovulation only occurs after copulation. The method of contraception by the use of the "safe period" depends on the assumption that ovulation only occurs 14 days before the next menstruation. If ovulation can take place as a result of intercourse the "safe period" may not be very good as a method of prevention of conception. There is some evidence that in fact ovulation can be made to take place before the expected time as a result of coitus. This can only happen if coitus is practised in the follicular phase of the ovarian cycle. Once the corpus luteum has formed it is unlikely that further ovulation can occur in that particular cycle. The evidence for "forced" ovulation comes from German sources where soldiers' leave dates were correlated with their wives' menstrual cycles. Conception occurred more readily than would have been expected on the assumptions implicit in "safe period" calculations.

Chapter IX

FERTILIZATION AND BIRTH CONTROL

Fertilization is the process of fusion of a spermatozoon with the ovum. It takes place in one or other Fallopian tube. By the time that the ovum is in the tube the first polar body has been extruded, that is the uneven division of the cytoplasm of the oöcytes has resulted in the large ovum proper and the polar body. Surrounding both these cells is the apparently structureless perivitelline space. Chemically it may be of mucoid type. Attached to the surface of this membrane there are still a few granulosa cells, the remains of the cumulus oöphorus which clings to the ovum as it is shed from the ovary. It used to be thought that the ovum was shed into the peritoneal cavity and secondarily was picked up by the Fallopian tube. Animal experiments with windows let into the abdominal wall have shown that in fact the fimbriae of the tube become adherent to the surface of the ovary over the site of ovulation, so that the ovum and its surrounding cells make their way immediately into the lumen of the tube. The ovum is transported along the tube partly by the ciliated epithelium of its lining and partly by peristaltic action of its muscular coats. Both mechanisms show their maximal activity at the time of ovulation and are under the influence of the high output of oestrogen from the ovary at that time. During intercourse many spermatozoa, together with some seminal plasma, are transported to the tube, and many thousands of them may now surround the ovum. Of course, many millions of sperm never gain access to the uterus and are lost in the vagina. Others perish on their way through the cervix and uterine body. Those that do not fertilize the egg disintegrate and die without harm to the woman.

Only one spermatozoon bores it way through the perivitelline membrane or "space" and through the cell membrane of the ovum, which is sometimes called the vitelline membrane (vitellus = yolk). It is probable that the hyaluronidase of the seminal plasma and the lysozyme in the head of the spermatozoon are involved in this process. Once within the cytoplasm of the ovum no further spermatozoa are able to enter. The cause of this is not known but presumably it involves some change in the surface layer of the ovum. Inside the cytoplasm the sperm quickly loses its tail, so that only the nuclear material of the head remains. At this point the ovum may extrude the second polar body if it has not already

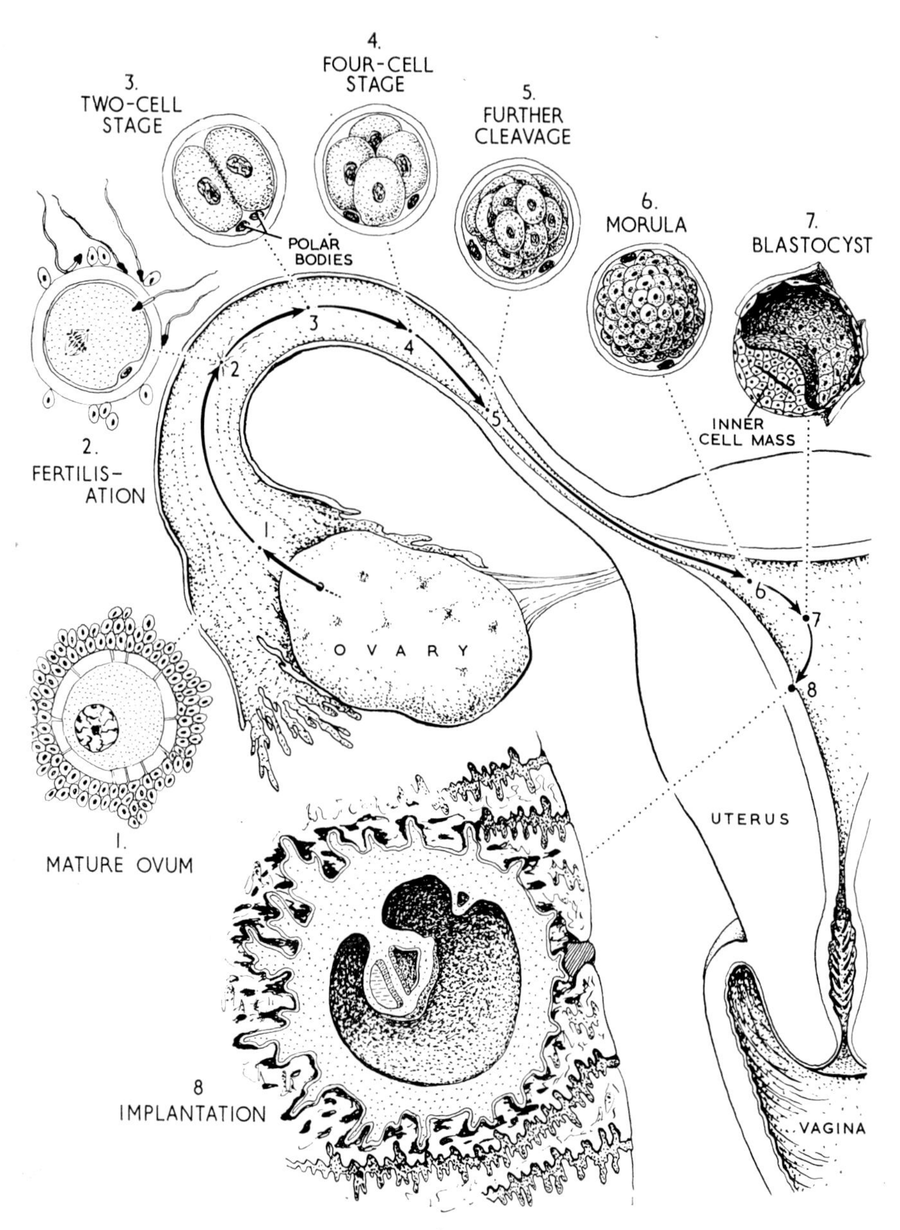

FIG. 9.1. Ovulation, fertilization, cleavage and embedding. Redrawn from *Expecting a Baby*. B.B.C. Publications, London (1967).

done so. The first polar body which was extruded earlier may also divide again so that within the perivitelline membrane there may now be the ovum and three polar bodies. Often there are only two polar bodies since the first one may not divide again.

The head of the spermatozoon makes its way through the cytoplasm of the ovum and fuses with the nucleus of the ovum. Each gamete contains the haploid number of chromosomes (23) and with fusion the diploid number (46) is restored and a new individual is on the way. It will be remembered that the diploid number is 22 + X for the ovum and either 22 + X or 22 + Y for the sperm. Depending on which kind of sperm fertilizes the ovum the sex of the child will be male or female.

CLEAVAGE

Very soon after fertilization the egg begins to divide and soon consists of a tight ball of cells, the morula (morula = mulberry). Although the mitotic activity is intense the total volume of the morula is only slightly larger than the original fertilized egg. That is the cytoplasm of each cell is not greatly increased. This is essential since the lumen of the Fallopian tube narrows as the uterus is approached. If the morula were unduly large it would be unable to negotiate the interstitial part of the tube where it goes through the uterine wall. An ectopic pregnancy would be the likely result. The metabolic activity involved in such rapid cell division must be very great and animal work suggests that it is done by anaerobic glycolysis so that oxygen is not required, or if it is, in only small quantities. The waste products of the process may be utilized further or they may diffuse out of the ball of cells into the Fallopian tube secretions. These come from the secretory cells of the tube which are interspersed with ciliated and "peg" cells. It is believed that the peg cells may be exhausted secretory cells. (See Fig. 4.17.)

Secretory activity is at its height at ovulation and is due to oestrogenic stimulation. The secretions of the tube may have other parts to play in reproductive processes but they are still obscure. They presumably form a stream which is helpful in conveying the ovum to the uterus, and recently it has been suggested that they may coat the egg with a protein membrane called the oölemma which is a necessary prerequisite for the ovum to be able to embed in the endometrium.

Animal experiments suggest that the time taken for the fertilized egg to reach the uterus varies from about 3 days to 7 days. Operations designed to unblock Fallopian tubes and restore fertility often shorten the tubes and even though it may be demonstrated that they are patent, pregnancy may not ensue. One explanation for this failure is that the ovum may reach the uterine cavity too prematurely and so not be able to embed. Another reason is that the epithelium of the tubes may be so destroyed by the inflammatory process which blocked them that an

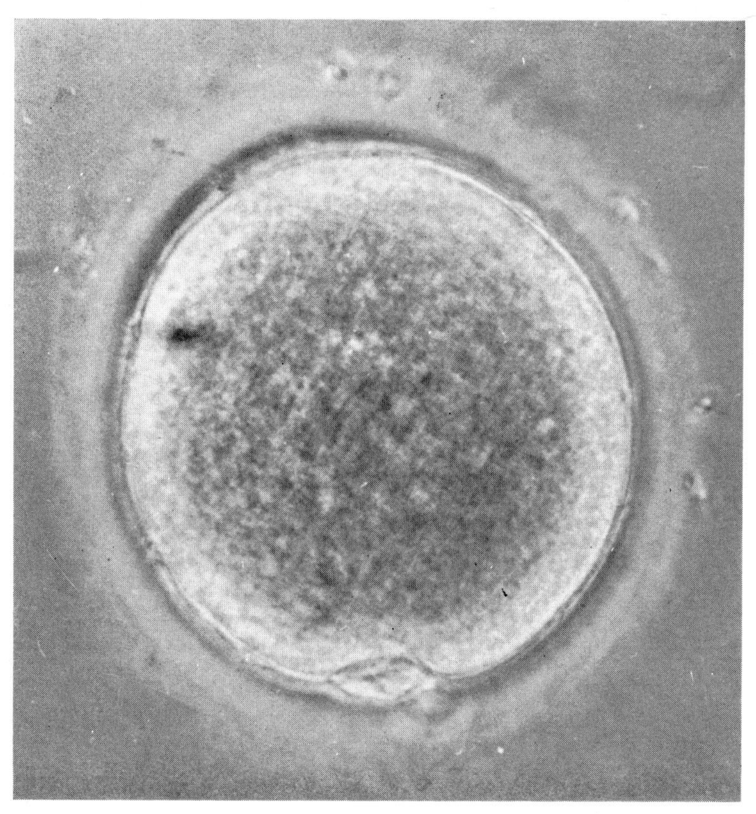

FIG. 9.2. An unfertilized egg; the zona pellucida can be seen around it, the first polar body is clearly seen, and there are spermatozoa on the outside of the zona (× 1,600). (*By courtesy of Dr. R. G. Edwards, Physiological Laboratory, Cambridge.*)

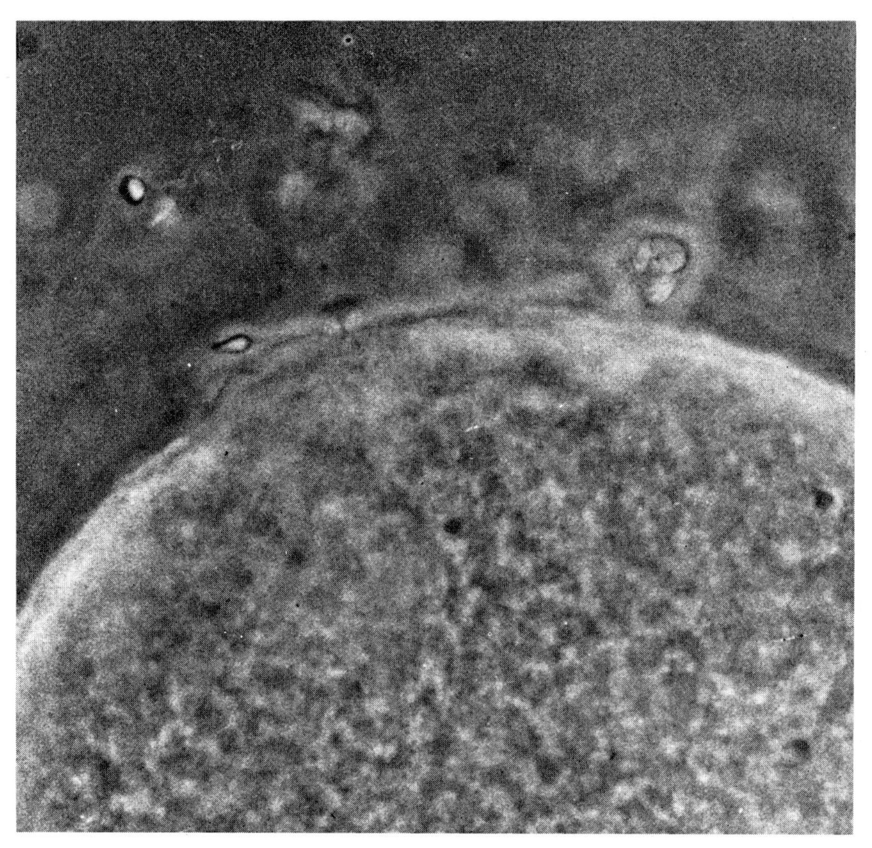

FIG. 9.3. The zona pellucida is more indistinct. A spermatozoa is in the perivitelline space. The cytoplasm extends towards the sperm-head (× 2,500). (*By courtesy of Dr. R. G. Edwards, Physiological Laboratory, Cambridge.*)

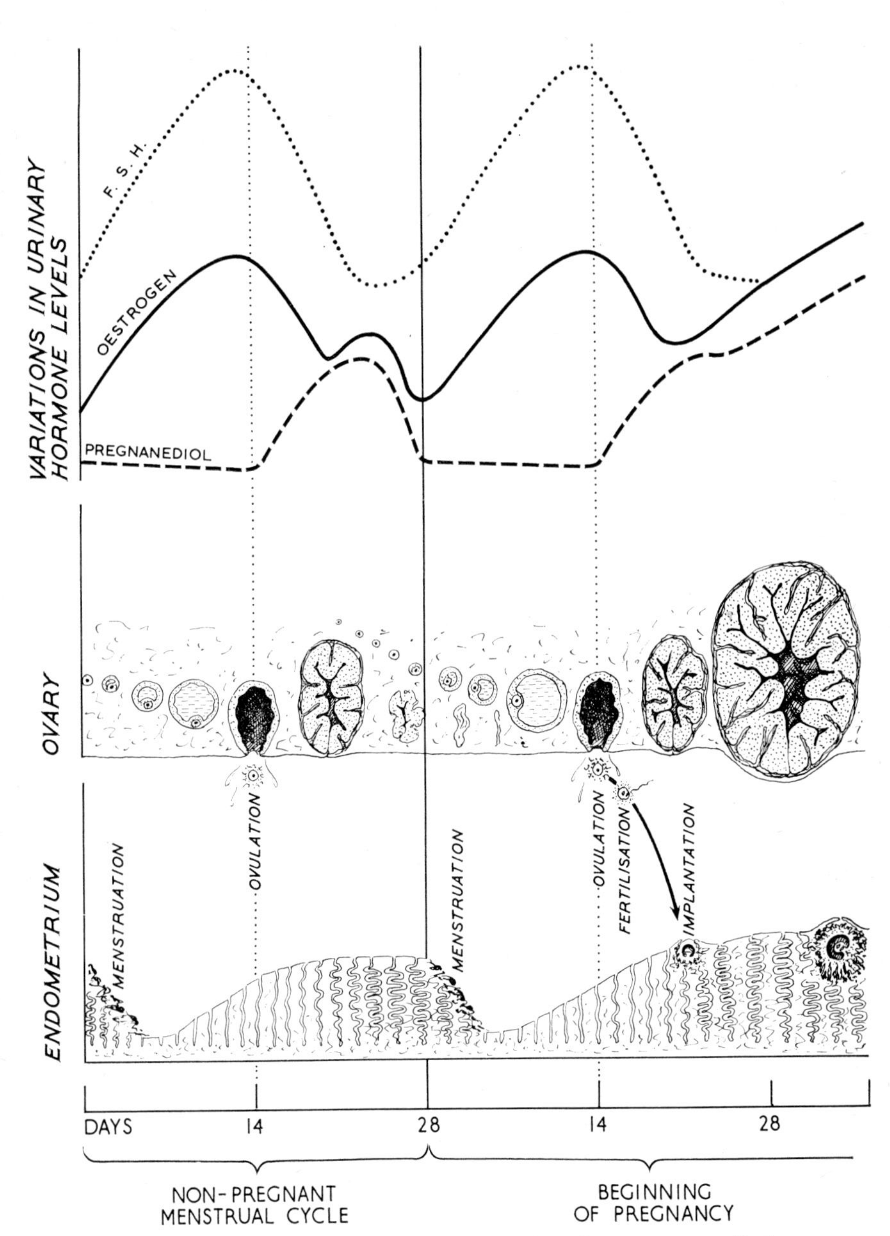

FIG. 9.4. Events of the menstrual cycle and early pregnancy. Redrawn from *Expecting a Baby*. B.B.C. Publications, London (1967).

oölemma cannot be put upon the egg and again it will be unable to embed. Alternatively spermatozoa may not be able to survive in a damaged tube or the ovum may be incapable of fertilization.

By the time the morula has reached the uterine cavity liquefaction of some of its central cells has taken place. There is now an outer envelope of cells which is the trophoblast and a ball of cells at one pole within the trophoblast which is the inner cell mass. This change is of immense importance, for the trophoblast will form the chorion and later the placenta for the nourishment of the baby, and the inner cell mass will differentiate and grow to form first the embryo and later the fetus. This stage of development is called the blastocyst.

It is the blastocyst which embeds in the endometrium. Whilst the blastocyst is within the uterine cavity it is possible that it is nourished in part by the secretions of the endometrial glands. In some animals with the so-called epithelio-chorial placenta the whole nourishment of the products of conception is derived directly from these secretions, the finger-like processes of the placenta dipping down into the glands and not actually invading the endometrium at any point. However, this is not so in the human, where the placenta is haemo-chorial; that is it erodes the endometrium so that its trophoblastic cells come to lie within the maternal blood stream.

The blastocyst adheres to the surface cells of the endometrium, a process mediated by some "stickiness" of the adjacent cell walls. Soon the inner cell mass, the embryonic pole of the blastocyst, erodes the surface of the endometrium, probably by the action of enzymes secreted by the trophoblast. The embedding takes place between two glands and very soon the trophoblast has established contact with the maternal blood, whilst the endometrium has healed over the site of entry of the blastocyst so that in effect the ovum is now excluded from the cavity of the uterus and lies wholly within the endometrium.

The later details of the formation of the chorion and subsequently the placenta and of the differentiation and growth of the embryo must be looked for in textbooks of embryology. Although much of the matter found there will be of descriptive nature, experimental embryology is making great strides and the control of the processes of embryogenesis and of placentation are of fundamental interest.

TROPHOBLAST

At first the trophoblast consists of two layers of cells. Nearer to the embryo is the cytotrophoblast and outside this is the syncytiotrophoblast. It is probable that the syncytium arises from the cytotrophoblast. It has been seen above that the trophoblast is soon bathed in maternal blood. When it develops a core of loose fibrous tissue and a fetal circulation runs through this the chorion is formed. It is obvious that the fertilized egg must make its presence felt for it has to redirect the maternal organism to its own growth and survival. In particular it must prevent menstruation and since this will occur about 14 days after ovulation it has to influence the mother before this time is up. In fact the trophoblast is in contact with maternal blood about 11 days after ovulation. A fetal circulation is perhaps not present until about 28 days after ovulation when the heart chambers can be seen to be distended. The maintenance of early pregnancy is therefore dependent upon the trophoblast, which although it is a fetal tissue must be nourished by maternal blood. The maternal blood is the sole nutritive support of the placenta throughout pregnancy.

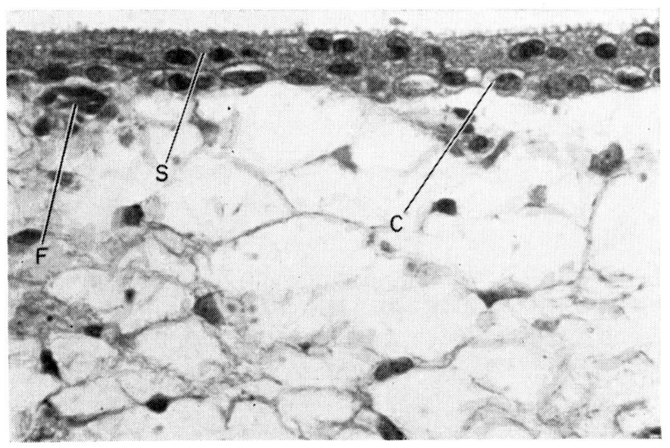

S = Syncytiotrophoblast. C = Cytotrophoblast.
F = Fetal capillary.

FIG. 9.5. Chorionic villus. This villus was from the tenth week of gestation. There are two layers of covering trophoblast. The outer syncytiotrophoblast has a microvillous surface and a finely vacuolated cytoplasm. The inner layer of cytotrophoblast is continuous at this stage of gestation but later it becomes interrupted. The villous core consists of mucous connective tissue containing fetal capillaries, fibroblasts and macrophages.

By histochemical methods it has been shown that the very early trophoblast contains RNA, glycogen and alkaline phosphatase, but its really important contribution to the welfare of the conceptus is the hormone chorionic gonadotrophin (HCG). This is probably produced by the cytotrophoblast which is mainly in evidence in early pregnancy, but tends to get relatively less as compared with syncytiotrophoblast later in pregnancy. Recent work suggests that the cytotrophoblast is more likely to flourish in a somewhat anoxic environment, whereas syncytiotrophoblast tends to succumb when there is an inadequate supply of oxygen. This is borne out by clinical experience, for when the placenta has had a diminished blood supply there tends to be a relative excess of cytotrophoblast, and in some circumstances (erythroblastosis fetalis) this increase in cytotrophoblast can be correlated with an increased excretion of HCG. Tissue culture experiments have also suggested that the cytotrophoblast is the site of production of HCG.

HCG probably acts upon the pituitary gland. It will be remembered that the basic cycle is as follows:

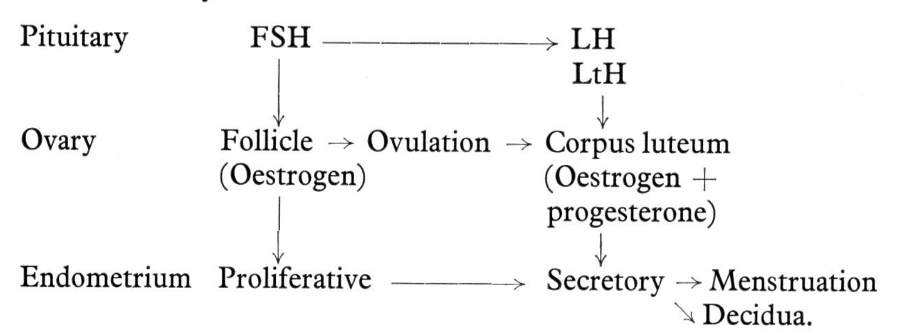

There are many feedback mechanisms in this cycle which maintain regular menstruation. HCG coming from the cytotrophoblast drives the right half of this cycle. That is it makes the pituitary continue with the production of LH and LtH so that the corpus luteum is maintained and its output of progesterone maintains the secretory endometrium and prevents it from breaking down in menstruation, which would destroy the embedded conceptus. In fact the increasing activity of the corpus luteum causes further changes in the secretory endometrium and converts it to the decidua. The secretory activity of the glands increases and the stroma becomes more obvious, the decidual cells becoming more prominent and their cellular outlines clearer. They seem to be loaded with nutritive materials. This serves to emphasize that the process of embedding depends upon the interaction of the trophoblast with the decidua.

A point of great interest is why the trophoblast stops its invasion once it has breached the capillary walls and comes to lie in the maternal blood. So far there is no explanation of this, but its interest lies in the control of differentiation and growth for it seems to be absent in most cases of cancer. Another factor to be considered is why the conceptus is not rejected by the mother. Usually any foreign protein injected into an animal causes the production of antibodies and if a graft is made this will usually be rejected. But this does not happen to the conceptus, half of whose genes are derived from the father and so the conceptus would be expected to produce proteins at least partly dissimilar from those of the mother. Later in life it can be shown that the offspring do in fact produce proteins which are foreign to the mother but at least in pregnancy there is some "immunological tolerance" of the conceptus by the mother. As pregnancy proceeds this tolerance is diminished as is well known in Rhesus incompatibility between the mother and the fetus. There is a suggestion that the trophoblast is covered by a layer of mucoid-like material which prevents the action of antibodies on the cell walls of the embedded placenta.

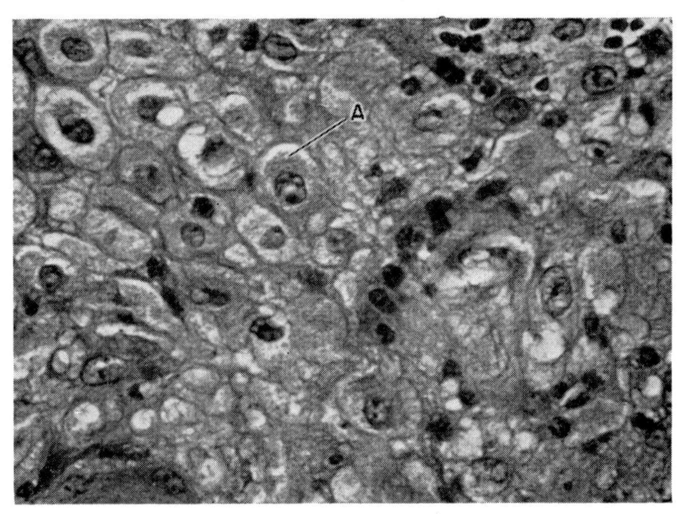

A = Abundant cytoplasm of decidual cell.

FIG. 9.6. Decidua. During pregnancy the stromal cells of the endometrium develop abundant cytoplasm, rich in glycogen, to form decidua. No intercellular ground substance is apparent between these decidual cells giving the stroma a compact appearance.

BIRTH CONTROL

With a greatly rising world population this is now a topic of major importance. In 1965 the world population was of the order of 3,300 million and it is expected to be about 6–7,000 million by A.D. 2000. Moreover, in the advanced countries, couples feel the need to restrict the size of their families for a variety of reasons. It is felt by many that unless the number of births throughout the world is regulated, the population will outrun the resources needed to maintain a reasonable standard of life. The subject raises ethical problems of a serious nature but whatever ethical beliefs may be held it is incumbent upon all educated people and particularly biologists and doctors to know something of the methods of contraception.

The word contraception has just been deliberately introduced because it raises the issue of what is meant by conception. In fact the word "conception" is so relatively vague that it is better not to use it. To some it means simply fertilization. To others it means all the processes from fertilization up to embedding. This may seem to be splitting hairs but it is important in discussing ethical considerations.

The methods of birth control can be grouped under various headings:

1. The suppression of the formation of gametes.
2. The prevention of fertilization.
3. The prevention of embedding.
4. Abortion of the embedded conceptus.

The inhibition of ovulation has recently become possible through the use of the so-called "contraceptive pill". This consists usually of an

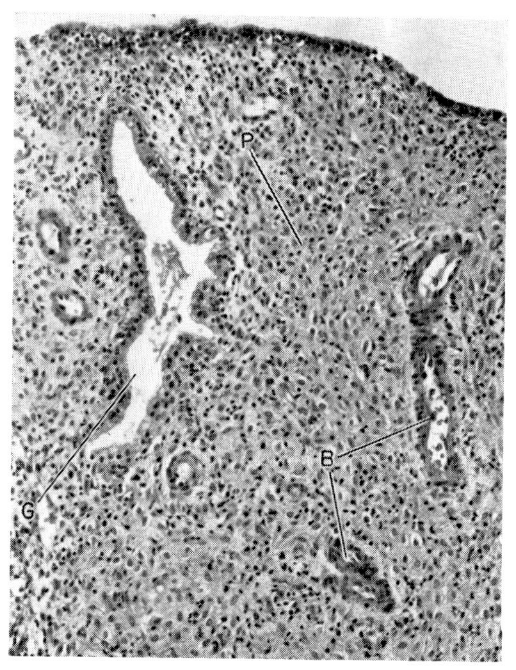

P = Pseudo-decidual stromal cells.
G = Gland. B = Blood Vessels.

FIG. 9.7. Endometrium under the influence of a contraceptive pill. The stromal cells have abundant cytoplasm and resemble decidua. The blood vessels have relatively thick walls but the gland is simple, without the tortuosity expected in the late secretory phase of the menstrual cycle or during pregnancy.

oestrogen combined with a progestagen. There are many varieties and combinations of these drugs on the market at present. They are taken orally from the 5th to the 25th day of the cycle, the first day of menstruation being day 1. They act on the pituitary to suppress the formation of FSH and LH. This is similar to the usual action of oestrogen and progesterone coming from the corpus luteum in the normal menstrual cycle. The body is therefore deprived of the two hormones which normally cause ovulation. In addition, however, these pills of combined hormones also affect the endometrium. They tend to reduce the amount of glandular tissue and relatively increase the stroma, so that even if an egg is fertilized it is possible that it would be unable to embed since there would be no proper decidua. Also the pills probably produce a change in the chemical and physical characteristics of the cervical mucus so that spermatozoa may be unable to get into the uterine cavity and so into the Fallopian tube. Finally there may be a direct action of the pills on the ovary itself suppressing ovulation. It will be seen that although these combined pills have been grouped under the heading of suppression of gamete formation they might also come under the headings of prevention of the meeting of gametes and prevention of embedding.

Experiments have also been conducted into the suppression of spermatogenesis and some success has attended the use of certain diamines. However, they have not found acceptance, since they are incompatible with the taking of alcoholic drinks. Large injections of testosterone may also suppress the formation of spermatozoa, but this is an inconvenient method and very expensive. A point of interest with both these methods is that it takes about two months from the beginning of treatment fully to reduce the sperm content of the seminal fluid. Also it takes about two months for the sperm count to return to normal after the cessation of treatment. This gives some indication of the cycle of production of spermatozoa in the normal testis.

Fertilization cannot occur without intercourse and perhaps abstinence from intercourse is the safest method of birth control but it is quite unacceptable. A widely used method is to try to predict the time of ovulation so that intercourse may be avoided at that time. This is the so-called "safe period" method. It depends upon the assumption that the life of the corpus luteum is almost always 14 days. Therefore ovulation will take place 14 days before the next period. Few women's cycles are exactly regular, but by careful observation the range of variation of the periods can be established. For argument assume that the cycles in a given woman vary from 26 to 30 days. If the life of the corpus luteum really is 14 days, and it usually is, then ovulation occurs 14 days before this, that is day 12 to day 16 of the cycle. Spermatozoa are probably capable of fertilizing the ovum for 24 hours after they have been deposited in the genital tract of the female. Therefore coitus on the 11th day of the cycle might fertilize an ovum shed the following day. Similarly the ovum is probably capable of fertilization for 24 to 36 hours after it has been shed from the ovary. Therefore coitus on day 17 might fertilize an ovum ovulated the day before. The woman might therefore be fertile from day 11 to day 17. Since the predictions are inevitably inexact a further day is included at either end so that the safe period is all those days outside days 10 to 18 of the cycle. As a matter of social preference intercourse is not usually practised during the time of menstruation so that the time when intercourse is interdicted may be quite long and some couples find it impossible to regulate their sexual activity in this way. Also it has been suggested earlier that intercourse may in some women actually cause ovulation in the first half of the cycle by some neurohormonal reflex, just as in rabbits and ferrets. This, if true, makes the "safe period" very unsafe. Despite its physiological drawbacks it is relatively free from ethical objections and has been accepted by the Roman Catholic Church as allowable for the spacing of children within families.

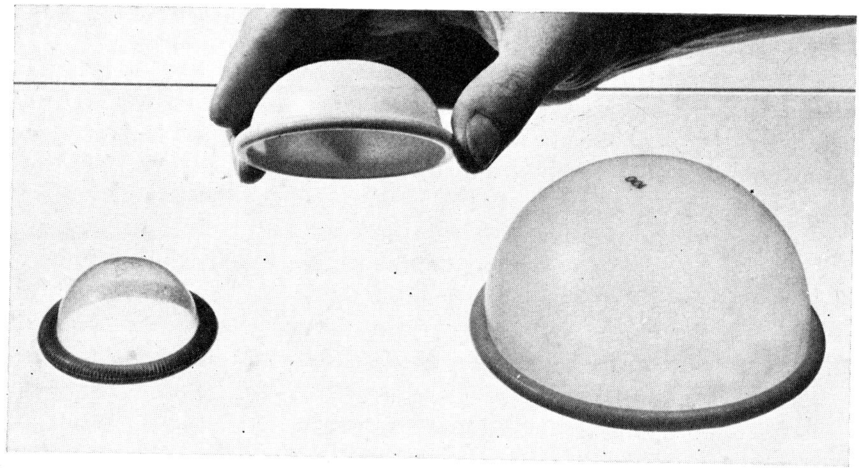

FIG. 9.8. Types of vaginal occlusive pessaries.

Which? supplement on Contraceptives (published by Consumers' Association, 14 Buckingham Street, London W.C.2.)

It is possible that future research may be able to pin-point the day of ovulation more exactly and then a safe period may be very safe. The temperature rise which occurs with ovulation is unfortunately not accurate enough for the prevention of fertilization, since it is probably due to the rising output of progesterone from the corpus luteum. This probably causes the temperature to rise some time after ovulation has occurred and so the egg may already be fertilized before the temperature goes up. Recent experiments have been directed to trying to detect the presence of LH (luteinizing hormone) by simple immunological methods in the urine or saliva. This might be an index that ovulation was imminent. Future research might be centred on a pill to be taken after coitus to prevent fertilization.

Most methods of birth control depend on the prevention of fertilization by interposing a barrier between the gametes. Most drastically the Fallopian tubes may be tied or removed, or the vasa may be similarly treated. More usually the male may wear a sheath or condom to contain the seminal fluid at ejaculation. Alternatively the female may wear a

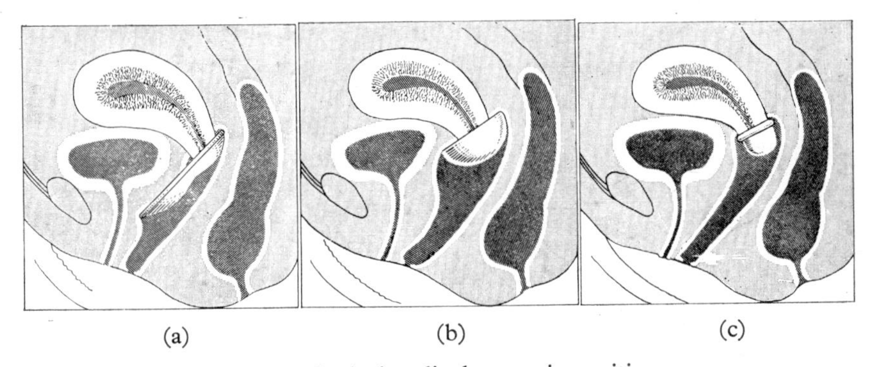

FIG. 9.9. Occlusive diaphragms in position.
Which? supplement on Contraceptives (published by Consumers' Association, 14 Buckingham Street, London W.C.2.)

rubber diaphragm stretched on a watch spring across the vault of the vagina. More rarely a small rubber cap is placed directly over the cervix. These last two female methods are usually combined with the use of a spermicidal cream whose base is often phenyl mercuric acetate. In this way it is hoped that any spermatozoa which pass the barrier may be killed by the spermicide. A somewhat less efficacious barrier may be formed by

FIG. 9.10. Application of spermicidal cream to Dutch cap.
Which? supplement on Contraceptives (published by Consumers' Association, 14 Buckingham Street, London W.C.2.)

using a foaming pessary which is inserted into the vagina. The foam entraps the sperm and also the pessary contains a spermicide.

The methods most often used today are those of the condom and the Dutch cap (the rubber diaphragm) combined with a spermicide. Rather less commonly used are the safe period and the foaming pessaries. Also in common use is the method of "Withdrawal" in which the penis is removed from the vagina before ejaculation has occurred. It is very undesirable since it usually prevents the woman from having an orgasm which would seem to be psychologically beneficial. Of more practical importance is the fact that fluid containing motile spermatozoa emerge from the penile urethra before overt ejaculation. Therefore the woman may become pregnant even though the withdrawal is apparently carried out exactly at the right time.

Recently introduced have been the intra-uterine contraceptive devices (I.U.C.D.) made of plastic. These are of various shapes and being pliable

FIG. 9.11. Intra-uterine contraceptive devices.

Which? supplement on Contraceptives (published by Consumers' Association, 14 Buckingham Street, London W.C.2.)

may be introduced into the cavity of the uterus through a special introducer without anaesthaesia. In general they are suitable for those women who have had children and so in whom the cervical canal is rather more lax than in the nulliparous patient. It was at first thought that these devices acted by preventing embedding of the fertilized egg or perhaps by causing an abortion of the conceptus after it had embedded. Both these actions are possible. Recently it has been suggested that by distending the uterus they may somehow prevent ovulation, though there is little evidence to support this. Alternatively it may be that the distension of the uterus may increase tubal motility so that the zygote is hurried through the tube and reaches the endometrium before it has the capacity to embed.

Abortion has been widely practised in all communities as a method of birth control, though ethically there may be objections to it. A miscarriage may be procured by such doubtful methods as hot baths, gin, or ergometrine which makes the uterus contract. For certainty, an instrument is usually passed through the cervix to disturb the embedded conceptus. Often women will procure abortions on themselves by inserting an enema nozzle through the cervix and forcing soapy or antiseptic solutions into the uterus. This is a thoroughly dangerous practice and

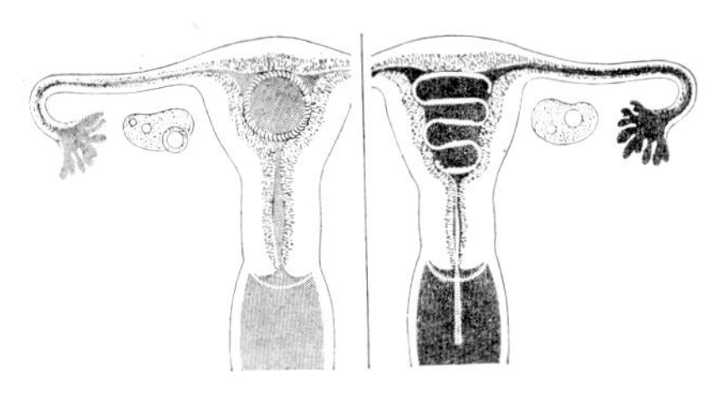

FIG. 9.12. Graafenberg ring and Lippe's loop in position.

Which? supplement on Contraceptives (published by Consumers' Association, 14 Buckingham Street, London W.C.2.)

may cause air embolus or the fluid may reach the peritoneal cavity through the Fallopian tubes and cause peritonitis.

It will be realized that the methods of birth control used in any community depend on cultural, psychological and social factors as well as upon economics and physiology.

The efficacy of birth control methods is measured in units of 100 woman years. This basic unit is of one hundred women using a particular form of birth control through one year, which really means about 1200 ovulations assuming that all are menstruating approximately once per month. The following table suggests the efficacy of the various methods,

Method	*Pregnancies per* 100 *woman years*
No attempt made to restrict conception	40 to 50
Simple douche after intercourse	31
Safe period	24
Spermicidal jelly alone	20
Coitus interruptus (withdrawal of the penis just before ejaculation)	18
Condom	14
Diaphragm (with or without spermicide)	12
Intra-uterine device	2?
Contraceptive pill	0–1

though it must be realized that some of the failures are not due to any inherent defect in the method but more in the way in which the method is used. Moreover, the more sophisticated the technique the more likely is it to fail among the uneducated, and therefore these advanced methods will more likely be used by the intelligent in whose hands any method is often effective.

SEX RATIO

It has been known for many years that more male than female babies are born. The sex ratio at birth in England and Wales is about 106 males for every 100 females. Since it has become possible to determine the sex of cells from the presence or absence of sex chromatin under the nuclear

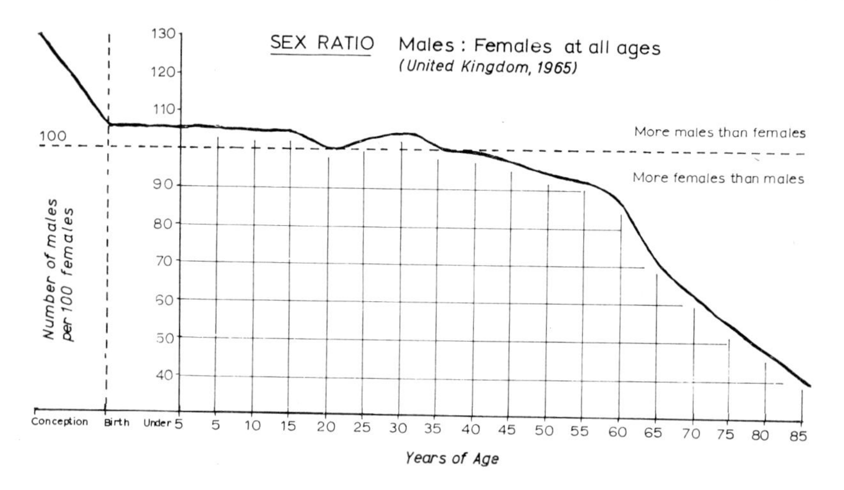

FIG. 9.13.

membrane (the Barr body) it has been found that the sex ratio at fertilization is possibly of the order of 120 to 130 males to 100 females. The estimates are derived from knowledge of the nuclear sex of abortuses and from the sex ratio of stillbirths and live births. It seems likely, therefore, that more males are aborted and die *in utero* than females so that the ratio is reduced from 130 : 100 at fertilization to 106 : 100 at birth. The reasons for this are complex but may be associated with immunological factors. It is known that the mortality for males is greater at all ages after birth and at maturity the sex ratio is about at parity, i.e. 100 : 100 due to the loss of males during the growth period. Of recent years this has changed as Fig. 9.13 shows, because more males are now living to reach maturity. In old age the ratio is still further reduced and may be of the order of 80 : 100 or even less.

The sex ratio differs in different parts of the world but the reasons for this are unknown. Also it is well attested that during wartime the sex ratio at birth rises and in the recent war it was about 108 : 100 in England and Wales.

The evidence suggests that the Y bearing spermatozoon has a better chance of fertilizing the ovum than the X bearing spermatozoa. The different types of sperm must be produced in equal numbers by the processes of mitosis and meiosis. It may be that Y chromosome bearing spermatozoa are more motile and active than X bearing ones, and with such a small cell as the spermatozoon the extra nuclear material may be relatively heavy in an X bearer, though this hypothesis seems unlikely when it is remembered that the sperm may be actively transported.

TWINNING AND OTHER MULTIPLE CONCEPTIONS

In populations of European descent the incidence of twins is about 1 in every 90 births. Early in pregnancy the ratio is higher because there is no doubt of the increased tendency to abortion when twins are present in the uterus. Under the stress of wartime conditions the twinning ratio falls but the reasons for this are not known. In the people of West African origin the incidence of twins is much higher and may be of the order of 1 in 50 or so.

Unlike twins result from the separate fertilization of two ova. The genetic complement of each of them is therefore different and they will be no more alike than any other two members of a family. Quite obviously they may be male and female. The incidence of such unlike twins increases with the increasing age of the mother and with the numbers of children she has had before. This suggests that there is a tendency for double ovulations to occur in older women. It may be too that herein lies a partial explanation of the lowered twinning ratio in wartime since the mothers then tend to be younger.

Like twins must come from the fertilization of only one egg and one sperm since the genetic material each receives is almost identical. Obviously they must be of the same sex. The formation of two separate individuals who are alike must come after fertilization and the totipotential cells of the morula must split into two halves. The mechanism of this process is unknown but it is genetically based. Some animals always produce like twins or like quadruplets, and identical twinning sometimes runs in a family. There may be a paternal effect since some men may sire twins in two different women.

Multiple births with more than two offspring arise from variants of the processes just described. Triplets may arise by multiple ovulation and fertilization, or from like twins and one separate fertilization and occasionally they may all be alike when the morula splits into three.

Chapter X

PREGNANCY

When the trophoblast has invaded the maternal endometrium and has made contact with the maternal blood stream by eroding into her capillaries it is able to begin to influence the maternal physiology, for the trophoblast is an endocrine organ. The endocrine secretions redirect much of the mother's physiology so that she fully nourishes the embryo and fetus within her uterus. It has already been pointed out that the trophoblast has to take over the control of the hypothalamic-pituitary-ovarian axis so that the endometrium may be maintained and converted into decidua (Chapter IX), and it must do this before the corpus luteum begins to break down about 10 to 12 days after ovulation. By about the 7th day after ovulation the trophoblast is within the maternal blood vessels and is producing chorionic gonadotrophin.

CHORIONIC GONADOTROPHIN (HCG)

The abbreviation HCG means human chorionic gonadotrophin. It is similar to the pituitary gonadotrophins, and is a glycoprotein containing mainly galactose as its carbohydrate constituent. In its physiological actions it is similar to LH and LtH. Although it is probably an over-simplification its action is best understood as being a taking over of the second half of the hypothalamic-pituitary-ovarian cycle. That is, once it is being secreted into the maternal blood stream it drives the hypothalamus to make the adenohypophysis continue to produce its LH and LtH. These two maintain the corpus luteum in the ovary and make it produce progesterone. This hormone continues to act upon the endometrium and so prevents it breaking down in menstruation. Moreover it causes the secretory changes in the endometrium to be intensified and so the embedded ovum is made secure (Fig 9.4.).

HCG is probably secreted by the cells of the cytotrophoblast. It seems to have been produced in tissue culture from these cells. The maximum amount of hormone is produced when these cells are most in evidence, and if under pathological conditions the cytotrophoblast increases (as in hydrops foetalis) the output of HCG increases.

HCG is excreted in the urine and is there the basis of the well-known pregnancy tests.

PREGNANCY TESTS

These are all performed on the urine. Usually an early morning specimen of the urine is used since then it is most concentrated. There are two main varieties of test, the biological and the immunological. In the first, animals are used. In the second the test is a kind of chemical reaction. In the biological tests the urine is injected into various test animals and the effect on the gonads is subsequently determined. The first animal to be so used was the mouse and the test was introduced by Aschheim and Zondek

Fig. 10.1. Immunological pregnancy test. This is the Pregnosticon test (Organon Laboratories Ltd). The test is based on an antigen-antibody reaction. The antigen is human chorionic gonadotrophin carried on specially sensitized red blood cells. Addition of rabbit anti-human chorionic gonadotrophin causes alteration of the sedimentation pattern of the erythrocytes. This is prevented by excess of antigen. The test urine is mixed with the antiserum and the sensitized red cells added to the mixture in an ampoule which acts as a test tube in the special rack. The reaction is allowed to stand for at least 2 hours. The test can be read in the angled mirror at the base of the stand. A clearly-defined brown ring indicates a positive result while a diffuse yellow-brown sedimentation indicates a negative result.

in 1927. The test took 5 days to complete and needed 10 mice to be sacrificed. Another animal was the rabbit in the Friedman test. Only one animal was needed and the answer came in 24 hours. In both mice and rabbits the end-point of the test was the appearance of reddish spots on the ovaries suggesting luteinization. The next animal used was the South African clawed toad (Xenopus laevis). Here the injection of urine containing HCG caused ovulation, which could easily be recognized as the ova were passed through the cloaca. The reaction took place about 18 hours after the injection. Later still the male toad (Bufo bufo) was used. The influence of the hormone on the gonad was recognized by the appearance of spermatozoa in the cloaca. They were discovered by aspirating the secretions in the cloaca and examining them under the microscope. The test took 3–5 hours to perform. In all these biological tests it was essential to see that the animals were under the influence of no physiological stimuli which might make them produce gametes. They had to be immature when mice and rabbits were used. Also all the test animals had to be kept away from others of the opposite sex so that they would have no sexual stimulation, and indeed since the sight of an animal of even the same sex might have been a stimulus they had to be isolated. Moreover, since urine is a foreign substance, there was a high mortality amongst the animals injected with it. Because of the costs and the difficulties the biological tests have almost been abandoned at least in routine laboratories and replaced by the immunological tests.

Immunological pregnancy tests depend on the demonstration of an interaction between HCG and the anti-hormone, anti-HCG prepared by the injection of HCG into some animal. The principle is to take the urine of the supposedly pregnant woman and incubate it with anti-HCG. The resultant fluid then has to be tested to see if any reaction has taken place. That is there are two phases to the test. First the incubation and secondly the indicator test. The indicators are of two main varieties. One is a suspension of latex particles coated with HCG, the other is red cells coated with HCG. Let it be assumed that the woman is pregnant and that the urine therefore contains HCG. When this is incubated with anti-HCG the result is a "neutral" solution. When it is added to the indicator nothing happens. If on the other hand the woman is not pregnant and the urine does not contain HCG the result is a solution of anti-HCG. When this is added to latex particles coated with HCG the result is to cause the particles to coalesce and form a precipitate. If the indicator is sheep's red cells and complement is added the result is to cause lysis of the cells which liberate their haemoglobin, which is easily recognized by its pink colour in solution. There are other variants on this immunological test but they only differ in the details of the indicator.

THE EXCRETION CURVE OF HCG

The graph shows the general form of the excretion of HCG. It will be seen that the maximal excretion is early in pregnancy with a peak at about 10 weeks or 70 days. Thereafter the amount in the urine declines quite rapidly to reach relatively low levels at about the 14th to 16th weeks, that is about 100 to 120 days. The excretion stays at this low level until term. The amount of HCG in the serum parallels this curve. The amount of HCG produced seems to depend upon the number of cells of cytotrophoblast in the placenta. It is known that by the standards of light microscopy the number of these cells increases rapidly in early pregnancy

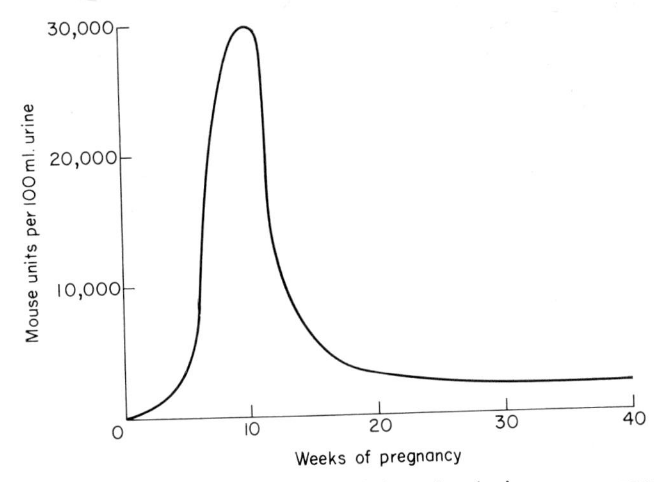

FIG. 10.2a. Concentration of HCG in *urine* during pregnancy.

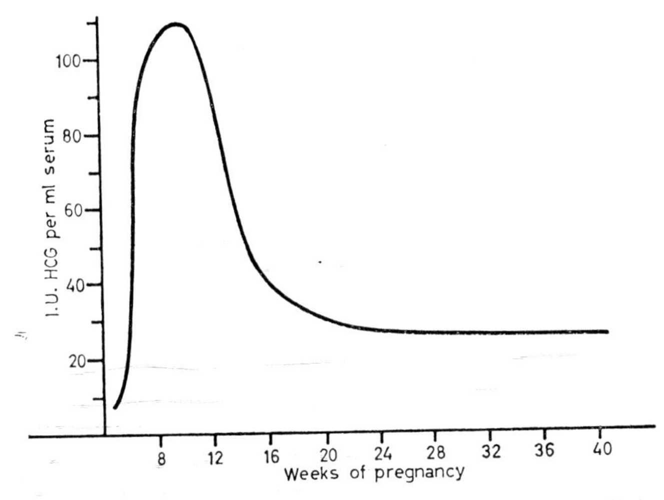

FIG. 10.2b. Concentration of chorionic gonadotrophin (HCG) in *serum* during pregnancy. From the data of Mishell, Wide & Gemzell (1963). Figures 10.2a and 10.2b are from *The Physiology of Human Pregnancy*, F. E. Hytten & I. Leitch, Blackwell (1964).

and then the number decreases until they can scarcely be seen at term. From this and from other evidence it is believed that the cytotrophoblast is the source of the hormone.

Occasionally because of some aberration of development the trophoblast proliferates greatly in the condition known as hydatidiform mole, and then no fetus is formed. With such massive growth of the trophoblast there is a very high excretion of HCG and this is used diagnostically, since the total amount of hormone excreted may be several hundred times higher with hydatidiform mole than in normal pregnancy. The same pregnancy tests are used as previously outlined but the test urine is diluted 100 or more times and if the test is positive this shows the very high excretion of HCG. However, the test needs care in interpretation and must be taken in conjunction with other clinical factors. Moreover it will be realized that the output of HCG even in normal pregnancy at about 70 days may be so very large that a pregnancy test may be positive even though the urine is much diluted. A further pitfall in the use of the pregnancy test is that the output of HCG may be so low very early after embedding and after the 16th week of pregnancy that the test may be negative.

The pregnancy test is really a test for HCG. Provided that this is known together with the shape of the excretion curve serious mistakes in interpretation of the test will not arise. Like all other laboratory tests it must be interpreted in the light of other findings which are predominantly clinical. Very few single tests of any kind have absolute validity in the clinicial situation.

The form of the excretion curve suggests that HCG is of great importance in the maintenance of early pregnancy but that its value in the economy of pregnancy declines later. It seems to have only the one function of maintaining the corpus luteum indirectly through the hypothalamus and adenohypohysis. It has not so far been shown to have any other effects though its exact mode of action has still to be demonstrated. Because of its luteinizing action it has been used therapeutically in conjunction with FSH to induce ovulation in some women in whom that function has failed. The general principle has been to give (say) ten daily doses of FSH and then 3 or 4 days of FSH and HCG combined. The results so far have been very good and tend to confirm the essential rightness of the views of the pituitary-ovarian cycle which have been outlined in Chapters II and V, that is that ovulation requires both the follicle-stimulating hormone and a luteinizing hormone. It is difficult to obtain LH directly from the pituitaries of recently dead people and so HCG has been substituted for it. HCG can of course be obtained in large quantity from the urine of pregnant women.

Early pregnancy seems to need the presence of progesterone which at first is supplied by the corpus luteum. Later the placenta is able to produce this hormone and the corpus luteum slowly regresses. The need for HCG therefore also disappears and this is shown in the excretion curve. In many animals the corpus luteum maintains the pregnancy till term and if the corpus luteum is surgically removed before that time then abortion takes place. This does not hold in Man where removal of the corpus luteum after about the 12th to 14th week has no direct effect on the pregnancy.

PROGESTERONE

This hormone has a most important part to play in the menstrual cycle (see Chapters II, III and IV) and also in the metabolism of many steroids especially in the adrenal glands. It is also produced by the placenta, probably by the syncytiotrophoblast. In Man it is essential for the maintenance of early pregnancy, for it is the hormone which causes changes in the endometrium such that the fertilized ovum can embed. For about the first three months of pregnancy it is secreted mainly by the corpus luteum, indirectly under the influence of HCG secreted from the cytotrophoblast. After three months the corpus luteum regresses but the secretion of progesterone goes on rising, to fall precipitately at the delivery of the placenta at term. It is for these reasons that the placenta is believed to be the major source of progesterone during pregnancy.

It has been estimated that the placenta produces about 30 mgm. of progesterone at the 12th week of pregnancy, about 40 mgm. at the 18th week and 200 mgm. or more up to 300 mgm. in the last few weeks of pregnancy. Despite these large amounts the content of the blood is of the order of 0·14 μg per ml. It is, therefore, very difficult to estimate in this fluid and cannot yet be used for routine clinical purposes. Injections of radioactive progesterone show that it very rapidly disappears from the blood stream and is distributed throughout the tissues, about one-third to one half being found in the fat depots. This explains its low concentration in blood.

The main known place of progesterone metabolism is the liver. Through a series of steps it is broken down to pregnanediol and other less important metabolites. Pregnanediol is excreted in the urine, and its quantity there is such that it can fairly readily be estimated. The estimations are performed on the amount of urine excreted in 24 hours.

THE PREGNANEDIOL EXCRETION CURVE IN PREGNANCY

The graph shows the general shape of the curve of excretion in pregnancy. It will be noted that the amounts excreted rise from a level of about 10 mgm. in 24 hours at about 10 weeks of pregnancy to about 40 mgm. at 38 weeks. Thereafter the level drops slightly till delivery. It will be noted too that there is a wide range of levels excreted in normal pregnancy so that the peak excretion may be between 20 and 70 mgm.

The fall-off in the excretion, and presumably the production, of progesterone at the end of pregnancy has some interest for it is believed by some that this fall may in part be responsible for the onset of labour. There is little doubt that progesterone has an effect on uterine muscle, and in general this is in the direction of damping down its contractions. That is it may act as a brake on the myometrium. It will later be seen that the excretion of oestrogens continues to rise steadily till the onset of labour and it has been suggested that the altered proportions of these two hormones may take the "brake" off the uterine muscle and apply the "accelerator" so that the uterus contracts and expels its contents to the outside world. This is a first approximation in the theory of the cause of the beginning of labour which will be taken up in more detail later.

Pregnanediol excretion can be used as an estimate of placental function. Especially in hypertension the placenta does not function well and the fetus may die *in utero* or be born small and may die shortly after birth

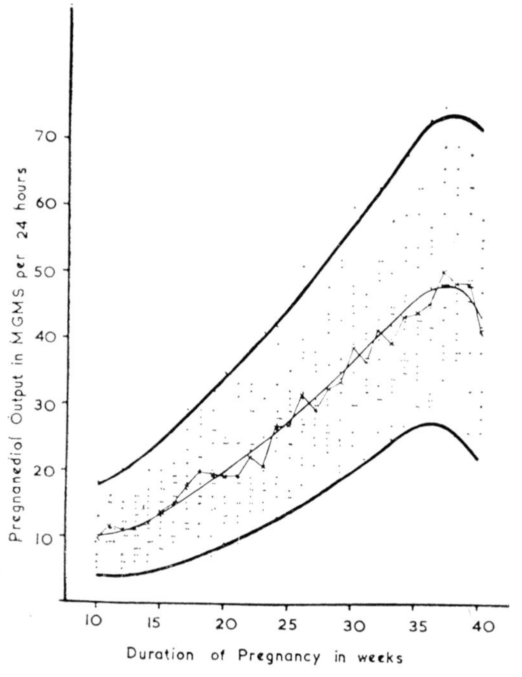

FIG. 10.3. Pregnanediol excretion curve in pregnancy. From C. S. Russell, M. G. Coyle, *Jl. Obst. Gynaec. Brit. Emp.* **64,** 649, (1957).

since it has been starved of essential nutrients. Placental function will be dealt with later, but if the pregnanediol excretion measured on several occasions is low for the time in pregnancy then it may be taken as a guide to placental efficiency, though it must be interpreted in the light of the clinical findings such as the blood pressure and the estimated size of the fetus.

ACTIONS OF PROGESTERONE IN PREGNANCY

The effects of progesterone are widespread, and it is responsible for much of the redirection of maternal physiology.

It has been seen that the endometrium is converted from a secretory pattern to that of decidua so that the fertilized ovum may embed. In addition it is the progesterone, acting as always synergistically with oestrogen, which is responsible for the growth of the uterine body to accommodate the growing products of conception. This action on the myometrium probably involves two processes, the first of which is the actual growth of the muscle fibres and the second the alteration of the pattern of myometrial contractions. In the early months of pregnancy the uterus grows more rapidly than the products of conception and the gestational sac does not fill the whole uterine cavity until about 12 weeks of pregnancy. The growth of the muscle is almost certainly due to progesterone. It is certain that the muscle cells of the uterus grow in size, but it is not known for certain whether there is an actual increase in the numbers of muscle fibres, but it is likely that this does occur.

The cervical mucus forms a filter preventing infection rising up from the vagina into the cavity of the uterus. If infection does involve the cavity in pregnancy the fetus dies. In Chapter IV it was pointed out that the character of the cervical mucus changes during the menstrual cycle, it being crystal clear and runny when the cervical glands are under the influence of oestrogen but tacky when under the influence of progesterone. The mucus is like this during the whole of pregnancy and forms a plug in the cervical canal, which is expelled at the beginning of labour. It is probable of course that the plug is undergoing constant renewal during pregnancy, and this is suggested clinically by the fact that the discharge from the vagina increases greatly during pregnancy.

The vaginal epithelium shows the effects of progesterone by a relative reduction in the numbers of superficial cells obtained in smear preparations. A typical count would be 0/90/10, meaning that in a count of 100 cells there would be no basal cells, 90 intermediate cells and 10 superficial cells. Any reduction in the relative number of intermediate cells and increase in the superficial ones would suggest an imbalance as between oestrogen and progesterone. A fall in progesterone output usually bodes ill for the success of the pregnancy so vaginal cytology has been used as an index for prognosis especially in abortions and in suspected placental insufficiency. It has the merit that it can be used in almost any routine laboratory whereas hormone excretion studies can be performed in few.

Fallopian tube motility probably decreases during pregnancy and its epithelium which is normally columnar becomes more cubical and it shows no evidence of secretion as it does at ovulation. These are probably hormonal effects and perhaps mainly due to progesterone.

The breasts are prepared for their function of lactation by oestrogens and progesterone acting in concert. It is probable that the growth of the ducts is mainly under the influence of oestrogens and the growth of glandular tissue under the influence of both the hormones.

Lactation after delivery depends on the pituitary secreting both prolactin (which is probably the same hormone as LtH, luteotrophin) and oxytocin. The stimulus to the production of prolactin is probably the withdrawal of oestrogen and progesterone consequent upon the delivery of the placenta. Thus these two hormones keep the anterior pituitary in check throughout pregnancy. In addition this involves the suppression of the formation of FSH for Graafian follicle development and ovulation are prevented during pregnancy.

Like many other steroids progesterone has an effect on kidney tubular function. In general it causes a relative increase in the excretion of sodium, but this is in large measure over-ridden by aldosterone and desoxycorticosterone. It will be seen later that the sodium and water balance of pregnancy is of great importance and it may be that in this context the action of progesterone could be significant.

Apart from its effects on uterine muscle progesterone acts on other smooth muscle. There is a general sluggishness of the alimentary musculature, especially early in pregnancy, and this shows itself clinically as a tendency to constipation. Stomach emptying time tends to be slow too. The veins, especially of the legs, increase in diameter and may become varicose. Some of this effect is believed to be due to progesterone. The ureters often dilate in pregnancy and though some of this may be due to pressure from the growing uterus, again progesterone contributes to the enlargement of their diameters. *In vitro* the response of the heart to digitalis preparations can be modified by progesterone, and digitalis also affects uterine contractions *in vitro*. These observations have no known clinical significance.

Other effects of the hormone are on respiration, for progesterone can lower the pCO_2, and early in pregnancy the output of progesterone raises the body temperature by about 0·6 to 0·8°F. This effect wears off at about 12 to 16 weeks of pregnancy.

Virtually nothing is known of the cellular effects of progesterone but it has been suggested that it influences enzymatic reactions requiring TPN. But it should be obvious from the above examples of its action that its effects are very widespread affecting cellular function at a fundamental level, but exactly how will involve much further research. In non-pregnant women progesterone diminishes the oestrogen uptake of the nuclei of endometrial cells.

OESTROGENS

The three main oestrogens of pregnancy are the same as those met with in the menstrual cycle, that is oestradiol, oestrone and oestriol. It should be realized, however, that there are smaller quantities of other oestrogens to be found but these have not yet been shown to be of physiological importance. Like progesterone it seems that the site of production of oestrogens is the placenta and the cells responsible are those of the syncytiotrophoblast. The level of oestrogen production rises throughout pregnancy and in general parallels the mass of placental tissue, and when the placenta is delivered the amount of oestrogen excreted shows a precipitate fall. Since there is no macroscopic or microscopic evidence of activity in the ovaries at the time of delivery it is assumed that the placenta must be the source of the hormones.

The level of total oestrogens in the blood rises from about 1 μg per 100 ml. at 12 weeks to about 17 μg at term. These are too small to be of value in the clinical field as yet. The oestrogens are probably widely distributed in the tissues, and indeed can be demonstrated by autoradiographic studies to penetrate the cells. In some areas the oestrogen is mainly in the cytoplasm and in others mainly in the nucleus.

The main site of metabolism of the oestrogens is the liver. This organ conjugates oestradiol and oestrone with glucuronic acid, or breaks them down and metabolizes them to compounds with greater solubility in water. These more soluble compounds are partly excreted through the bile and into the gut and partly they are excreted from the blood through the kidney and into the urine. Much of the oestrogen that is excreted into the gut is reabsorbed and passes to the liver again. Finally about 10 per cent of the oestrogens is excreted in the faeces and about 65 per cent in the urine. The rest of an injected dose is "lost" and this gives an indication of the fact that there is still much to be known about this sex steroid.

THE EXCRETION CURVE OF OESTROGENS IN PREGNANCY

The graph shows the excretion curve of the various oestrogens during pregnancy. It will be remembered that during the menstrual cycle the output of oestrogens is measurable in μg, there being a peak of about 60 or so μg at the time of ovulation. During pregnancy the output is measured in mgm., rising to a figure of the order of 35 mgm. This gives an indication of the vast amounts of oestrogens secreted by the placenta.

An unexplained feature of the excretion of oestrogens in pregnancy is the enormously increased amount of oestriol as compared with the other two main oestrogens. In the menstrual cycle the ratio of oestrone: oestradiol: oestriol in the urine is 1·0 : 0·3 : 1·1. At the end of pregnancy this has changed to 1·0 : 0·4 : 22.

As with progesterone the high output of the oestrogens falls precipitately as soon as the placenta is delivered.

Oestrogen excretion in pregnancy has been used as a measure of placental efficiency, and so of the welfare of the fetus. The techniques can be difficult, though they have recently been simplified and so they are becoming widely used routinely. One of the more interesting features of oestrogen curves is that they seem to be related to the fetus rather than to the placenta, so that in a case where the fetus dies the oestrogen, and particularly the oestriol, excretions fall before those of pregnanediol. The rôle of the fetus in oestrogen metabolism will be left till later.

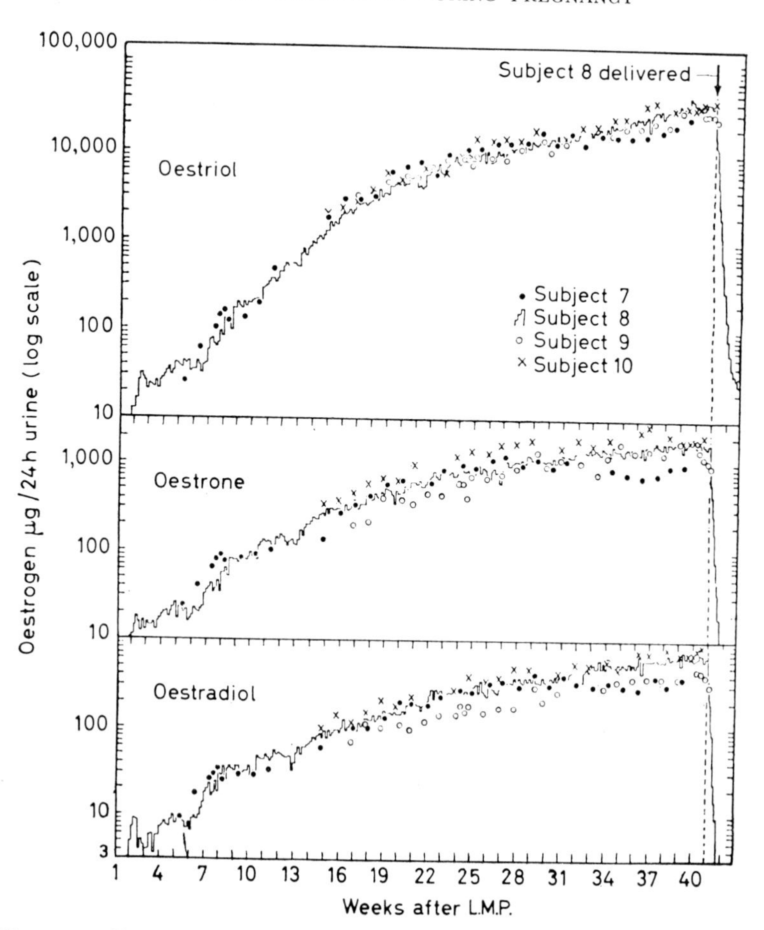

FIG. 10.4. From *Modern Trends in Human Reproductive Physiology*, ed. H. M. Carey. Chapter IV by J. B. Brown. Butterworths (1963).

The place of oestrogens in body economy is similar to that detailed in the section on progesterone. Progesterone cannot act at all until oestrogen has primed a particular tissue. There is synergism between the two sex steroids. The two acting together are responsible for the changes in the decidua, in the myometrium both in muscle growth and contractile

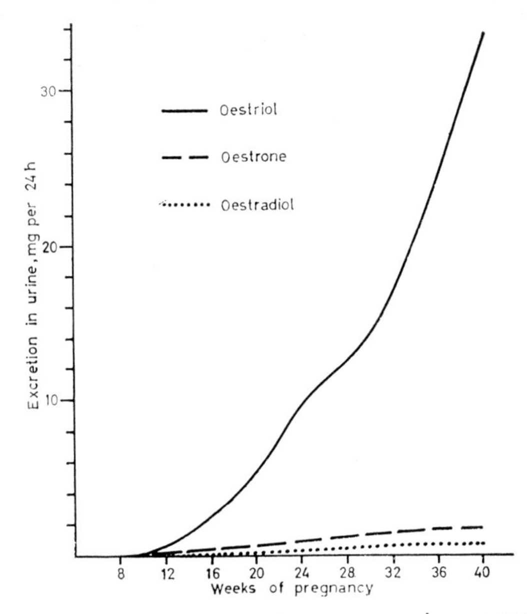

FIG. 10.5. Excretion of the three major oestrogens in pregnancy (Brown, 1956). *The Physiology of Human Pregnancy*, F. E. Hytten & I. Leitch, Blackwell, (1964).

activity, in the alteration of cervical mucus and in the vaginal epithelial cytology. Moreover they are needed together to bring about the changes in the Fallopian tubes and the breasts, as well as in modifying the activity of the pituitary and the kidneys and in changing fluid and electrolyte balance. Little is yet known of the action of oestrogens at cellular level but there is evidence that they affect the synthesis of proteins and so probably of enzymes.

Chapter XI

PLACENTAL FUNCTION

Up to this point we have looked at the growth of a girl from her birth and followed her through the establishment of her ovarian, menstrual and pituitary cycles so that she may arrive at a time when she is physiologically able to conceive and bear a child. The processes of sexual intercourse, fertilization, embedding of the ovum and early differentiation have been followed. From here the mother and her products of conception live together in symbiosis for the length of pregnancy. Each modifies the other, though each to some extent must live his or her life independently. We are concerned mainly with the points at which they interact. The only real point of anatomical interaction is at the placenta, but through the mediation of this remarkable organ both the fetus and the mother have to some extent to be controlled. For the next part of this book the maternal adaptations to her products of conception will be considered. Later will come the fetal adaptations to its life within the uterus.

At the outset it must be realized that the fertilized ovum consists of two main parts, that is the fetus proper and its apparatus designed for adaptation to the uterine environment. This apparatus consists of the placenta, the umbilical cord, the chorion and amnion and the liquor amnii. These all increase greatly in amount and size during the nine months of pregnancy so that taken together they may come to weigh something of the order of 10 or 11 lb. But this growth is controlled and organized and this is one of the phenomenal characteristics of the whole process. To accommodate this massive proliferation from two microscopic cells to a fully formed baby and its secundines requires adjustment of almost every system in the mother's body. Thus the fetus and its secundines need to grow and for this they need nourishment. They need to be able to respire and to excrete waste-products. Much else is also needed to supply the needs of all kinds of metabolism. It is to subserve these ends of the fetus that the mother is so changed in her physiology. The changes are vastly important both for the individual mothers and also for the race as a whole, for the species must in large measure depend on the efficiency of its processes of reproduction. Whilst for present purposes the emphasis is to be laid on the physiological processes of adaptation it must not be lost sight of that adaptations are needed by the individual in the emotional, mental and sociological spheres, and society too has to respond to the needs of new lives. Just as physiological adaptations may be smooth, or they may fail giving rise to physical disorder, so may psycho-social adaptations be failures or successes and these may have repercussions for society and its mental health, and also psycho-social failure of adaptation may have its physiological effects. The life of a woman, and especially of a pregnant woman, is an interconnected nexus. Any abstraction, such as that of physiology, from the integrated whole is inevitably only a part of the truth. It must be with this realization in mind that the abstraction can usefully be made.

The chorion is the essential structure through which the fetus obtains its nutrition. At the placental site the chorionic membrane is thickened to

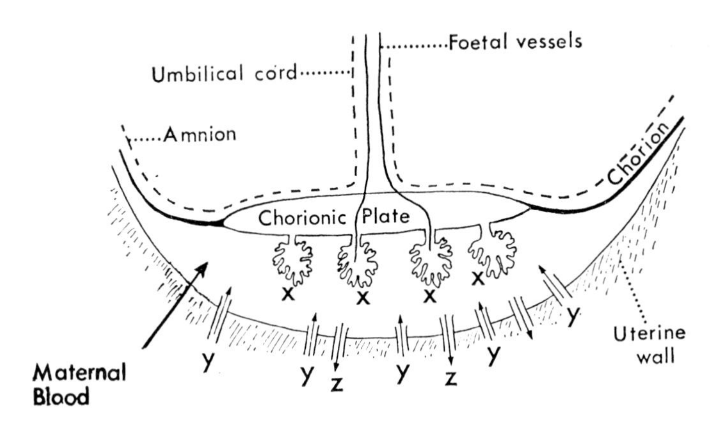

FIG. 11.1. From *An Introduction to Gynaecology and Obstetrics* (Rhodes, 1967) Lloyd-Luke.

form the chorionic plate, which may be looked upon as the skeletal framework of the placenta. The maternal or uterine surface of the chorionic plate is thrown up into folds which are covered with trophoblastic epithelium. These folds increase the effective exchanging surface between the maternal blood stream and the fetal blood to about 10 square metres. The chorionic plate is held in position against the lining of the mother's uterus (decidua lying on myometrium) by the chorionic membrane at the edges of the chorionic plate and by villi which anchor themselves to the decidua. The chorionic plate therefore seals off a space, the chorio-decidual space, into which maternal blood is poured through several arterioles. From here the blood is returned to the maternal circulation by way of venules.

Into the fetal side of the placenta comes the fetal blood conveyed along the umbilical arteries, and the umbilical vein takes it away again. The fetal arteries break up into capillaries as elsewhere in the circulation, but on the maternal side there are no capillaries since the maternal blood lake is in effect one vast capillary. The maternal blood flows all round the villi, that is the folds of the chorion and trophoblast, and on the other side of this area the fetal blood is flowing through capillaries so that there is opportunity for the exchange of materials across the barrier without the two blood streams intermingling.

From this brief description, which is only intended to make later explanations intelligible, it will be realized that there are three main parts to the placenta, viz.: (i) the maternal input of blood, (ii) the parenchyma of the placenta itself, that is the chorion plus trophoblast, and (iii) the fetal input of blood. On these three finally depends the efficiency of the placenta in doing its job which is to nourish (in its widest sense) the fetus, and to control the mother so that she supplies the materials with which the placenta can see to its care of the fetus.

The functions of the placenta are, in broadest outline, nutrition, respiration and excretion for the fetus. In addition it produces the hormones HCG, oestrogens and progesterone. For all this its raw materials are in the first instance brought to it by the maternal blood. Therefore the first consideration must be of the changes in that blood. Later must come the circulatory adaptations to see that that blood reaches the site. Following this it has to be seen how the mother obtains the necessary increases in oxygen and nutriments to supply her fetus and how she copes with the excretory products that it produces.

Chapter XII

THE BLOOD VOLUME

THE BLOOD VOLUME

During pregnancy a great deal of new maternal tissue is manufactured, especially in the uterus and the breasts. These areas increase the size of the vascular bed. Also there is redistribution of blood, most evident being an increase in the flow of blood through the skin. Theoretically it might be possible to use the same amount of blood in pregnancy as was used by the non-pregnant women by shunting it from one area to another. In fact, however, the blood volume increases during pregnancy to fill the enlarged vascular bed.

There are several ways in which blood volume can be measured. In essence they all depend on the injection of a known amount of a substance, allowing it time to mix with the blood, then estimating by how much the substance has been diluted. Then, for instance, if 10 ml. of the test substance has been diluted 1000 times, that is to 0·01 ml. in 10 ml. of blood, then it has been distributed in 1000 ml. It is obvious that any substance injected must be easily recognizable in small amounts after it has been diluted. Moreover, it is obvious that the test substance must not, under the conditions of the experiment, escape outside the vascular tree nor must it form pools anywhere within the vascular bed. Since the blood is made up of plasma and cells it is possible to associate the "indicator" with either the fluid or the cells. Both methods have been used and the best results are obtained when both are used together. However, this has not yet often been done in pregnancy. If only plasma or only cells are used, then the haematocrit reading is needed to obtain the percentage of cells within the total amount of blood. This brief survey of the principles of the method obscures a lot of technical difficulty and also difficulties in interpretation.

Evans Blue is a dye which has been much used to measure the circulating plasma volume. It is measured colorimetrically before and after mixing. Alternatively red cells can be "tagged" with ^{32}P or ^{51}Cr and sometimes radio-isotopes of iron. Having obtained the total dilution of the substance the proportion of plasma and of red cells contributing to the dilution are estimated from the haematocrit. There are pitfalls in this too.

There is much biological variation between women as regards the blood volume. Nevertheless, it is valuable to have some idea of the average of a group of normal people. This has been done for pregnancy by the Aberdeen workers, who have made many of the most meaningful physiological observations. Realizing that there are great variations round the averages they may be given as follows for blood volume.

Non-pregnant	Plasma	2600 ml.
	Red cell mass	1400 ml.

Increase over non-pregnant	10 weeks	20 weeks	30 weeks	40 weeks
Plasma	50 ml.	550 ml.	1150 ml.	1000 ml.
Red cell mass	—	50 ml.	150 ml.	250 ml.

Points to note about these figures are that the total increase of blood volume is of the order of 25 to 30 per cent, but that the percentage increase of plasma is almost 50 per cent whilst that of the red cells is only 18 per cent. This shows that there is a much greater increase in plasma

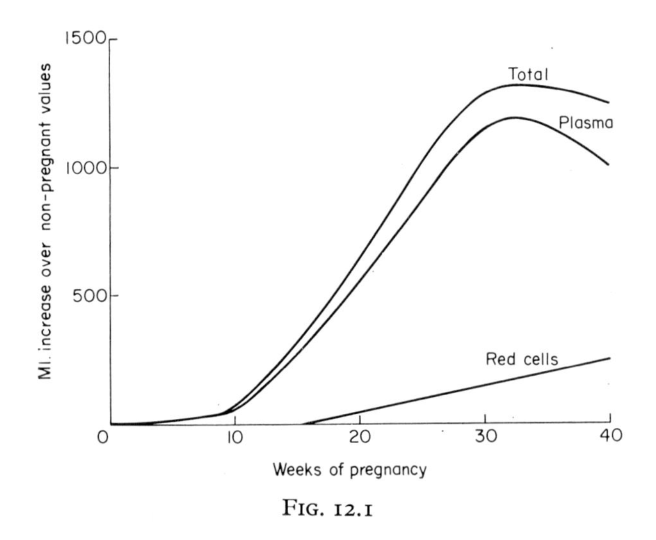

FIG. 12.1

than in cells and this is a fundamental feature of pregnancy. Moreover the maximal increase of plasma is found at about 34 weeks when it amounts to 1300 ml. The graph drawn from these figures demonstrates the points, and also that whilst the plasma volume tends to decline from 34 weeks to 40 weeks, the red cell volume still goes on slightly increasing. The changing blood volume is important since it will be seen later that its changes are paralleled by cardiac output and by certain changes in the extra-cellular blood volume.

THE RED CELL VOLUME AND HAEMOGLOBIN

The red cell total volume goes on increasing until term, that is for 40 weeks from the last menstrual period. On average the pregnant woman at term will have 85 more grams of haemoglobin than her non-pregnant self. The non-pregnant red cell mass is about 1400 ml. and the rise in pregnancy reaches about 250 ml. or an increase of about 18 per cent. Because the plasma volume increases by very nearly 50 per cent it will be realized that the red cells are diluted in pregnancy. The ordinary methods of measurement of haemoglobin determine only how much haemoglobin there is in 100 ml. of total blood. Therefore, inevitably in pregnancy, the level of haemoglobin as estimated by these usual methods must fall. This has led to the concept that there is a "physiological anaemia of pregnancy". But it will be realized that although the relative amount of haemoglobin per 100 ml. of blood has fallen the actual total amount of haemoglobin in the body has increased. Therefore, there can only be a "physiological anaemia" in one sense, but not in another. However, clinicians have been impressed by the worse results for the mother if she becomes grossly anaemic in the relative sense, and so it is now usual to give supplements of iron to all pregnant women. When this is done it is found that the total circulating haemoglobin may rise by about 30 per cent rather than by the 18 per cent if the woman does not receive iron. This fact has been used as evidence that the "physiological anaemia" is a misnomer. However, this may be taking a too simplified view for it can be shown that iron therapy is a stimulant to the bone marrow and even in non-pregnant patients, who are not anaemic by any standards, it will cause a rise in haemoglobin level of the blood. Perhaps this is all an argument in semantics only, which stems from the fact that it is difficult to define exactly what is meant by the word anaemia. In practice it seems to have been of value to give iron therapy to all pregnant women, and even folic acid as well, in certain clinics.

It has been suggested that the relative dilution of the blood in pregnancy serves a useful purpose in lowering the viscosity and so allowing of more efficient perfusion of the placenta. There is no doubt that the viscosity of the blood is lowered from a non-pregnant value of about 4·61 to 3·84 or less when the blood is compared with distilled water. There is no evidence for or against this concept of easier perfusion.

Other effects of haemodilution in pregnancy are seen in the lowered haematocrit reading which falls from 40–42 per cent of cells in the non-pregnant woman to about 34 per cent in the pregnant. Also it can be shown that the percentage of water in the blood rises from about 91 per cent to about 92 per cent, and that electrolyte levels and protein levels fall. For instance the sodium level falls by about 4 mEq. per litre from a non-pregnant level of 143·3 mEq. All the other electrolyte levels fall too but not usually by as much as this. Protein levels fall from a level of just above 7 g. per 100 ml. to just under 6 g. per 100 ml.

The relative changes in haemoglobin are the result of the dilution by the plasma and also by the amount of production of haemoglobin. In similar fashion the levels of electrolytes and protein are not fully explicable only in terms of dilution. There are changes in the physiological control of these substances as well as simple dilution. This is evidenced by the fact that the electrolytes do not all have their concentrations reduced to the same degree, and as far as the proteins are concerned the albumin fraction falls a great deal more than the globulin. For fuller details of these changes, which are off the present main theme it is suggested that the reader should consult more advanced books such as that of Hytten and Leitch on the *Physiology of Human Pregnancy*. (1964).

THE IRON REQUIREMENTS OF PREGNANCY

It is well known that women hover on the verge of anaemia throughout their lives mainly because of menstruation. The total haemoglobin mass holds about 2500 mgm. of iron and there are stores of about another 1000 mgm. in the liver, spleen and bone marrow. This 3500 mgm. is

TABLE XII.1.

	Iron	*Mgm.*
Stores	Haemoglobin mass	2500
	Liver, spleen, bone marrow	1000
		3500
Losses	Daily	1·0–1·5
	Menstruation	10–30
	Fetus and placenta	500
	Blood loss at delivery	180
	Lactation	180
Intake	Absorbed from food daily	1·0–1·5
	Iron therapy must supply in pregnancy and lactation	*c.* 1000

depleted daily by losses of about 1–1·5 mgm. but they are replenished by the 10–15 mgm. taken in daily in the normal diet. However, of this 10–15 mgm., only about 10 per cent is absorbed into the bloodstream so that the intake only just balances the output. But women lose in addition anything from 10 to 30 mgm. of iron with each menstrual period, and during reproduction they have to find about 500 mgm. for the fetus and placenta, and 180 mgm. for the blood loss after delivery. Therefore, unless the pregnant woman receives iron supplements, the 750 or so mgm. she has to supply for reproductive needs may almost deplete her iron stores. Added to this must also be another 180 mgm. needed for lactation. If before pregnancy her iron stores are full she may just be able to supply the reproductive needs and maintain her own haemoglobin level, but if her stores have been previously depleted by heavy menstrual periods she must inevitably become anaemic. The depletion of her iron stores will show itself in a low level of serum iron even before the haemoglobin level falls. The normal level of serum iron varies from laboratory to laboratory but is of the order of 70 to 160 μg. per 100 ml. of blood.

Another possible factor affecting iron metabolism in pregnancy might be an alteration in absorption of the mineral from the gut. However, there is no direct evidence of an increase or decrease in such absorption, though since there is an increased demand for iron in pregnancy the uptake does tend to increase, but this appears to be no different from the condition pertaining in the non-pregnant since the uptake of iron seems to depend upon demand.

In similar fashion to the demands for iron there are demands for folic acid, another important factor in the manufacture of haemoglobin. Megaloblastic anaemia is not at all uncommon, and biochemical tests show that the level of serum folate falls during pregnancy and that abnormal metabolites of histidine, namely formimino-glutamic acid, FIGLU, appear in the urine when folic acid is deficient. This evidence has led to many clinicians giving all their patients folic acid at least during the last three months of pregnancy. Surveys of large numbers of pregnant women have shown that when they are given both iron and folic acid supplements the numbers of them who reach term with a haemoglobin under 10 g. (about 70 per cent) is very much reduced. The amounts of the supplements of iron and folic acid are 400–600 mgm. of ferrous sulphate daily and 10–15 mgm. of folic acid weekly for the last 12 or more weeks of pregnancy.

THE PLASMA AND TISSUE FLUID VOLUMES

It has been seen that the plasma volume increases by about 1250 ml. in pregnancy, over the normal non-pregnant value of 2600 ml., an increase of almost 50 per cent. In first pregnancies the increase is less than in later ones, and after the first pregnancy the plasma volume may increase by 1500 ml. However, it will be remembered that the maximum increase in the plasma volume is at about 34 weeks of pregnancy and that thereafter it declines to 40 weeks by about 200 ml. Since the haemoglobin rises steadily to term it is obvious that the maximum dilution of the blood occurs at 34 weeks. These changes in the fluid content of the blood are intriguing and have not so far been fully explained but the possible physiological mechanisms for the changes must now be considered. They are of especial importance since changes in fluid balance may be very significant in the pathology of pregnancy.

The increased blood volume of pregnancy is mainly in the plasma, which is mainly made up of water, so the explanation of the increase is concerned essentially with the retention of water in pregnancy. The amount of water in the body must depend on intake, on output and on the distribution of water.

The intake of water is by fluid which is drunk and by fluid contained in apparently solid food. In addition there is also some water derived from the oxidation of fat. In a civilized community the amount of food and fluid taken in is largely determined by habit and desire and almost never is it determined by the availability of supplies. In general, more is delivered to the metabolic processes of the body than are strictly required. As far as fluid is concerned the controlling mechanism for the seeking of water is thirst and for food it is hunger, but these are not of extreme degree in advanced communities. They will not be further examined here. For the construction of a 24-hour balance sheet of water the following is usually given:

Intake	
As fluid	1500 ml.
In food	1000 ml.
From oxidation	300 ml.
	2800 ml.

For clinical purposes it is usually assumed that the intake in health should be of the order of 3 litres per day.

The output of fluid is insensible, due to losses from lungs and skin and a little is lost in the faeces. The final adjustment of water loss is made by the kidneys. Under extreme conditions there may be sweating, but in temperate climates this will not amount to a great deal and will soon be made up by drinking. The following scheme shows the second half of the fluid balance sheet:

Output		
Insensible loss	Skin	800 ml.
	Lungs	400 ml.
	Faeces	100 ml.
	Urine	1500 ml.
		2800 ml.

The important thing to note about this output is that the insensible loss is obligatory and little can alter it, simply because it is the expression of a wet warm body in air, which is less wet and less warm. The final regulator of the body water is therefore essentially the volume of urine and this depends on renal physiology. As a generalization, keeping in mind the reservations just suggested, it can be said that the increased water volume of pregnancy, which is such a characteristic feature of that state, is due to alterations in renal physiology. The essential problem is then an understanding of the factors which alter renal mechanisms. It will be seen later that glomerular filtration rate, renal plasma flow and renal tubular function are all changed in pregnancy. The tubular function is altered by the action of some hormones upon the cells of the tubules, including aldosterone, adrenal corticosteroids and perhaps sex steroids. Also the anti-diuretic hormone may be involved as well as other factors. They will be considered later.

For the moment it is to be accepted that during pregnancy there is a retention of water, some of which is held in the plasma. It is obvious that water is needed for incorporation into growing tissues such as those of the fetus, the placenta, the umbilical cord, the uterus and the breasts and in addition there is the water of the liquor amnii. Apart from the uterus and the breasts these tissues and fluids may be looked upon as not directly involved with the maternal circulation. Estimates, both direct and indirect, have suggested that the total body water increase in pregnancy is about 7–8 litres. Of this about 5–6 litres is extracellular and this is apportioned between the "mother" and her "products of conception". The products of conception, that is fetus, placenta and liquor accounts for about 2·5 litres of the E.C.F. and the rest is distributed in the maternal tissues proper. This leaves about 3·5 litres to be accounted for. About 0·5 litre is held in the enlarged uterus and breasts and the 3 litres remaining is apportioned between the plasma and the tissue fluid in all parts of the body.

It will be realized that these are all very rough estimates and that there is very wide variability between women. The following is a summary:

Total body water increase	8 litres
Extra-cellular fluid volume increase	6 litres
E.C.F. volume in products of conception	2·5 litres
E.C.F. volume in uterus and breasts	0·5 litres
E.C.F. volume in plasma and tissue fluid	3·0 litres

It is already shown that the plasma volume increases by about 1·5 litres, therefore it is easy to see that the tissue fluid spaces must contain about 1·5 litres of extra fluid during pregnancy.

Since the plasma is part of the E.C.F. and is in equilibrium with the rest of the E.C.F. in the tissue spaces it will have become obvious that the plasma volume could not increase without an accompanying increase in the E.C.F. of the tissue fluid. It has been seen that the initiating factor in this increase of fluid is alteration in renal physiology.

The plasma and the tissue fluid are in dynamic equilibrium and some of the factors involved in this are summarized in the diagram below. The fluid interchange is dependent on capillary function. The pressure at the arterial end of the capillary is about 32 mm. Hg and at the venous end it has fallen to 12 mm. Hg. Since the osmotic pressure of the proteins of the plasma is about 25 mm. Hg and they are imprisoned within the vascular tree there is a net force driving fluid out at the arterial end (32 — 25 mm. Hg) and a net force taking fluid back into the capillary (25 — 12 mm. Hg) at the venous end. This is a very over-simplified scheme for it leaves out of account the movements both of electrolytes and of protein and there may be factors affecting the permeability of the capillary wall,

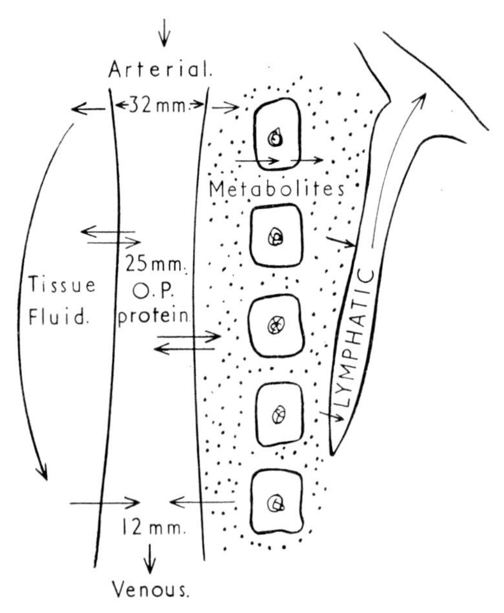

FIG. 12.2. From *Rhodes' Fluid Balance in Obstetrics.* (1960).

such as the pre- and post-capillary sphincters, and also altering osmotic pressure in the tissue fluid. Nevertheless, it is an essentially correct view of the situation. The magnitude of the capillary exchanges of water may be understood when it is remembered that about 8 litres of fluid is poured into and out of the alimentary tract every day and that the glomeruli filter about 170 litres per day, of which about 168·5 litres is reabsorbed into the bloodstream. Under these circumstances even slight changes in rates of flow of fluid into and out of various body compartments can cause great disturbance very quickly. This may be very important in pregnancy since there are many pathological disorders of pregnancy in which fluid balance plays a very large part.

Factors which may disturb the fluid equilibria of pregnancy are those involving the kidney and all its controlling mechanisms, and also those which alter distribution. That is, there may be alterations in the total amount of water in the body and there may be changes in the distribution of it as between the intra-cellular fluid and the extra-cellular fluid. It is not fully known where the extra water that is in the tissue fluid is stored, but clinically almost all women in pregnancy hover on the brink of oedema of the legs. It is known that the venous pressure in the legs is raised and it may be that this is a major factor in storing water there. This is an index that local hydrostatic pressure determines where blood and water will pool. Hypertension is common in pregnancy and this may raise the arterial pressure at the proximal end of the capillaries. The osmotic pressure of the proteins falls during pregnancy, though it is surprising that even in the rare congenital cases where there is no albumin, oedema is not seen. Metabolism may so increase locally that the osmotic pressure of the tissue fluid may rise and perhaps on occasion the capillary walls may be damaged. The fluid balance of pregnancy both general and local is determined by a multiplicity of factors most of which are not very well understood and all of which must be accorded due weight in considering the problem.

The burden of the present chapter has been that the blood volume increases in pregnancy, partly to fill the enlarged vascular tree. The volume increase is partly of cells and partly of plasma, but the increase in plasma outweighs that of the red cells, with resultant effects in the usual parameters which measure blood changes. The stimuli to the bone marrow are not known and the stimuli for the increase in the plasma volume are ill-understood. The major alterations in plasma volume must result in the first instance from changes in renal physiology such that more water is held in the body. Thereafter the extra water has to be distributed as between the various body compartments by other mechanisms. Inevitably the increase in plasma volume must have repercussions on the tissue fluid space since plasma and tissue fluid are parts of the same extra-cellular space.

The next chapter is concerned with the circulatory adjustments required to distribute the increased blood volume round the body and in particular to see that the placental site receives a sufficient supply of blood so that the products of conception may flourish and survive.

Chapter XIII

BLOOD PRESSURE AND DISTRIBUTION

During pregnancy the body weight increases as a result of growth of the products of conception, the uterus, the breasts and the increases in the blood and tissue fluid volumes and the storage of fat. These changes cause resultant adaptations in the cardio-vascular system. The vital organs to which a circulation must be maintained at all times are the brain, heart, liver and kidneys. In pregnancy another vital area is added, that of the uterus and its contained products of conception.

CARDIAC OUTPUT

The demands for an increased flow of blood are met mainly by increasing the cardiac output. In the average non-pregnant woman this output is about 4·5 litres per minute. At the eighth month of pregnancy this has risen to about 5·5 litres. At various times in pregnancy the increase in output is as follows:

Non-pregnant	4·5 litres/minute
3–4 months of pregnancy	5·5 ,,
Mid-pregnancy	6·0 ,,
8 months of pregnancy	5·5 ,,

It will be seen that if a graph were plotted the output rises to a peak in the middle of pregnancy and then slowly declines thereafter though it still remains one litre per minute above the non-pregnant values. In general shape the curve follows that for the blood volume, but there is no adequate explanation of why either the cardiac output or the blood volume should fall from a peak as the time for delivery approaches, though recently it has been shown that the fall-off in cardiac output only occurs when the mother is on her back, with the uterus obstructing the venous return through the inferior vena cava, so diminishing the venous return to the heart.

Cardiac output depends on the heart rate and on the output of the ventricles at each beat, that is the stroke volume. It is known that the resting pulse rate increases in pregnancy and representative figures are:

Non-pregnant	70 beats per minute
Early pregnancy	78 beats per minute
End of pregnancy	85 beats per minute

The heart rate may therefore increase by about 15 beats per minute and the increase seems to be progressive. The stroke volume of the heart in pregnancy has not been measured but if the heart rate increases by 15 beats per minute and the output is up by 1·5 litres per minute it would suggest an increase in the stroke volume. If 70 beats per minute will eject 4500 ml. then the stroke volume will be about 64 ml. If 85 beats per minute eject 6000 ml. then the stroke volume must have increased to

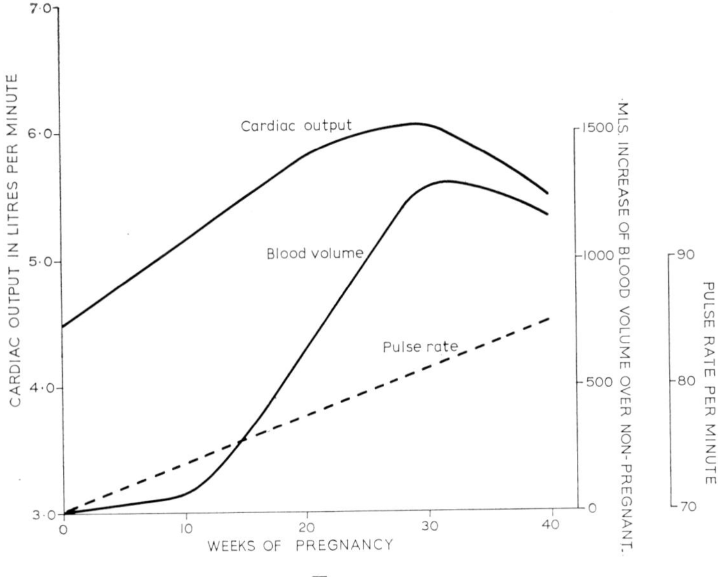

FIG. 13.1.

about 71 ml., that is an increase of about 10 per cent over the non-pregnant value. It will be noted that the pulse rate increase is more of the order of 20 per cent. Thus most of the increased cardiac output of pregnancy is due to an increase in heart rate and rather less to an increase in stroke volume. Since it seems that the cardiac output decreases towards the end of pregnancy whilst the pulse rate does not fall the change is presumably due to a fall in the stroke volume to very nearly non-pregnant levels. On a similar calculation to that just made it would be about 65 ml. Looking at the whole course of pregnancy it would appear that in the early months of pregnancy the stroke volume rises rapidly to a peak and then declines whilst the pulse rate slowly increases, so that the two mechanisms of increasing the cardiac output have varying importance at the extremes of pregnancy.

These physiological changes have clinical importance in the management of patients with cardiac disease in pregnancy. The maximum load on the heart seems to be somewhere about the 30th week of pregnancy and therefore this is the most likely time for cardiac failure to occur, which is borne out in practice.

A further unusual feature of pregnancy is the arterio-venous oxygen difference. Surprisingly this is less in the first six months or so of pregnancy than in the non-pregnant woman. Only at term is the arterio-venous oxygen difference greater. The following figures may be representative:

Arterio-venous oxygen difference:

Non-pregnant	46 ml. oxygen per litre
First 3 months of pregnancy	42 ml. ,, ,, ,,
Second 3 months of pregnancy	44 ml. ,, ,, ,,
Third 3 months of pregnancy	48 ml. ,, ,, ,,

This shows that in early pregnancy there is comparatively less uptake of oxygen at the periphery, even when the products of conception are

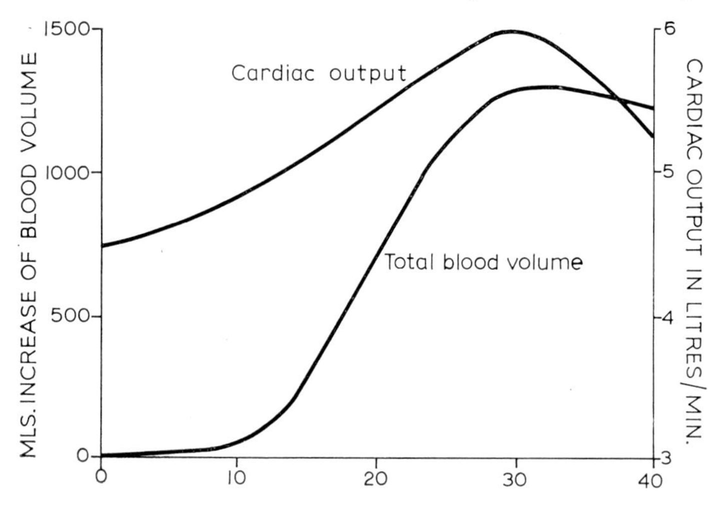

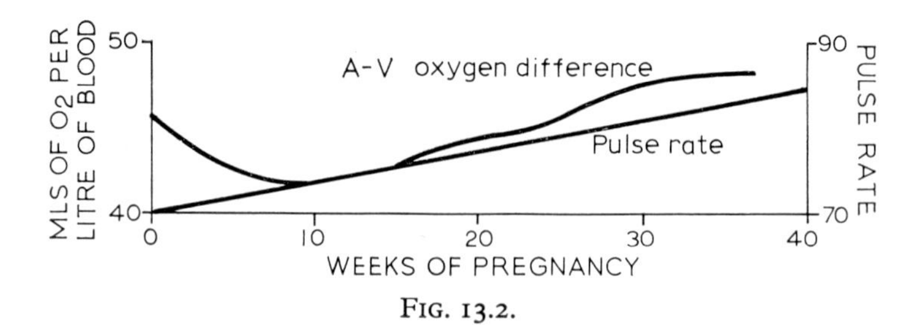

FIG. 13.2.

growing at a fast rate. It suggests that the circulation has increased to a greater extent than is absolutely necessary for the delivery of oxygen to the periphery, and might seem to be an adaptation allowing for a degree of safety for the fetus should the circulation partially fail. In late pregnancy as the cardiac output is falling the peripheral uptake of oxygen increases.

THE ARTERIAL BLOOD PRESSURE

The blood pressure in pregnancy is of inordinate interest to the obstetrician because there is a condition of "toxaemia of pregnancy" or pre-eclampsia which is very common and is characterized by a rise in blood pressure together with oedema and often proteinuria. Toxaemia of pregnancy is badly named since there is no evidence of a toxin being present. The importance of pre-eclampsia, and also of hypertension not due to toxaemia, is that it is a disorder which kills large numbers of babies and a small number of mothers. Unfortunately very little is known of its causes so that treatment has to be empirical, but even so it can be shown that the treatments available have had some success, but it would be more satisfactory if it could be made possible to understand the reasons why treatment was valuable. The disorder of pre-eclampsia is not predictable at an early stage in pregnancy but the worst effects of the disorder in terms of maternal and fetal death can to a large extent be offset if minor rises in blood pressure can be found at the earliest possible moment so that treatment may be instituted. To be able to do this the patient has to attend her doctor many times during pregnancy and one of the major aims of this ante-natal care is in fact to detect any undue rises in blood pressure. Therefore it is imperative that the blood pressure be recorded at every visit of the patient to her doctor.

Blood pressure recordings are made with the sphygmomanometer, but this instrument has some drawbacks. It requires care in use in seeing that the cuff is applied evenly and that it be inflated rapidly when it is applied. If the rate of inflation is slow it causes some venous congestion which in its turn may cause constriction of the capillaries and arterioles. Such constriction leads to a local rise in the arterial blood pressure which is being recorded by the stethoscope placed just below the cuff.

If after pumping up the cuff to a level above that of the blood pressure the mercury column is allowed to fall too rapidly then there will be much observer error since there is delay in correlating the sight of the upper level of the mercury column against the scale and the sounds as heard through the stethoscope. On the other hand the tight cuff can be uncomfortable and if the patient is nervous she may cause a rise in her blood pressure by her emotional changes. It is well known too that observers display what is called digit preference in that they prefer to record readings ending in noughts or fives but not the intermediate numbers. As between individual observers there are also differences since some will tend to prefer slightly higher numbers and some the rather lower ones. That is, in their digit preference some correct upwards from the true reading and some correct downwards, so that if a "true" reading is 83 mm. Hg some will call it 85 and some 80.

The business end of the stethoscope must be placed directly over the brachial artery so that the sounds emanating from the partially occluded vessel may be heard as soon as they appear. However, even when the sounds are heard, there is still some difficulty in interpretation of what they mean. There are several of these Korotkoff sounds and they have been given as:

1. A sudden clear sound is heard.
2. The sound takes on a softer murmuring note.
3. The sound becomes louder and tapping in quality.
4. The sound suddenly becomes softer.
5. Finally the sound disappears.

All are agreed that the first sound is a good representative of the systolic pressure. The major problem is to decide what sound represents the diastolic pressure, and it is probable that clinically this is more important than the systolic. It has been customary to read the diastolic pressure as the sound change between phases 3 and 4, that is at the sudden change in note. However there has recently been a move to take the entire disappearance of the sounds as representing the diastolic pressure. But a very little experience shows that sometimes the note does not disappear until the mercury column is at about 10 or 20 mm. and this obviously cannot be the true reading. The dilemma is still unresolved.

There is no absolute standard against which the sphygmomanometer readings can be calibrated but it would seem likely that intra-arterial readings taken through a needle inserted into the artery may be more exact than those taken with the sphygmomanometer. When the two types of reading are compared it is found that the sphygmomanometer may cause systolic readings to be about 2–3 mm. Hg. too low and the diastolic to be about 8 mm. Hg. too high. The error in a single measurement of the blood pressure with the sphygmomanometer may be as much as $\pm$ 8 mm. Hg. This makes for clinical difficulties since therapeutic action has to be taken on differences of about 10 mm. Hg. in the diastolic pressure as recorded at weekly intervals. As a matter of practical politics, however, the drawbacks of casual readings of the blood pressure have to be accepted and acted upon as seems reasonable in the light of the total clinical situation.

Further errors can be introduced into the measurement of blood pressure by altering the size of the inflatable cuff and by increasing circumference of the arm occasioned by fat. Nevertheless it is *changes* in the reading of the blood pressure that matter and provided that it is recorded under the same conditions at each visit of the patient the major errors should tend to cancel out.

In normal pregnancy the blood pressure in the first few months is similar to that for the non-pregnant woman. In the middle three months, however, the blood pressure tends to fall on an average by about 3 to 5 mm. Hg. But sometimes the blood pressure drop may be of the order of 20–30 mm. Hg. though the patients seem not to suffer at all from this. In the last three months of pregnancy the blood pressure slowly rises again until it comes back to the normal non-pregnant level.

In general the blood pressure is the product of the cardiac output and the peripheral resistance. The peripheral resistance is mainly due to arteriolar tone, the arterioles being under the control of the vasomotor centre in the floor of the fourth ventricle. The diastolic pressure is a

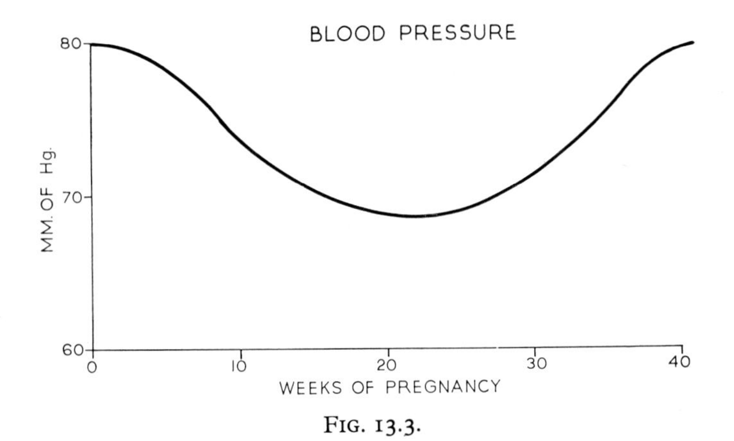

FIG. 13.3.

measure of the constant load on the heart and indirectly is a measure of the continuing arteriolar tone at the end of diastole. If there were no peripheral resistance the blood pressure would fall away to zero after each systolic ejection.

The cardiac output gradually rises to a peak during pregnancy at about thirty weeks from the last menstrual period and thereafter it declines and the increase is in part dependent on an increased stroke volume and in part on an increase in pulse rate. Since the blood pressure is at its lowest from about the 12th to 26th week of pregnancy it is obvious that the peripheral resistance must have decreased. The most likely explanation of this is that new areas of circulation have been opened up in the uterus. It has been suggested that the circulation through the placenta is so increased that it acts rather like an arterio-venous fistula, one of the effects of which is to lower the blood pressure. In addition an arterio-venous fistula increases the pulse pressure, that is the difference between the systolic and diastolic pressures and such an increase in pulse pressure is found in pregnancy too. Also there is little doubt that the skin vascular bed increases during pregnancy to allow of the dissipation of the extra heat arising from the growth of the uterus and its contents. As well as these changes in the size of the vascular bed it is possible that arteriolar tone diminishes and also to be taken into account is the decreased viscosity of the blood which is a feature of pregnancy. It falls from a non-pregnant level of 4·61 in relation to distilled water to 3·84 at about 22–28 weeks.

In the last three months of pregnancy when the blood pressure is rising again the cardiac output is falling. Therefore the peripheral resistance is increasing. The skin circulation probably goes on rising till term so that presumably there is an increase in arteriolar tone elsewhere and this may be in the utero-placental area but there is no direct evidence of this. However, it is known that the rate of growth of the fetus slows down as term approaches and this may be dependent on a falling blood flow to the placenta. Like so many of the other physiological phenomena of pregnancy this one is not understood. It seems almost that a peak of efficiency in physiological function is reached some weeks before labour supervenes and that thereafter efficiency declines, but this may not be true and the problems of interpretation of the changes remain.

The control of the arteriolar tone is mainly the function of the vasomotor centre, but besides this nervous mechanism there is also a humoral one which is usually deemed to be less important. The main regulators of the blood pressure seem to be the afferent endings in the aorta and in the carotid sinus, but in addition afferents reach the vasomotor centre from the respiratory centre and from other areas of the body. Also carbon dioxide and oxygen changes in the blood may affect the centre directly. During pregnancy there is in fact some lowering of the alveolar CO_2 and the partial pressure of oxygen is little changed. Of the humoral factors that might be expected to mediate the blood pressure changes of pregnancy none has been demonstrated to have any effect though attempts have been made to evaluate the effects of the renin-angiotensin mechanisms and of those of adrenaline and nor-adrenaline as well as others such as 5-hydroxy-tryptamine.

THE VENOUS PRESSURE

In the legs the venous pressure rises during pregnancy. In the femoral vein in the supine patient the pressure is of the order of 9 cm. of water in the non-pregnant. This may rise threefold to 27 cm. of water in the pregnant woman at term. It can also be shown by radioisotopic methods that the flow rate of venous blood in the legs is halved in pregnancy. In early pregnancy the foot to groin time is about 17 seconds whilst later in pregnancy it is about 35 seconds. The cause of these changes in the legs is the growing uterus in the abdomen.

It has been shown radiographically by injection of radio-opaque dyes that when a pregnant woman at term lies down her large uterus entirely occludes the inferior vena cava as high as the level of the renal veins. The blood has to find its way round the block by anastomotic channels running in the posterior and anterior walls of the abdomen. In a small percentage of women the return of blood to the heart is so impaired that the cardiac output falls and the woman faints. This is called the *supine hypotensive syndrome*. It is immediately reversed by placing the woman on her side so that the pressure of the uterus is removed from the vena cava. In addition to the direct pressure effect of the uterus there is also the fact that a greatly increased flow of blood goes through the internal iliac vein. This flow competes for the common iliac vein with the flow returning from the legs through the external iliac vein. An interesting point con-

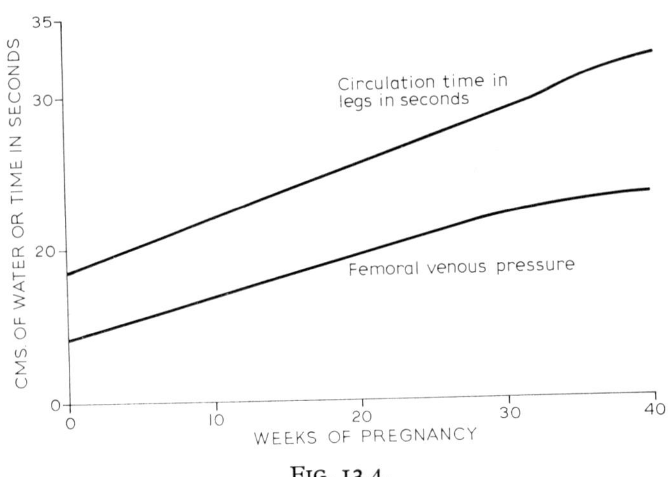

FIG. 13.4.

firming this is that when the placenta is (say) on the right side of the uterus the venous pressure in the right leg will be between 2 and 3 cm. of water higher than in the left. It might be expected that in such a case the major venous return from the uterus would be on the right side.

In prolonged standing in lordosis many pregnant women will pass protein in the urine. Ureteric catheterization shows that the protein comes mainly from the left kidney. It is probable that this occurs because the left renal vein crosses the abdominal aorta just below the superior mesenteric artery. In lordosis the vein may be pressed upon in this situation so raising the venous pressure in the left kidney.

Apart from the legs and left kidney there are no other known changes in venous pressure. The pulmonary blood pressure is the same in pregnancy as in the non-pregnant woman and in the upper limbs there is no change in circulation time.

THE DISTRIBUTION OF THE BLOOD

The major physiological changes of pregnancy result in an alteration in the distribution of blood as compared with the non-pregnant state. These alterations include the uterus obviously, but also the skin, kidneys, liver, brain and lungs.

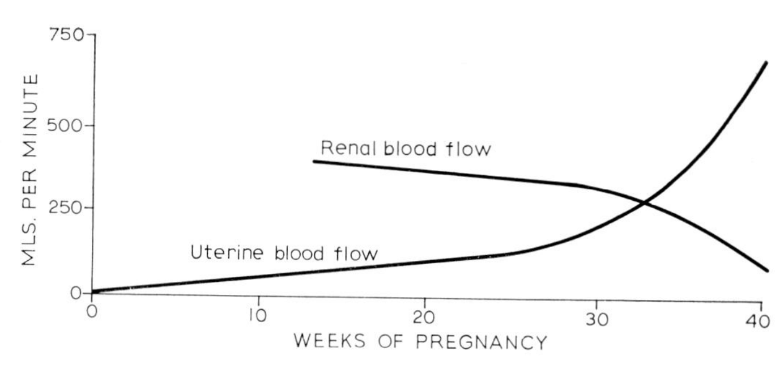

FIG. 13.5.

The measurement of uterine blood flow in the human is very difficult and various methods have been tried, but all involve many assumptions so that the figures to be presented cannot be taken necessarily to be very accurate. Nevertheless the following figures may be helpful:

Uterine blood flow.

10 weeks	50 ml. per minute
20 weeks	100 ml. per minute
30 weeks	200 ml. per minute
40 weeks	700 ml. per minute

The figures given for the earlier part of pregnancy are probably very approximate and if anything probably very much on the low side. There seems to be much more agreement that the blood flow at term is about 700–800 ml. per minute.

The renal blood flow seems to be anomalous since it is maximal in early pregnancy and gradually falls throughout pregnancy. The flow in the non-pregnant woman may be of the order of 800 ml. per minute. The increase over this figure in pregnancy is as follows:

Increased renal blood flow.

13 weeks	400 ml. per minute
32 weeks	300 ml. per minute
40 weeks	100 ml. per minute

It should be noted that the uterus and kidneys between them take

1200–1500 ml. of the total cardiac output of 5500 ml. per minute at term, that is a percentage of about 25.

It is well known clinically that the blood flow through the skin in pregnancy is increased, for the skin feels warm even in the coldest weather, and pregnant women find that they need fewer clothes to keep warm when they are pregnant. The increased skin flow is an adaptation to the presence of the growing products of conception which produce heat which must be dissipated if the woman's temperature is not to rise unduly. The surface temperature of the skin can be measured under standard conditions of ambient temperature and one estimate showed a rise from the non-pregnant level of about 22°C in the fingers to 34°C at term. One estimate of the increased blood flow through the skin at term is of 500 ml. per minute.

Concomitant with the increased flow through the skin there is probably a total increased flow through the limbs. That is there is an increased flow through muscle, and it may well be almost doubled. A calculated estimate is that the flow through the forearms and legs may rise from 130 ml. per minute to 230 ml. per minute a month before term, but nearer term the flow through the legs may decrease, and this is presumably an effect of the increased venous pressure which has been discussed.

Pulmonary blood flow must parallel the cardiac output and therefore it rises, but the pulmonary vascular tree is so accommodating that there is no rise in venous pressure. Because of the increased cardiac output the coronary circulation must also increase but it has not been measured. It has been suggested that liver blood flow does not increase or decrease in pregnancy and that brain blood flow may increase a little, but these only serve to show that much more can be known about blood flow in pregnancy and that reasonably firm estimates are only available for the uterus, kidneys and skin.

Chapter XIV

RESPIRATION

Respiratory function in pregnancy is of special importance since the life of the fetus depends primarily upon its oxygen supply. In addition it has to be rid of its production of carbon dioxide. The mother must adapt her respiration both to supply the additional oxygen needed by her fetus and to excrete its production of carbon dioxide.

It has long been thought that the lower ribs flare outwards during pregnancy and that the transverse diameter of the chest increases. The subcostal angle has been measured and found to increase from a non-pregnant value of 68° to 103° in late pregnancy. The maximum transverse diameter of the chest also increases by about 2 cm. Radiography shows that the diaphragm rises by about 4 cm. from its non-pregnant position. It was thought that these changes arose from the increasing size of the uterus, but it is now thought that the anatomical changes in the chest begin before the size of the uterus can have any great effect.

Comparatively few direct measurements of respiratory function have been made, but those that have suggest that the respiratory rate does not appreciably rise at rest. The vital capacity does not increase. This is the total amount of air that can be expelled from the chest by forced expiration after a maximal inspiration. The tidal volume, however, does increase. This is the amount of air inspired or expired with each breath. Immediately after delivery, i.e. in the puerperium, the tidal volume was found to be 487 ml., which is to be compared with a tidal volume before delivery of 678 ml., the difference being 191 ml. or a percentage increase of 39. The minute ventilation is the tidal volume multiplied by the respiratory rate per minute. Since the respiration rate scarcely rises and yet the minute ventilation increases as shown by the foregoing figures it is obvious that the mother's respiration increases in depth but not in rate as pregnancy proceeds. In fact the tidal volume or depth of respiration increases slowly from the beginning of pregnancy to its end and shows no falling off as the cardiac output does. The amount of air flowing to the alveoli increases so that gaseous exchange between the air and the pulmonary blood can be increased.

The vital capacity is made up of expiratory reserve, tidal volume and inspiratory reserve. Expiratory reserve is that amount of air which can be expelled from the lungs after normal quiet respiration. Inspiratory reserve is that amount of air which can be taken into the lungs after a quiet inspiration. Since vital capacity does not change in pregnancy the tidal volume must increase at the expense of both the expiratory and inspiratory reserves.

OXYGEN CONSUMPTION

More studies are needed of oxygen consumption but those that have been made suggest that at term the mother is using between 50 and 60 ml. more oxygen per minute than she is in the non-pregnant state. At a respiration rate of 16 per minute this amounts to about 3 to 4 ml. per breath or about 3·3 litres per hour.

Oxygen is needed for the products of conception and the breasts and also for the increased work by the heart, lungs and possibly the kidneys. Hytten and Leitch (1964) have made an ingenious calculation and suggest the following figures:

Increment of oxygen consumption in ml./minute

	Weeks of pregnancy.			
	10	20	30	40
Fetus	0	1·1	5·5	12·0
Placenta	0	0·5	2·2	3·7
Uterine muscle	0·5	2·2	3·0	3·3
Breasts	0·1	0·6	1·2	1·4
Cardiac output	4·5	6·8	6·8	3·4
Respiration	0·5	1·8	3·0	3·8
Total	5·6	13·0	21·7	27·6

From this it will be seen that of the total increased oxygen uptake of pregnancy at term about half of the observed extra uptake of 50 ml. per minute is not accounted for. Of the 27·6 ml. per minute accounted for, the products of conception take the lion's share, 20·4 ml. per minute if the breasts are also included. It has been shown that the 27·6 ml. per minute increment is 14 per cent above the non-pregnant values, and this should be compared with the 18 per cent increase in the oxygen carrying capacity of the blood. In this is seen the reserve that so often shows itself in the physiological changes of pregnancy so that the fetus is sheltered from serious insults that may be visited upon the mother.

CARBON DIOXIDE

The output of carbon dioxide is almost impossible to measure accurately since it is so variable from minute to minute. This is because the amount

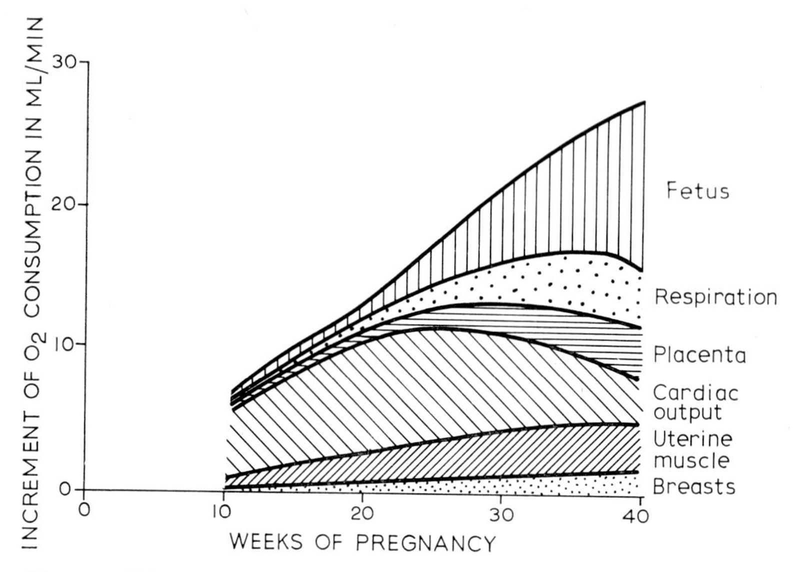

FIG. 14.1. Distribution of the increased oxygen consumption of pregnancy. Re-drawn from Hytten, F. E. & Leitch, I. *The Physiology of Human Pregnancy*. Blackwell, Oxford (1964).

put out is only that amount not kept in the blood by the storage mechanisms and which is not involved in acid-base equilibrium. However, the relatively deep breathing of pregnancy does lower the alveolar CO_2 and some time ago it was shown that the percentage of carbon dioxide in expired air was 3·31 in the non-pregnant and this was lowered to 2·83 in late pregnancy. The partial pressure of carbon dioxide in the non-pregnant has been found to be 37·3 mm. Hg in the expired air and it had fallen to 30·9 mm. Hg by the end of the sixth month of pregnancy and stayed at that level till term. The arterial pCO_2 falls *pari passu* with the alveolar partial pressure. It is of interest that the pCO_2 falls towards the end of the menstrual cycle when the body is under the influence of progesterone and it may be that the carbon dioxide changes of pregnancy are due to this hormone also.

As far as respiration goes there seems to be no difficulty on the part of the mother in supplying her fetus and other products of conception with oxygen and in getting rid of the carbon dioxide that they produce. However, she is likely to become breathless on exertion more easily than when she is not pregnant. The mechanism of this dyspnoea remains obscure.

Chapter XV

RENAL PHYSIOLOGY

It needs no imagination to realize that the growth of the products of conception, and especially of the fetus, involves metabolic changes of no mean order, and that the results of these changes must involve the production of waste materials in greater amounts than occur in the non-pregnant woman. It would be a surprise, therefore, if there were no changes in renal physiology in pregnancy. Moreover the kidneys are a major factor in the maintenance of the constancy of the internal environment which it must be presumed is of enormous importance for the survival of offspring. One of the most obvious changes biochemically in pregnancy is the retention of water and minerals and in this the kidneys have a key role.

Kidney function depends especially on (i) blood flow, which determines the substances on which it has to work, (ii) the glomerular filtration rate which determines the sort of filtrate which will be delivered to the tubules, and (iii) tubular function in reabsorbing materials into the blood which have been delivered to the lumen of the nephron by glomerular filtration.

Tubular reabsorption probably does not change during pregnancy in any qualitative way, though as will be seen later it must change quantitatively. Sodium is reabsorbed from the glomerular filtrate mainly in the proximal tubule, and much of the water is absorbed in the same situation. Bicarbonate, phosphate, potassium and glucose are also absorbed in the proximal convoluted tubule. In the loop of Henle and in the distal convoluted tubule further reabsorption of sodium occurs and although the intimate mechanisms are complex, chloride is also absorbed and the urine acidified here, and there are further changes in bicarbonate excretion. The tubules are active in the secretion of ammonium ions and there may be secretion of other substances by the tubules. These basic changes are fully dealt with in textbooks of physiology and need no further elaboration here.

Renal function may be measured by the renal blood flow, renal plasma flow and glomerular filtration rate. It should be noted that all these measures are related to a standard surface area of 1·73 sq. metres. Surface area is calculated from a complex formula whose functions are height and weight. Weight changes quite rapidly in pregnancy so that the formula may not give an accurate assessment of true surface area in pregnancy but the convention is followed so that comparisons between the pregnant and non-pregnant state may be made. Whatever else may be said about this, it at least means that the data are subject to the same mathematical calculations and to that extent are therefore more comparable with one another.

The estimates of renal blood flow and of glomerular filtration rate are all based in Man on the concept of clearance. Clearance of a substance is given by the formula $\frac{UV}{P}$ where U is the concentration of the substance in urine in mgm. per 100 ml., where V is the volume of urine secreted in ml. per minute, and where P is the concentration of the substance in plasma in mgm. per 100 ml. The theoretical background of clearance is the same for pregnancy as for the non-pregnant woman and for present purposes is not important. From these theoretical considerations the renal plasma flow (RPF) is measured by the excretion of para-amino-hippuric acid (PAH), and the glomerular filtration rate (GFR) from the inulin clearance.

The renal plasma flow is raised in pregnancy by about 45 per cent. In the non-pregnant the figure is about 500–725 ml. per minute and in pregnancy these results are increased by about 225 ml. per minute. But strangely the flow is maximal in early pregnancy and falls as pregnancy proceeds until at about term the increased RPF amounts only to about 100 ml. per minute. This is curiously like the stroke volume of the heart which is maximal also when one would expect the least load. There is no explanation as yet for either phenomenon.

The total renal blood flow is calculated from the RPF by the formula $RBF = \frac{RPF}{\text{1-haematocrit reading}}$. This is the correction needed because blood consists both of plasma and of cells. The RBF has been found to be raised above the non-pregnant level which is of the order of 885 ml. per minute. In the second three months of pregnancy it is 1200 ml. per minute and at 34 weeks it is about 1100 ml. per minute. Since the calculation is dependent on the values for the RPF the RBF also falls towards term. Even so it is worth noting that it forms about 20 per cent of the cardiac output at that time, and more before it.

The GFR is also raised in pregnancy but by about 60 per cent as compared with the rise in RPF of 45 per cent. Moreover the GFR does not fall towards the end of pregnancy as the RPF does. In the non-pregnant patient the GFR is about 90 ml. per minute and during pregnancy it rises to 145 ml. per minute.

Since the blood flow to the kidneys is falling as pregnancy proceeds and the GFR remains the same the filtration mechanism at the glomeruli must improve. This may be expressed by the calculation of the filtration fraction (FF). The formula for this is $FF = \frac{GFR}{RPF}$. In the non-pregnant this is 0·18 and it rises to 0·20 in the middle of pregnancy and is 0·25 in the last few weeks. The effective force for filtration at the glomerulus is the arterial blood pressure minus the osmotic pressure of the plasma proteins. It is known that the O.P. of the proteins falls during pregnancy and it is probable that this is the mechanism whereby increasing filtration occurs in the face of the falling blood flow.

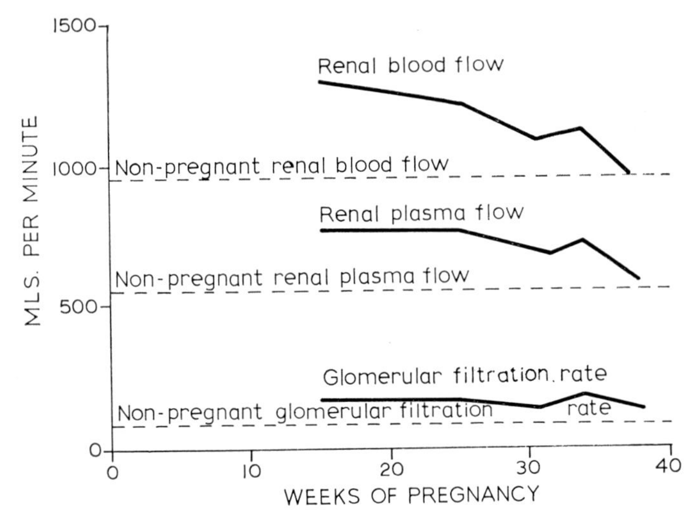

FIG. 15.1. Data for graph from Hytten & Leitch (1964). *Physiology of Human Pregnancy*. Blackwell. pp. 117, 118 & 119.

Mean week Pregnancy	RPF/1·73 m²	RBF/1·73 m²	GFR/1·73 m²
15·4	769	1280	157
25·2	761	1216	152
30·6	680	1098	147
33·7	716	1116	163
37·8	589	941	146
Late puerperium	549	965	97

R.P.F. = clearance using P.A.H.
G.F.R. = clearance using inulin.

These results for RPF and GFR perhaps mean little in themselves but they form one part of the substratum for the biochemical changes of pregnancy especially in water and mineral balance. To take a simple example the blood urea tends to be lower in pregnancy and in health it is not expected that the blood urea will be above 20–25 mgm. per cent, and more amino-acids are lost in the urine in pregnancy than at other times—

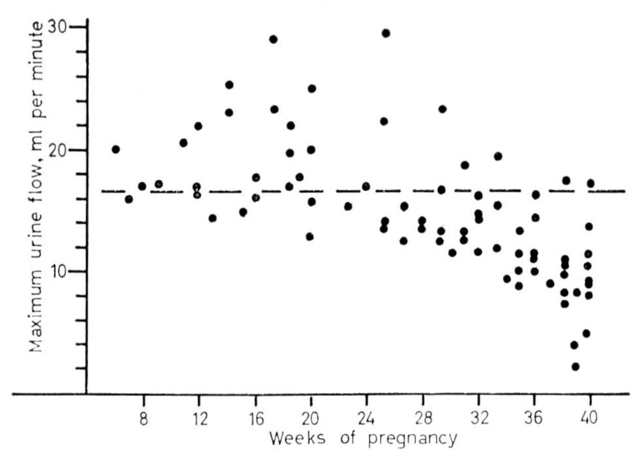

FIG. 15.2. The maximum rate of urine flow after drinking one litre of water in pregnancy (Hytten and Klopper, 1963).

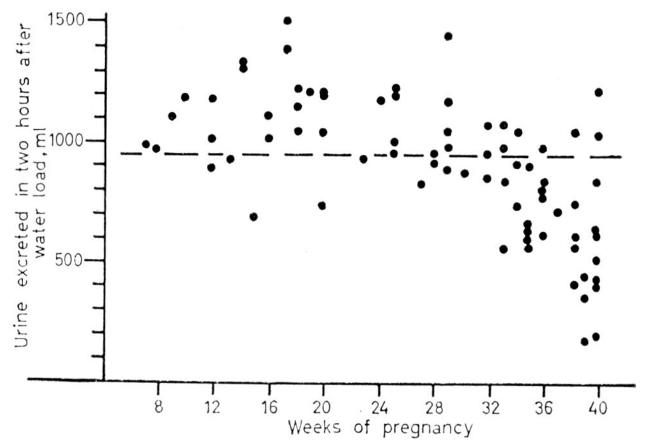

FIG. 15.3. The volume of urine passed in two hours after drinking one litre of water in pregnancy (Hytten and Klopper, 1963). The original figures are by F. E. Hytten, and A. I. Klopper, (1963), *Jl. Obst. Gynaec. Brit. Cwlth.* **70,** 811.

a curious fact. However, it will be realized that the biochemical changes as seen in the blood are the resultant of many factors of which the kidneys are only one. Also there are major changes in water metabolism since during pregnancy about 8 litres have to be retained and along with this there is much retention of sodium. One ingenious simple experiment has been done to show that the kidney handling of water diminishes in success as pregnancy nears term. Pregnant women were given 1 litre of water to drink and then the urine excreted was measured in amount at 15-minute intervals for the next $3\frac{1}{2}$ hours. When the amounts excreted per minute were plotted against the length of pregnancy it immediately became obvious that the ability to excrete a water load diminished markedly as pregnancy went on and whereas under the conditions of the experiment a woman might excrete 20 ml. urine per minute at 16 weeks of pregnancy, at 40 weeks she might only manage to excrete about 10 ml. of urine per minute.

In hypertension, that scourge of pregnancy, whether it is mild or severe and whatever its cause, the renal plasma flow falls and there are changes in the GFR and the FF, though these may be comparatively less than in the blood flow.

Apart from these changes in renal physiology it has been known for a long time that the ureters and kidney pelves dilate certainly by the middle of pregnancy. Before the use of radiography was restricted during pregnancy because of the possibly deleterious effects of X-rays, the dilatation of the pelves and ureters was evident on all intravenous pyelograms. The cause of the dilatation has been variously ascribed to pressure of the growing uterus on the ureters at the pelvic brim and on the muscular relaxant effect of progesterone. Recently the concept of dilatation of the renal pelves and ureters as a normal phenomenon in pregnancy has been called into question.

Chapter XVI

OTHER PHYSIOLOGICAL CHANGES

The major changes in maternal physiology are those already considered, namely in the blood, the cardio-vascular system, the respiratory system and the excretory system. In addition to these there are others which seem of apparently relatively less importance, but this may be only because it is not yet possible to see their significance.

Bones are the area of storage of calcium and there is no doubt that calcium is required for the growth of the fetus and products of conception as well as for growth in other areas such as the breasts, and moreover, it is needed for the maintenance of proper muscular irritability in voluntary muscle, in the heart and, especially in this context, in the uterus. Moreover, there is need of calcium for the proper function of nervous tissue. It has been suggested that there is decalcification of the bones in pregnancy and this may be so under conditions of adversity such as are unlikely to be seen in this or other advanced countries. Certainly pregnancy seems to make the state of oesteomalacia worse. This disorder is seen in some eastern countries and consists of gross demineralization of bones so that they become soft and bend easily under the strain of weight-bearing or other forces. In the childbearing woman this can be very serious since the pelvis may be greatly distorted, which leads to difficulties in labour. The level of calcium in the blood does not change during pregnancy and remains between about 9 and 11 mgm. per cent, nor does there seem to be any difference in the ionizable and non-ionizable fractions.

In many animals, especially the guinea-pig, it is well known that some *joints* become more freely mobile due to the softening of ligaments. In these animals the hormone relaxin has been implicated, and it is probably a secretion of the ovary. However, its role in human physiology has not been established and it may not even be produced at all in women. Nevertheless, there is good clinical evidence of the relaxation of the pelvic joints in pregnancy, which sometimes may be so great as to cause incapacity. Especially is the symphysis pubis involved and when this shows undue mobility, which can be demonstrated radiologically, there is also demonstrable laxity of the sacro-iliac joints. In addition it is possible that the ligamentous laxity may be present in the lower joints of the lumbar spine.

Muscles, except those of the pelvic region, may not be much changed in their physiology in pregnancy, though many pregnant women say that they feel more tired than when they are not pregnant. However, it is unlikely that lassitude and fatigue are mainly a function of muscle only. Nerves, and especially the central nervous system, are at the base of the symptoms, and these areas are probably of more importance than the muscles themselves. In the pelvis the levator ani, and probably other muscles, becomes more relaxed. This has been shown in the past by radiographic methods when metal clips have been placed on the area of the levator just within the vagina and the level of the clips in relation to fixed bony points can be demonstrated. These observations showed a descent of the perineum of the order of ½ to 1 inch. The cause of this relaxation is probably the infiltration of the muscle with oedema fluid and perhaps other changes too. The effect may be to allow easier egress of the fetus at labour, but during pregnancy these muscular changes may occasion micturition symptoms such as stress-incontinence, in which a little urine is expelled from the urethra whenever the intra-abdominal pressure is suddenly raised as in coughing, sneezing, laughing or in the worst cases on walking.

The *nerves* peripherally show no obvious changes in their function, but there must be major changes in the *central nervous system*. There is not much direct evidence of this but it is probable that the metabolic changes are mediated through the hypothalamus at least, and in some women there are peculiar changes in appetite. Moreover, pregnancy is a highly emotional affair and every woman shows some psychological changes, especially in wide swings of mood. Much of this change is probably in response to social factors, but there is also little doubt that cerebral function changes under the impact of the stimulus of hormones. There is sound experimental evidence in animals for the effect of the sex steroids on sexual behaviour which is dependent on an intact cerebral cortex. In the puerperium too, after the baby is born, there are major changes of mood tending towards depression at the time when the hormone levels in the blood are rapidly falling as a result of the withdrawal of the placenta. Other evidences of change in the central nervous system are occasionally seen in the heightened senses of smell, vision and hearing.

The *skin* blood flow has already been dealt with, but in addition it shows pigmentation for which no satisfactory explanation has been found. Although the breasts are a skin appendage their physiology of lactation will be left till later. But quite early in pregnancy they show pigmentation of the areola immediately surrounding the nipple. As in all other areas the pigment is melanin. About the sixteenth week of pregnancy there is also a diffuse stippled pigmentation in the skin just beyond the areola, and this has been called the secondary areola. Surrounding the nipple there are several small glandular structures which are probably sebaceous, though there has been some argument as to whether they may be milk glands. Their purpose is believed to be the lubrication of the nipple and aerola with sebaceous material so that the skin here stays soft and supple ready for suckling. These small glands enlarge very early in pregnancy and are then named Montgomery's tubercles. They are presumably under the influence of the sex steroids. Having enlarged in one pregnancy they do not entirely disappear.

Apart from the increasing pigmentation of the breast skin the skin of the forehead, cheeks and abdomen also darken. In some women the pigmentation of the face is so great as to be a cause of social embarrassment, though usually it is slight. On the abdomen the main area of pigmentation is in the central line running from the xiphisternum to the symphysis and it is called the linea nigra.

Wherever the skin is stretched and especially over the abdomen, breasts and thighs the dermal fibres may rupture leaving wavy red scars which may be scanty or abundant. These are called *striae gravidarum*. At one time it was thought that these were only due to stretching, but it has been shown that although stretching is a factor the reason for the rupture is alteration in the tensile strength of the dermal fibres caused by the high output of adrenal corticosteroids in pregnancy. Certainly similar striae are seen in cases of adrenal hyperplasia especially in Cushing's disease. Moreover, in women with gross striae, there can often be demonstrated an impairment of glucose tolerance, which is probably due to high corticosteroid levels.

There is subjective evidence from many pregnant women that the apocrine glands of the axilla, and skin glands in the vulval area increase their activity, changing the scent arising from these places.

There is clinical evidence of alteration of function in the *alimentary system*. Early in pregnancy there is often constipation and a feeling of nausea is so commonplace as to be known to almost everyone. This feeling is probably due to lowered tone of the stomach musculature caused by the rising levels of sex steroids in the blood. The evidence for this statement is that women who are put on the "contraceptive pill", which contains oestrogens and a progestagen, experience the same feeling of nausea that they have had in pregnancy. Studies have been made of the secretion of the stomach during pregnancy and these have shown that acid secretion is about 40 per cent above its normal value in early pregnancy and then falls to only 50 per cent of its non-pregnant value at 24 weeks, and then astonishingly it returns to its high level of 40 per cent above the non-pregnant values at term. There is no satisfactory meaning to be given to these changes. However, it is an oddity that peptic ulcer symptoms are almost never seen in pregnant women and most peptic ulcers seem to undergo temporary healing. But oddly too there is a high incidence of heartburn in pregnancy and this is due to reflux of gastric contents through the cardiac sphincter, and this seems to be due to some impairment of whatever mechanism normally maintains the integrity of the junction between stomach and oesophagus. The *liver* must have much to do with the physiological adaptations of pregnancy, but nothing is solidly known of them.

There are obviously enormous changes taking place in the *genital system* itself but these will be left to a later chapter, and also the changes in all systems are mediated by changes in the *internal secretions* of various glands, and it is these which form the matter of the next chapter.

Chapter XVII

HORMONES

At implantation of the fertilized ovum a new endocrine gland is put into the maternal body. In Chapter X the role of the hormones HCG, progesterone and oestrogen, which are produced by the trophoblast have been considered. The last few chapters have shown something of the changes that occur in maternal physiology during pregnancy. The knowledge of these changes is still meagre but at least the knowledge provides some answers to the question of *what* occurs in pregnancy. The knowledge of *how* the changes are brought about is almost nil, but in seeking the answer to *how* in physiology it is natural to look at the endocrine system as a whole since so much of bodily economy is controlled by it, and at the same time it is not possible to divorce the central nervous system from the endocrine system in the control of the internal environment.

THE PITUITARY GLAND

Through its various trophic hormones the anterior pituitary controls much of the activity of other endocrine glands. HCG coming from the trophoblast probably has its main action on the pituitary, making luteinizing and luteotrophic hormones continue to be secreted so that the corpus luteum and thus the decidua of the uterus are kept in a suitable state to maintain the fertilized ovum in its implantation site. At the same time the secretion of FSH (follicle stimulating hormone) is kept in abeyance. At about the 12th to 14th weeks of pregnancy, in the human, the corpus luteum regresses and its secretion of progesterone and of oestrogen is taken over by the placenta itself. How this change is brought about is quite unknown and in fact almost nothing is known of how the placenta is controlled. It seems almost autonomous in its behaviour but there must be some controlling mechanisms. Perhaps these reside more in the fetus than in the mother and although the trophoblast may live on for a little time after the death of a fetus the trophoblast itself usually soon ceases to function. Where there is no fetus as in the neoplastic diseases of the placenta, hydatidiform mole and chorion epithelioma, the trophoblast runs wild. Whether this is because of the inherent autonomy of the trophoblast or whether it behaves wildly because of the lack of control by a fetus is not known, though these considerations show that the fetus and its placenta must be looked upon as an integrated whole.

The other trophic hormones of the pituitary are best considered along with their target organs.

THE THYROID GLAND

During pregnancy there is an increased uptake of oxygen (Chapter XIV) and therefore the Basal Metabolic Rate (BMR) is raised. Also the pulse rate is increased (Chapter XIII), and the skin blood flow rises (Chapter XIV). These changes might be taken as evidence of increased thyroid function. However, the serum cholesterol is raised during pregnancy from an average non-pregnant level of about 180 mgm. per 100 ml. serum to about 260 mgm. per 100 ml. at about 36 weeks. Thereafter the level falls a little. Such a rise in cholesterol levels occurs in hypothyroidism. Therefore it is not possible to infer anything about thyroid function from these changes.

However, an increase in thyroid function is probably demonstrated by the rise in protein-bound iodine (PBI) from 5 μg/100 ml. in the non-pregnant to 8–10 μg/100 ml. in the pregnant. The PBI is a measure of iodine bound to thyroxine, but even these blood levels are difficult to interpret as they are for many other hormones, since one effect of the rising level of oestrogen in the blood is to increase the plasma globulins which are therefore capable of holding larger amounts of protein-bound hormones. The difficulty is to know whether such bound hormone is physiologically active or whether it is simply a relatively inactive store of hormone to be called upon in emergency.

Iodine uptake by the thyroid gland in pregnancy is increased, but again this is not easy to interpret. Hytten and Leitch (1964) quote unpublished work by Aboul-Khair which seems to show that the kidneys, because of their changed physiology (Chapter XV) lose iodine rather freely. This reduces the plasma inorganic iodine to one half its normal value so that for the thyroid to function in pregnancy at non-pregnant levels demands that it should clear twice as much blood of iodine as it does in the non-pregnant state. Therefore as far as the thyroid is concerned there is a relative lack of iodine supplied to it during pregnancy. Not surprisingly this may show itself clinically as an enlargement of the gland, and in fact this has been observed in about one-third of all pregnant women. The role of thyroid stimulating hormone (TSH) in all this is not known since it has not been satisfactorily measured.

Nothing is yet known of thyrocalcitonin in pregnancy.

THE PARATHYROID GLANDS

Nothing of value can be said about these. Calcium metabolism is probably changed in pregnancy if only to meet the needs of the growing fetus but no obvious changes have been measured, although there tends to be some decalcification of the bones in some women.

THE ADRENAL GLANDS

Nothing of value is known of the secretions of the medulla, adrenaline and *nor*-adrenaline in pregnancy.

The zona glomerulosa of the adrenal cortex is probably responsible for the production of aldosterone, which secretion is not under the control of ACTH from the anterior pituitary. The levels of aldosterone in the blood and urine are both raised during pregnancy but the most convincing evidence is derived from the production rate of aldosterone measured by radio-isotopic methods. From a level of about 192 μg per day in the non-pregnant the level may be increased two- to five-fold.

The importance of aldosterone is its role in water and mineral metabolism which is greatly altered during pregnancy, and probably in the disease of pregnancy called pre-eclamptic toxaemia which at its worst can kill mothers and babies. In Chapter XII it was seen that the blood volume increases in pregnancy and that the blood comes to contain relatively more water than in the non-pregnant state. Normally the blood volume is held within fairly well-defined limits perhaps by the activity of "volume receptors" situated in the thorax. These receptors indirectly affect the secretion of antidiuretic hormone and of aldosterone, which in their turn affect renal function so that fluid and electrolytes are either excreted or retained according to the needs of the moment. In pregnancy it is obvious that the "homeostat", whatever its ultimate nature, is set at a higher level than normally so that the body can retain more salt and water without ill-effects. This may "explain" the high production rate of aldosterone though how the changes in the homeostatic systems are brought about remains obscure.

Cortisol (hydrocortisone), the major corticosteroid secretion coming probably from the zona fasciculata, is also increased in the blood in pregnancy. Its secretion is under the control of ACTH from the pars anterior but there is no definite evidence of a rise in the production of this pituitary hormone. The normal non-pregnant level of cortisone in the blood is about 3–11 μg./100 ml. plasma, and in pregnancy this may be doubled or trebled. Again the problem arises of the protein binding capacity of the plasma proteins which is increased in pregnancy by the action of oestrogen. So it is not known if this increased level of cortisone in the blood is physiologically active or whether it is just a store held in reserve, or whether it is an evolved device to prevent undue loss of steroid in the urine. However, production rates of cortisol in pregnancy are increased two- to three-fold in pregnancy and range from 20 to 40 mgm. per day.

The effects of cortisol are widespread as demonstrated by the clinical states of Addison's disease and of Cushing's disease. Protein, carbohydrate and mineral metabolism are in part under its control. All of these are changed in pregnancy. Mineral metabolism has been mentioned in connexion with aldosterone. Protein metabolism is altered to serve the needs of the growing fetus and amino-acid transfer to the fetus brings about changes in the blood levels of these substances in the mother. An often observed clinical change in the mother is decreased sugar tolerance in pregnancy amounting sometimes to the induction of a mild diabetic state which disappears after the pregnancy is over. Although insulin is involved in this it is almost certain that cortisol is too. In fact some observers have affected to see pregnancy as a "Cushingoid" state and have attributed many of the problems of pregnancy to an over-activity of the adrenal gland. The increasing trunk fat and the relatively thin limbs, together with the pink striae of the skin which are often seen in pregnancy are similar to the changes seen in Cushing's disease, but this is probably to over-simplify the physiological problems of pregnancy.

An interesting suggestion has been that the placenta may secrete either ACTH or corticosteroids or both. The placenta is believed to contain more corticosteroids than would be expected, but there is no proof that they are produced there and alternative possibilities are that they are stored there or that they may come from the fetus.

INSULIN

There has long been presumptive evidence of changes in insulin metabolism in the changed carbohydrate tolerance of pregnancy and of the worsening of diabetes mellitus in pregnancy but direct evidence in upsets of insulin secretion has not been demonstrated. Confusing the issue is that the absorption of glucose from the alimentary tract is changed so that absorption is speeded up, and this may give rise to a "lag storage" type of glucose tolerance curve. But insulin injections near term do not show such a fall of blood glucose as would be expected in the non-pregnant woman. Therefore, there seems to be a definite insulin resistance.

Apart from diabetes being made worse during pregnancy there is no doubt that diabetic women are more prone to develop the syndrome of pre-eclamptic toxaemia, hydramnios and to have bigger babies than normal. These may not be due to direct effects of over-secretion or under-secretion of hormones, but rather due to the changes which these alterations in endocrine function may bring about. It is tempting to suppose that the over-secretion of diabetogenic or growth hormone from the anterior pituitary is responsible for the increased weight of the babies of diabetic mothers but there is no evidence to support this hypothesis.

MELANOCYTE STIMULATING HORMONE (MSH)

This hormone has long been known to be active in Amphibia causing darkening of the skin under certain circumstances. It is secreted by the pars intermedia of the pituitary and chemically is similar to ACTH. It has recently been shown to be increased in the blood during pregnancy and may be responsible for the well-known pigmentary changes of pregnancy. Thus there may be pigmentation of the forehead, cheeks and virtually always there is increasing pigmentation of the nipple and areola of the breasts as well as of the central line of the abdomen which is called the linea nigra. There appears to be no useful purpose in this increased pigmentation and it has been suggested that MSH is a by-product of the metabolism of ACTH.

THE POSTERIOR LOBE OF THE PITUITARY

This secretes the two octapeptide hormones, oxytocin and anti-diuretic hormone (ADH), the last being also known as vasopressin. Oxytocin causes the uterus to contract and will be best dealt with under the heading of myometrial activity. It is almost certain that increased or at least changed secretion of ADH has a part to play in the economy of pregnancy but its role is not so far proven. It is known that there is an altered water balance in pregnancy mediated through alterations in renal physiology. Some measurements of ADH have suggested an increase in the blood; others have failed to detect it. The hormone has also immense theoretical interest since in its role as vasopressin it has an effect on the blood pressure. Raised blood pressure is an essential feature of the syndrome of pre-eclamptic toxaemia of pregnancy but no place has yet been assigned to vasopressin in its genesis. In addition there is probably an upset in water balance in pre-eclampsia in which ADH may be involved. Again the evidence as to whether it is present in excess in the syndrome is conflicting.

Chapter XVIII

THE PLACENTA

The reproductive cycle of a woman has been followed from birth through growth to maturity, the production of gametes, sexual intercourse and the coming together of male and female gametes in the process of fertilization. In Chapter IX the early cleavage of the fertilized ovum and its

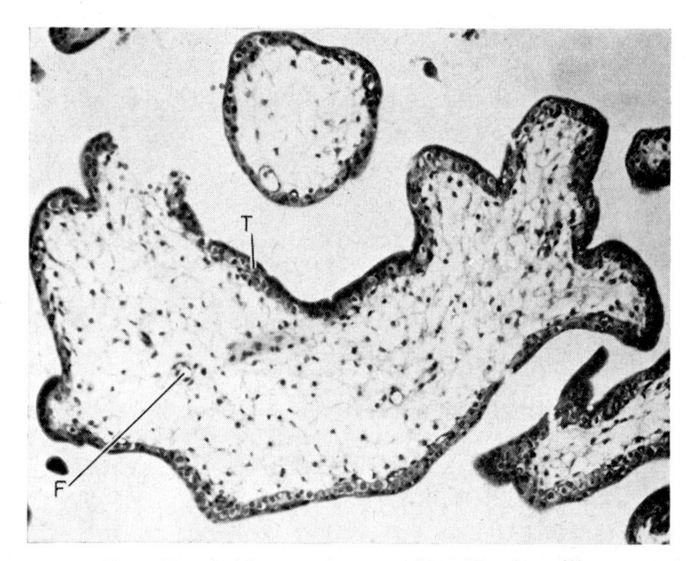

T = Trophoblast F = Fetal capillary

FIG. 18.1. Histology of early placenta. The chorionic villi are larger than at term and they are covered by a double layer of trophoblast, an inner cytotrophoblast and an outer syncytiotrophoblast. The core of the villus consists of mucous connective tissue. Fetal capillaries have developed in the villi by the twelfth week of pregnancy.

differentiation into the inner cell mass which forms the embryo and the trophoblast which nourishes the embryo was traced. At first the trophoblast consists of two layers, the cytotrophoblast and the syncytiotrophoblast. The cytotrophoblast nearest to the embryo has cellular walls and probably produces chorionic gonadotrophin. The syncytiotrophoblast does not have cellular boundaries. It is closest to the maternal decidua lining the walls of the uterus and probably produces enzymes which erode through the decidua so that ultimately this fetal epithelium comes to lie in maternal venous sinuses. It is from the maternal blood that the embryo and fetus derives all its nutrition during the nine months of its pre-natal existence. The two layers of the trophoblast are nourished from

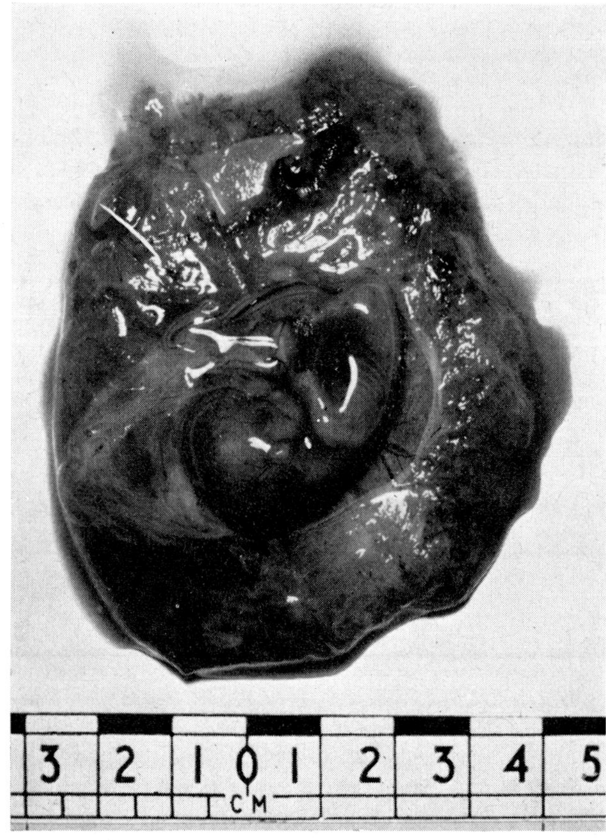

Fig. 18.2. Fetus of 9 weeks menstrual age covered with glistening amnion. Chorionic villi are seen especially at the upper edge of the photograph.

the maternal blood but for the embryo to take advantage of the trophoblast it has to establish its own vascular system running into the trophoblast.

The chorion is a fetal production consisting of a type of loose connective tissue. From the fetal side the chorion pushes its way into the trophoblast and everts it so that the chorion forms a core to the epithelium. Soon afterwards the chorion is invaded by fetal blood vessels and the two layers of trophoblast on their core of chorion containing fetal blood vessels are called villi. It is these villi which are the essential structures concerned with the interchange of materials between the mother and her fetus. Following a substance from the mother to the fetus it will be realized that it has to make its way from the maternal venous sinuses, through the syncytiotrophoblast, then through the cytotrophoblast, then through the connective tissue of the chorion and so into the fetal capillaries from where it can make its way along the umbilical vein into the body of the fetus.

For the first few weeks of life the embryo is completely surrounded by chorionic villi but about the 14th to 16th week the fetal circulation is established to only a part of the chorion and this part becomes the placental disc whilst the rest atrophies. To summarize, the process is fertilization, cleavage, differentiation into inner cell mass and trophoblast, embedding in the decidua, formation of chorionic villi with a fetal circulation, formation of the placenta with atrophy of other villi.

The placenta, or rather its chorionic villi, is truly remarkable. It invades the maternal bloodstream just so far and no further, and then with a most simple structure it becomes responsible for the nutrition, respiration and excretion of the fetus, and by its own endocrine secretions namely chorionic gonadotrophin, oestrogens and progesterone it becomes responsible for redirecting maternal physiology so that the fetus receives what it requires for differentiation and growth. After its brief span of life of nine months it is delivered after the fetus and, its job done, is consigned to the furnace.

The previous eight chapters have dealt with the maternal adaptations to the presence of the products of conception within the uterus. Now attention is to be turned to events within the uterus.

THE PLACENTA AT TERM

At term the placenta (Latin for "cake") is about 7 inches (17·5 cm.) in diameter and about 1 inch (2·5 cm.) thick. It weighs about 1 lb. (0·5 kg.). After delivery the surface which was attached to the uterine wall is seen to be shaggy and red and roughly broken up by grooves into areas about 2 inches (5 cm.) across. These are called cotyledons and consist of groups of villi tightly packed together. There are 15 to 20 of these cotyledons. Careful inspection of this surface shows that it is almost completely covered by a thin broken layer of greyish tissue. This is the superficial layer of the maternal decidua which comes away at birth leaving only a thin layer of decidua still attached to the muscular wall of the uterus. With a dissecting microscope it is possible to see some of the small maternal blood vessels, now broken across, which supplied the maternal blood to the spaces between the villi.

The fetal surface of the placenta shows the smooth and glistening amnion which was a complete sac surrounding the fetus and containing

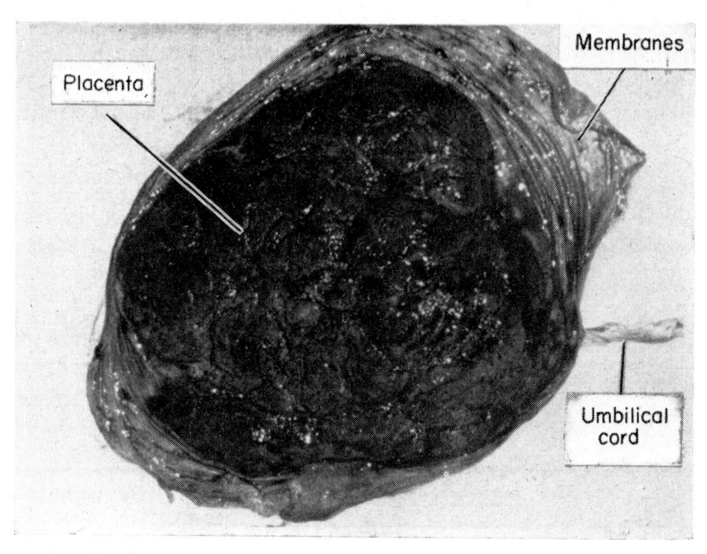

FIG. 18.3. Surface of placenta applied to uterine wall. Surface broken up into cotyledons.

fluid, the liquor amnii. Attached to this surface is the umbilical cord with its three vessels surrounded by Wharton's jelly. The amnion can easily be stripped off the placenta and for the moment is of no further importance. With the amnion removed the fetal vessels can be seen coursing in all directions over a red flattened area and this is the chorionic plate. At its edge and stuck to the outer surface of the amnion is the much attenuated

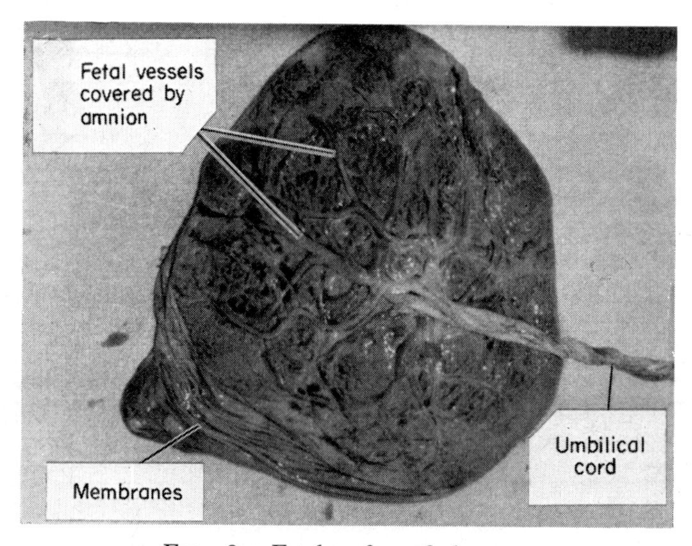

FIG. 18.4. Fetal surface of placenta.

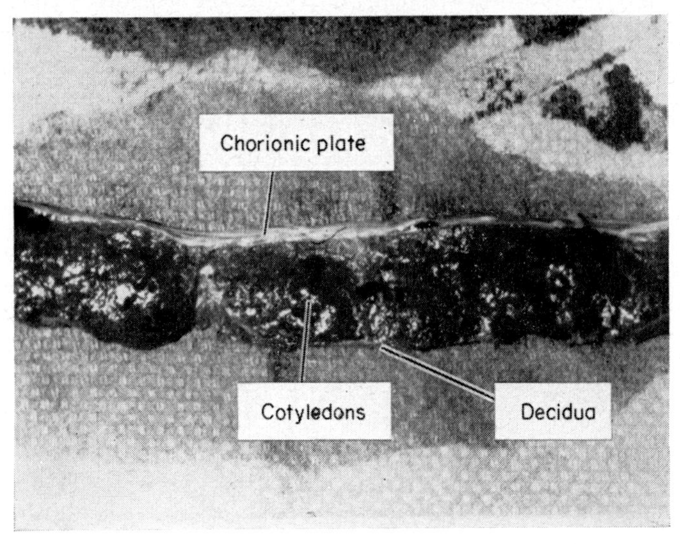

FIG. 18.5. Cut surface of placenta. Fetal surface above.

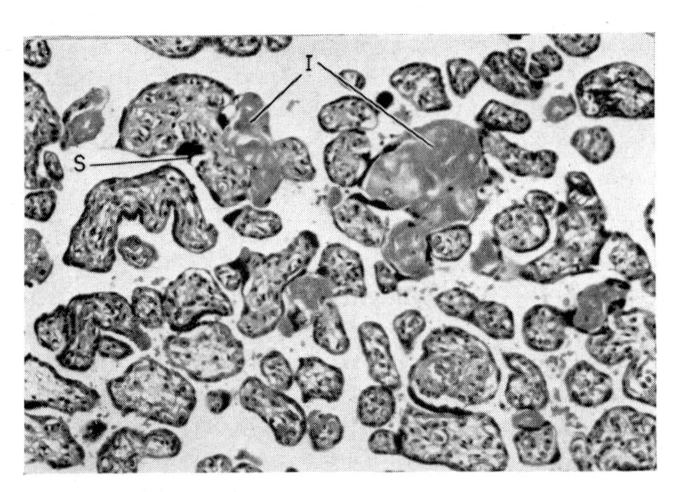

I = Intervillous fibrin. S = Syncytial knot.

FIG. 18.6. Histology of late placenta. The chorionic villi at term are small. For the most part the trophoblastic covering has atrophied to a single layer of syncytiotrophoblast, although isolated cytotrophoblast cells may be found by careful search. In some areas, trophoblast is deficient and here dark-staining fibrin is deposited on the villi. The villous core contains fetal blood vessels forming into peripheral sinusoids beneath the trophoblast. More collagen is apparent at term than in the earlier months of pregnancy.

chorion and this is all that remains of the masses of chorionic villi which surrounded the whole ovum up to the first 12 or 14 weeks of pregnancy. It is the chorionic plate which forms, as it were, the skeleton of the placenta. Its tissue is continuous with the connective tissue cores of the villi. The plate and its cores carry and support the fetal blood vessels.

THE TWO CIRCULATIONS OF THE PLACENTA

On the *maternal* side blood is delivered to the placenta mainly through the uterine arteries. The pressure at which it is delivered is not accurately known but is probably of the order of 60 to 80 mm. Hg. diastolic. The pressure within the amniotic cavity and within the maternal lake of blood in the placenta is about 10 to 15 mm. Hg. in the resting state. Therefore some mechanism must be available to lower the pressure across the uterine wall from the 60 mm. Hg. in the artery to 15 mm. Hg. in the intervillous space. This is achieved partly by the artery breaking up into many smaller tributaries in the muscle wall of the uterus and then as the arterioles traverse the decidua they become very coiled, a system associated with rapid pressure reductions over a short length of tube and they open into a lake of low resistance. Estimates have been made of the number of arterioles opening into the intervillous space and there are probably about 200 to 300 at term and rather smaller numbers earlier in pregnancy, though these estimates are based on only a few cases.

The rate of blood flow through the maternal side of the placenta has been investigated by injecting radioactive isotopes through the abdominal wall into the intervillous space and then tracing how fast the isotopes were removed from the site. Most estimates agree roughly and it is thought that the blood flow is of the order of 500 to 750 ml. of blood per minute or it may be expressed as 100 ml. blood/100 g. of placenta per minute. (Muscle is 2 ml./100 g. per minute.) This represents about 10 to 15 per cent of the cardiac output of the pregnant woman near term. A matter of some importance is that in some maternal diseases and especially in hypertension the blood flow to the maternal side of the placenta may be very much reduced and this may endanger the life of the fetus or if long continued may seriously interfere with its growth.

The exact mode of circulation of the blood within the intervillous space is not fully known for Man, though X-ray photographs show that it is discrete and bear out for Man the experimental findings of Elizabeth Ramsay in the rhesus monkey and other Primates. She injected indian ink into pregnant monkeys and then sectioned the uterus with the placenta *in situ*. The results show that from the mouths of the arterioles emerging from the decidua the blood spurts up to the chorionic plate under some pressure. From here the blood trickles slowly back over the surface of the villi and then is drained away from the decidual area by

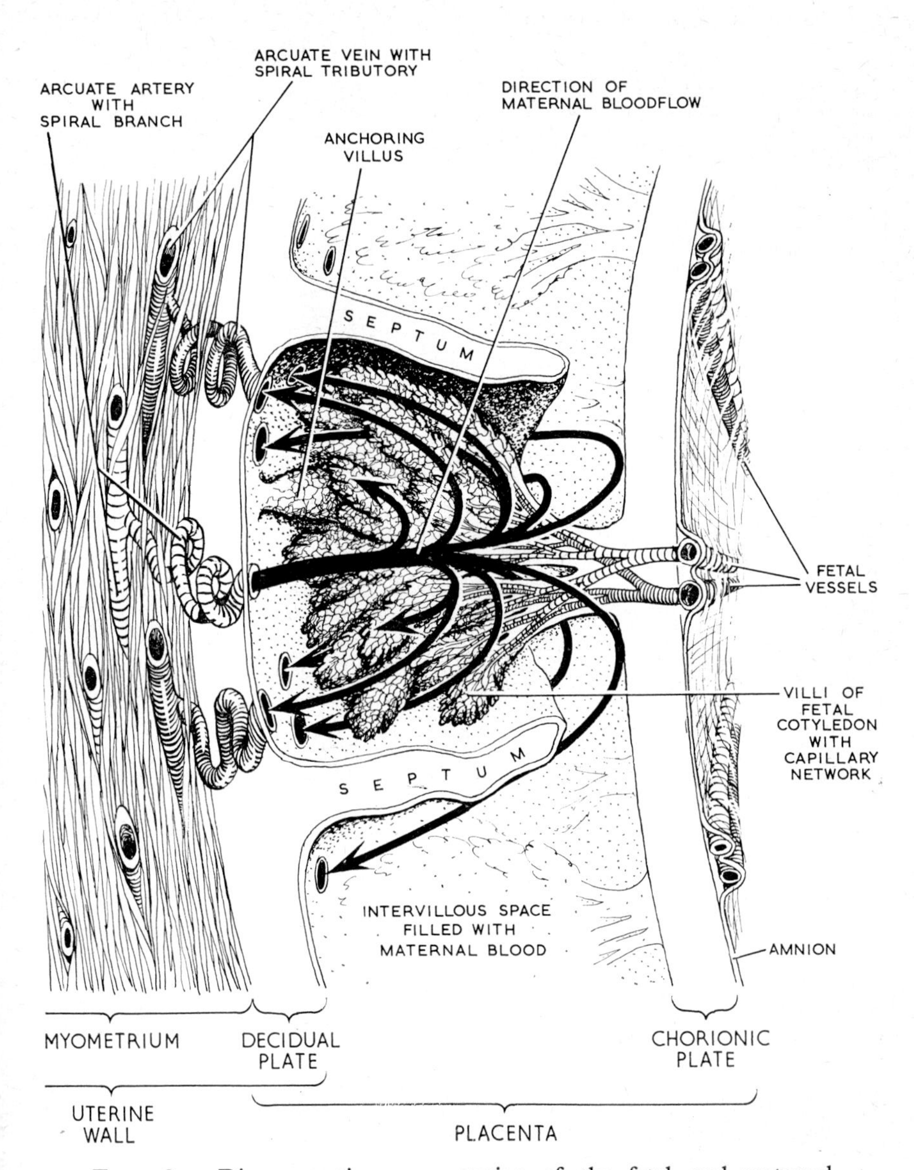

FIG. 18.7. Diagrammatic representation of the fetal and maternal circulations in the human placenta. Note that each villus is bathed by maternal blood of varying composition, intermediate between that of arterial and venous oxygen tension.

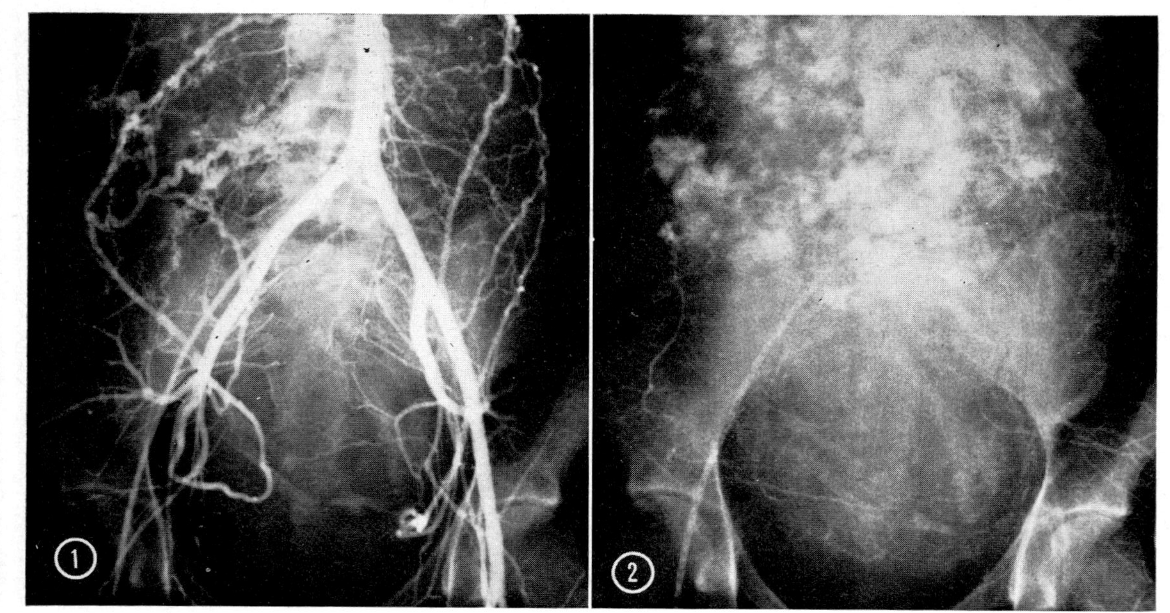

FIG. 18.8. The catheter, which is introduced into the right femoral artery and passed up beyond the bifurcation of the aorta, can be seen on all four radiographs.

(1) The radio-opaque dye has been injected and fills the aorta, both common iliac arteries and both internal iliac arteries, but not the right external iliac artery where the catheter is. The leash of vessels arising from the internal iliac arteries is clearly visible and the arteries coursing up the sides of the uterus are seen. Note the tortuosity of all the distal arteries.

(2) A few seconds later the dye has been carried away from the major vessels but still lingers in some of the minor ones. The fuzzy cotton-wool like areas show the dye in the intervillous spaces of the placenta.

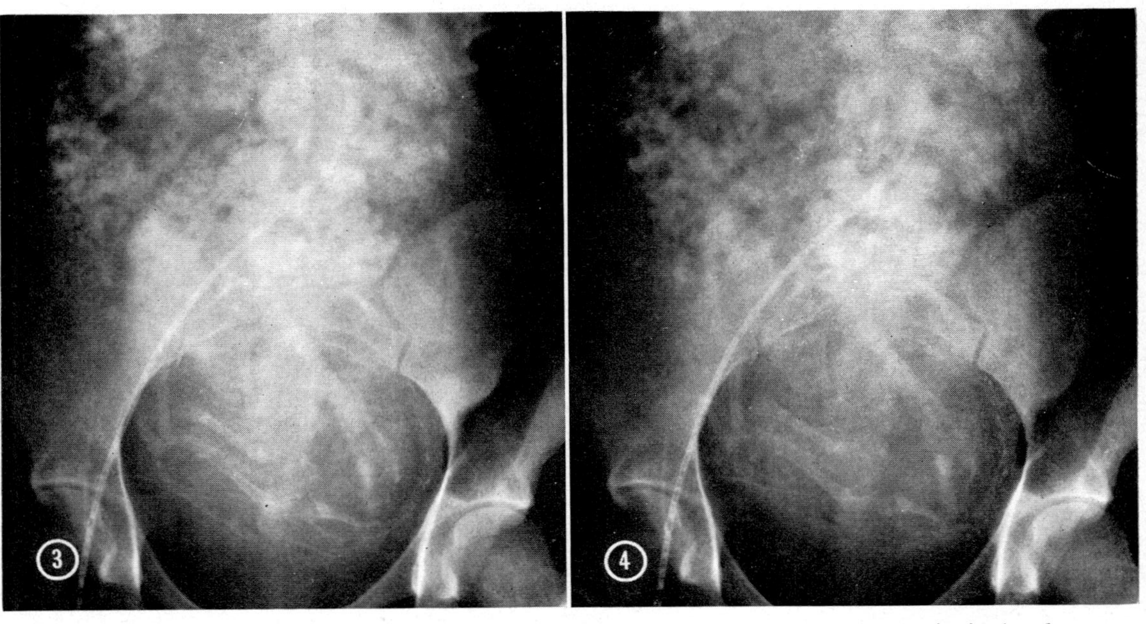

(*b*)

(FIG. 18.8). (3) A few seconds later still the dye has left the fine vessels but the radio-opacity in the placenta is still increasing. The X-ray shows that the placental circulation is broken up into relatively independent areas.

(4) The dye is beginning to drain away from the placental area, though the pattern is still discrete. The fetal skeleton shows distinctly in the maternal bony pelvis. The time covered by the four radiographs of Fig. 18.8 is about 10 seconds.

Fig. 18.8 (*a*) and (*b*). Series of photographs by permission of Dr. M. Lea Thomas, St. Thomas's Hospital.

many small venules which exceed in number the arterioles. The pressure in the venules is not known for certain but would be expected to be not more than, say 5 mm. Hg., and perhaps less, and might vary with respiration as it does in the inferior vena cava. For simplicity, the venules are not shown in the diagram Fig. 18.7, but they join up one with another as they do elsewhere and finally drain into the uterine and ovarian veins which form large plexuses draining respectively into the internal iliac veins and into the inferior vena cava, though on the left side the ovarian vein drains into the left renal vein.

The blood in the intervillous space is prevented from being lost to the vascular system by the adherence of the chorion to the decidua. This area is often of clinical importance since if the chorion is weakened at this point maternal blood may escape from the placental site and may be manifest as vaginal bleeding. Such bleeding can be severe and what is more it is evidence of impairment of the maternal placental circulation which may cause serious deprivation of the fetus.

On the *fetal* side of the placenta blood is conveyed by the two umbilical arteries. These reach the chorionic plate through the umbilical cord where they break up into several smaller vessels which gradually sink through the plate to supply each of the villi with an arteriole. Thereafter they break up still further to form capillaries lying under the trophoblastic covering of the villi. On the other side of the trophoblast is the maternal blood in the intervillous space.

The villi are collected into groups called cotyledons and figure 18.9 shows how the arterial system breaks up to supply the fetal placenta. The proximal one-third nearest to the chorionic plate is straight and then very rapidly breaks up into its constituent smaller vessels in the distal two-thirds. As elsewhere the capillaries join up into venules and these become veins and sinking back through the chorionic plate they are gathered up to form the umbilical vein which conveys the blood along the umbilical cord to the fetus.

The pressures within the fetal blood vessels of the placenta are not fully known for Man though they have been the subject of investigation in other animals. Measurements in the umbilical arteries of the newborn baby suggest that the pressure here is of the order of 70/45 mm. Hg. From the animal work it would seem likely that the pressures in the villi would be of the order of 30 mm. Hg. and somewhat less in the umbilical vein. In the newborn baby it has been computed that the cardiac output is of the order of 450 to 550 ml. per minute and it would seem that something rather less than this amount is flowing through the fetal side of the placenta each minute. The placenta must take the major share of the cardiac output since the shape of the fetus (small arms and legs, medium size trunk and big head) reduces the amount of blood needed for metabolism. In the sheep the placenta takes two-thirds of the cardiac output

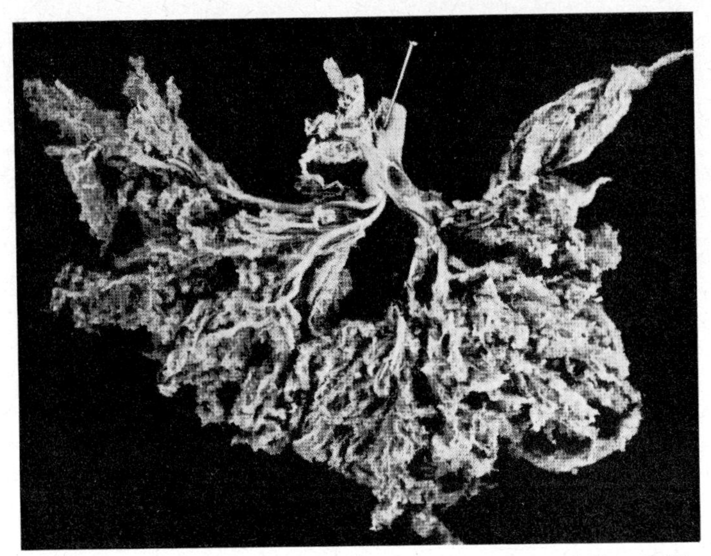

FIG. 18.9. (From C. A. Villee (1960) "The Placenta and Foetal Membranes". Williams and Wilkins.)

of about 500 ml./min. The liquor amnii reduces the need for the maintenance of body temperature and so for skin blood flow. Finally there is very little flow through the lungs until after birth.

THE GROWTH OF THE PLACENTA

At first the trophoblast completely surrounds the amniotic cavity. Not until the 14th to 16th week is the placenta formed as a disc, and the fetal and maternal circulations are fully functioning. Placental weights at various stages of gestation have been measured with the following results:

Length of gestation (From first day of last menstrual period)	*Placental weight in grams.*
10 weeks	20
20 "	170
30 "	430
40 "	650

The graph from these averaged figures shows that the placenta is growing steadily throughout the whole of pregnancy and further that the probable maximal phase of growth is between the 20th and 30th weeks and that thereafter the rate of growth tends to slow a little. This is in keeping with many other physiological features of pregnancy where there is a tailing off towards the end of pregnancy, e.g. growth of fetus, volume of liquor, steroid production, cardiac output, blood volume. The cause of these changes is obscure.

Apart from growth being measured crudely by weight, estimates have been made of placental volume and of its functioning surface area. This area is most important since it is the effective surface across which interchanges between the mother and her fetus occur. Aherne and Dunnill (*Morphometry of the Human Placenta in The Foetus and the New-Born,* British Medical Bulletin Volume 22, Number 1, January 1966, page 5) found the volume of the placenta at term to vary between 450 and 500 ml. Of this total volume it was shown that just over 20 per cent

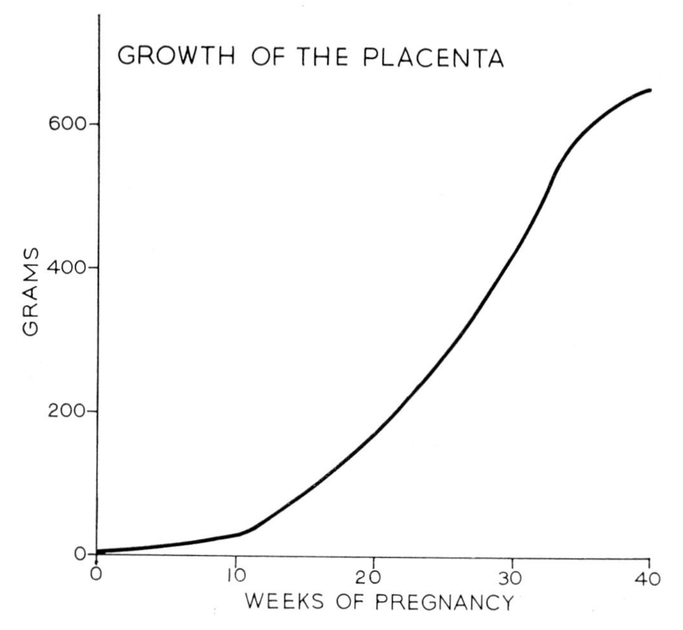

FIG. 18.10. Data from Hytten and Leitch (1964).

is supporting tissue with no function in the transport of materials between mother and fetus. The remaining 80 per cent or so is "parenchyma" and consists of the chorionic villi and the intervillous space where the maternal blood circulates. This space forms about 42 per cent of the "parenchyma" and in absolute terms this is about 140 ml. The remaining 58 per cent of the parenchyma consists of (i) fetal capillaries 12 per cent, (ii) connective tissue stroma 31 per cent, and (iii) trophoblast 15 per cent. These proportions stay the same from the 28th week of pregnancy to term. The surface area of the villi measured by a most erudite technique came to 11 square metres, at term, though others have found it to be slightly more. Of great clinical interest is the fact that this villous surface area is reduced in some maternal diseases such as hypertension with consequent effects on the growth of the fetus. The surface area of the villi in contact with the maternal blood bears a linear relationship with the placental volume and therefore the routine weighing of the placenta at delivery is of some value. The diameters of the villi are nearly all in the range of 20 to 60 μ.

PLACENTAL EFFICIENCY

It can be seen from the foregoing that placental efficiency depends on (1) the maternal input of blood, (2) the parenchyma of the placental villi, (3) the fetal input of blood. To summarize what has gone before there would seem to be a blood flow of about 600 ml. per minute of maternal blood through a space of about 140 ml. and it flows over a surface area of about 11 sq. metres. Coming in on the fetal side may be something of the order of 300 ml. blood per minute. Other factors are involved in the transfer of materials between the mother and the fetus but these three are basic to understanding. With them in mind it is possible to consider the exchange of various materials across the placental barrier and it is well to remember that the essential functions of the placenta are respiration, nutrition and excretion. In addition the placenta produces hormones and also protects the fetus from infection, partly by acting as a filter against organisms and partly by transferring antibodies from the mother to the fetus.

PLACENTAL TRANSFER

The transfer of substances between mother and fetus occurs in the parenchyma of the placenta which lies between the maternal and fetal fluids. Early in gestation the parenchyma has considerable depth, 25 μ, and at term it is still 5–7 μ thick, about equal to the diameter of a red cell. Transport is effected by the syncytiotrophoblast, which is now usually called "the placental membrane", to indicate the active nature of many of the mechanisms involved. The cells are rich in enzymes and mitochondria in order to carry out these functions. The high oxygen consumption of the placenta, about 10 ml./Kg./min., is twice that of the fetus itself. Relatively little is known of the rate of transport of substances in either direction across the membrane, but "net" transfer of the important constituents is readily measured by analysing fetuses which have died in the perinatal period.

RESPIRATION (O_2 AND CO_2 TRANSPORT)

Though the fetal heart may continue to beat for much longer periods than the adult when the oxygen supply is reduced, it cannot be too frequently emphasized that the oxygen utilization of the whole fetus is the same as the adult at rest, 5 ml./Kg./min. and that oxygen is, therefore, as necessary for survival in them both. The oxygen consumption of the fetus and placenta at term has been computed from blood flow measurements on either side of the placenta and either uterine or umbilical arterio-venous (A-V) differences. The oxygen consumption is about 25 ml./min., 15 ml. being required by the fetus and 10 ml. by the placenta. Any impairment of the maternal blood flow, upon which the existence of the fetus primarily depends, will create an oxygen debt in both placenta and fetus.

The course of the maternal blood flow in the chorionic space and its relation to the chorionic villi (see page 190) suggests that it would be impossible for the mean oxygen tension in the maternal spaces to be high,

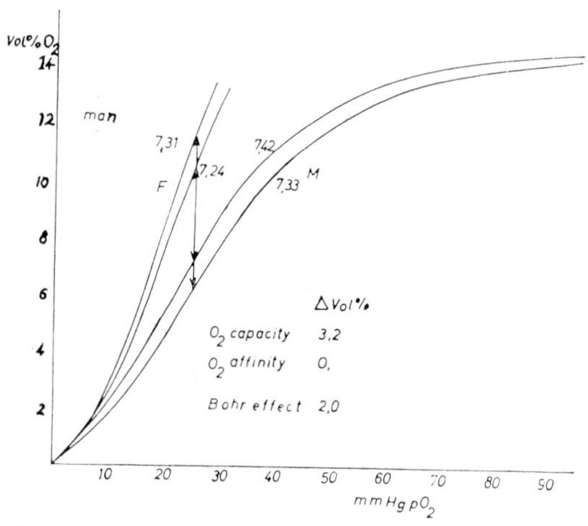

FIG. 18.11. From *Proc. Inter. Un. Phys. Sci. IV*. Diagrammatic representation of the human maternal and fetal oxygen dissociation curves. This shows the approximate importance of the oxygen capacity (3·2 vols $\%_{O_2}$) and the Bohr effect (2·0 vols $\%_{O_2}$) during oxygen transport from the maternal to the fetal blood.

and that in the umbilical vein must be correspondingly low. Each villus is bathed by maternal blood of varying composition intermediate between that of arterial and venous oxygen tension, a situation analagous to unequal ventilation perfusion in the lung (Fig. 18.7). The mean oxygen tension in the intervillous space is 40–50 mm. Hg., and in the fetal arterialized blood in the umbilical vein, it is 20–30 mm. Hg. The tension difference across the membrane is therefore about 20 mm. Hg. O_2 in comparison with 60 mm. Hg. O_2 across the alveolar membrane.

Oxygen transport across the placenta at these low oxygen tensions is effected by the high maternal and fetal placental blood flows, which have already been described (page 197), and the relative positions of the maternal and fetal blood oxygen dissociation curves. The figure 18.11

shows that the fetal dissociation curve lies to the left of the maternal, in spite of the lower pH of fetal blood. This means that at any given oxygen tension the fetal blood will contain more oxygen than will the maternal blood. The separation of the oxygen dissociation curves is not brought about by differences in oxygen affinities of the bloods in the human subject, for the differing oxygen affinities of the two haemoglobins is cancelled by their environment in the red cells. It is mainly due to the relative amounts of haemoglobin and, at term, the oxygen capacity of the fetal blood may be of the order of 20–25 vols. per cent and that of the mother's blood may fall to 15 vols. per cent. At the placental site the exchange of CO_2 causes an increase in tension in the maternal blood, but a decrease in the fetal. These changes in CO_2 tension will, by the Bohr effect, separate the curves even further, so increasing the oxygen content of the fetal blood and decreasing that of the maternal. Note, particularly, in the diagram that the combined influence of the oxygen capacity and Bohr effect results in a 6 vols. per cent O_2 content difference between the maternal and fetal bloods in the range 20–30 mm. Hg. pO_2, found in the umbilical vein blood. The A-V O_2 difference across both the maternal and fetal side of the placenta is about 6 vols. per cent, similar to that across the alveolar membrane.

The carbon dioxide tension difference across the placenta is almost the same as across the alveolar membrane, 6 mm. Hg., but may be as much as 10 mm. Hg. This gradient from the fetus to the mother is ensured by the relatively high fetal plasma bicarbonate and the reduction in the maternal level which occurs early in gestation. So that even at the same pH there will always be a higher tension in the fetal plasma.

NUTRITION (WATER AND ELECTROLYTES)

Studies with isotopically labelled molecules have demonstrated that water and sodium and potassium ions and many other electrolytes cross from the maternal to the fetal bloodstream and attain equilibrium rapidly, as they do between the maternal body fluid compartments. But little is known of these transport mechanisms nor of the control of the osmolarity of the fetal body fluids. The normal relationship between capillary pressures and colloidal osmotic pressures which govern fluid distribution in the adult body do not appear to operate across the placental membrane. The pressure in the fetal villi is about 30 mm. Hg. The pressure in the maternal sinuses is similar to the intra-uterine pressure and not usually more than 10 mm. Hg.; yet the colloidal osmotic pressure of the fetal plasma is not correspondingly high. It is well established that the peak rate of transfer from mother to fetus occurs at about eight months gestation: the increase is probably due to the increasing fetal and maternal blood flows, and the reduction in thickness of the membrane, but the reasons for subsequent decline are not known.

MINERALS

Higher concentrations of iron, calcium and iodine are found in fetal blood in comparison with the maternal, suggesting that an active process is involved in the transport across the placental membrane. There is, again, little information on the mechanisms involved: the active process would appear to be on the maternal side of the membrane because iron and phosphorus have been shown to be concentrated within the trophoblast during transport. Iron is of particular interest because a single injection into the maternal bloodstream is practically completely cleared into the fetus at term.

CARBOHYDRATE, FAT AND PROTEIN

The fetal blood glucose is usually slightly lower than the maternal level. The rate of transfer suggests a facilitating mechanism within the trophoblast in the direction from the maternal to fetal bloodstream: fetal blood sugar increases when the maternal levels are raised and decreases when

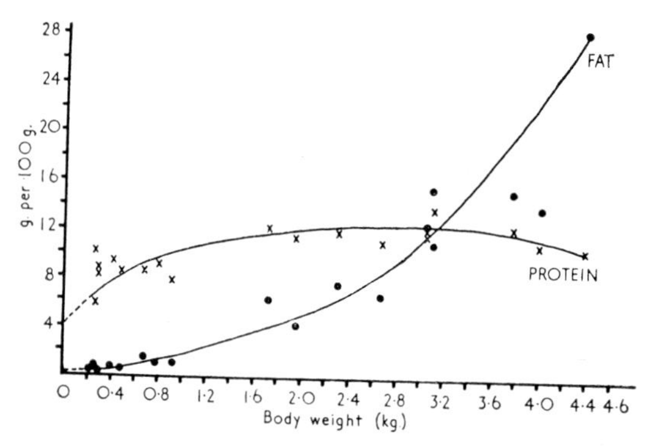

FIG. 18.12. Protein and Fat in the fetal body during development. Note the continuous laying down of protein and the sudden increment in fat deposition late in gestation. (*Archives of Diseases of Childhood*, **26**, (1951).)

they are lowered. Only small amounts of fructose are present in human fetal plasma. The relation of the placental glycogen to fetal blood glucose levels is not fully understood, but it is assumed that it forms a carbohydrate store in the same manner as that of adult liver glycogen. Placental glycogen decreases towards term as the fetal liver stores increase.

Cholesterol and triglycerides do not cross the placental membrane but free fatty acids pass readily. Fat is not laid down until late in gestation as shown in Fig. 18.12.

With the exception of the γ globulins proteins as such are not transferred from the maternal to the fetal bloodstream. Free amino-acids cross readily against a concentration gradient: the fetal: maternal plasma free amino-acid ratio is about 2:1 at term; early in gestation it is higher. As with iron and phosphorus the amino-acids are concentrated in the trophoblast during transport. A selective mechanism for the transfer of the natural L-isomers has been demonstrated. The mechanisms involved can be impaired by inhibitors of oxidative metabolism, but are uninfluenced by oestrogen, and progesterone and fetal requirements. The latter emphasizes the autonomous nature of the placenta. The free amino-acid levels in fetal plasma remain unimpaired in the post-mature period and are not disturbed when fetal growth is poor. The net retention of nitrogen is shown in Fig. 18.12.

VITAMINS AND HORMONES

The vitamin requirements of growth are high and the fetal plasma concentration of the water soluble vitamins is maintained at a higher level than in the maternal plasma. Special mechanisms for the transfer of riboflavin and Vitamin C have been demonstrated. It has been shown that thyroxine is transferred slowly but in physiological amounts from the maternal to the fetal bloodstream: tri-iodo thyronine is transferred more rapidly. It is possible that all the pituitary hormones are similarly transported. Steroid hormones are transported quite readily but the concentrations in the fetal plasma are always lower than in the maternal plasma.

PINOCYTOSIS AND 'LEAKAGE' ACROSS THE PLACENTAL MEMBRANE

Pinocytosis describes the closure of the endothelial villi on the maternal side of the membrane around vesicles of plasma, which migrate through the cytoplasm and discharge their contents at the opposite side of the cytotrophoblast. It has been suggested that maternal plasma proteins such as globulins might reach the fetal plasma this way.

Of greater practical importance is the probability that "leakage" may occur from the fetal to the maternal circulation through small defects in the membrane. Since the pressure in the chorionic villi is higher than in the maternal sinuses, leaks are most likely to be in the direction from fetus to mother. Fetal red cells, in particular, may be so transferred for the presence of fetal haemoglobin in the mother's blood can be demonstrated in 5 per cent of all obstetric cases. The maternal Rh antibody response may be due to such leakage. Recently, it has been shown that this response usually occurs after the birth of the first child and may, therefore, result from leakage through the large defects in the membrane occurring as the placenta separates at birth.

The rubella and smallpox viruses cross from the maternal to the fetal bloodstream, possibly through defects in the membrane caused by inflammation. The entry of the causal organism of syphilis is probably by the same route. Section of such syphilitic placentas after birth show the Treponemata flourishing in the parenchyma.

EXCRETION

Urea and Bilirubin. The preponderance of anabolism over catabolism in the growing fetus probably means that waste products are not formed in large amounts. The fetal kidney can produce urine by the third month of intra-uterine life, and this is excreted into the amniotic fluid. It is doubtful whether renal excretion plays any part in the regulation of the fetal *milieu interieur*; the placenta is responsible for maintaining intra-uterine homeostasis, and infants born with no kidneys do not have high blood ureas nor do they become acidotic. Fetal plasma urea rises gradually during gestation and at term umbilical A-V differences demonstrate diffusion across the placental membrane from the fetus to the mother.

Fetal plasma bilirubin is about 1 mg. per cent at term. It is nearly all in the free unconjugated form and is probably excreted slowly by diffusion through the placental membrane: the conjugated bilirubin in the maternal plasma does not cross into the fetal bloodstream.

DRUGS

The diffusion of drugs from the maternal to the fetal bloodstream depends upon a number of physical factors such as the concentration gradient, lipoid solubility, molecular weight and degree of ionization, which in turn is influenced by the pH of the body fluids. Biologically the transport is also determined by the maternal and fetal circulation rates both of which may be influenced by the presence of the pharmacological substance.

It cannot be assumed that the effect of a drug on the fetus is the same as in the adult, for little is known of the activity of the developing effector organs. Further, it is well established that many of the liver enzymes required to detoxicate drugs are deficient in the developing liver, and harmful concentrations may accumulate in the fetal bloodstream.

The sedatives and analgesics, alcohol, bromide and salicylates are known to cross the placenta in the human subject. Pethidine and morphine are also transferred and withdrawal symptoms may be observed in the newborn of morphine addicts. The anaesthetic gases, ether, nitrous oxide and cyclopropane are readily transferred. The barbiturates also reach the fetus swiftly. The placental transmission of d-Tubocurarine is negligible and it may therefore be used with safety to cause relaxation of maternal skeletal muscle during Caesarean section and other obstetric operations. The human fetal neuro-muscular junction is paralysed by d-Tubocurarine as early as 18 weeks of gestation.

Sulphonamides cross the placental membrane and the antibiotics, streptomycin, penicillin and the tetracyclines quickly reach therapeutic levels in the cord blood. Dindevan (an anti-coagulant), the antithyroid drugs and iodine all easily pass to the fetus.

The sympathomimetic amines and other pressor substances probably do not reach the fetal circulation in significant quantities. In animal experiments both pressor and depressor substances cause fetal hypoxia and bradycardia by reducing the maternal placental circulation rate, due to the vasoconstriction of the pressor agents or the hypotension of the depressors.

Chapter XIX

FETAL GROWTH

The placenta is a mainly fetal organ which dips, as it were, into the maternal bloodstream. By its secretions it is able to influence the mother's physiology so that it is redirected to take account of the products of conception growing within the uterus. Also from its place in the maternal bloodstream the placenta is able to take up those nutrients which the fetus requires for its differentiation and growth. The process of differentiation is a complex one to which textbooks of embryology are devoted. Little is known of the physiology of this stage of development, and it will not be further considered here, though it is of immense importance in its own right and because studies of this early phase are of value in the understanding of teratology. At present for the physiologist and doctor the time of growth of the fetus is more rewarding. Very roughly differentiation of all the fetal organs is nearly complete at about the 12th to 14th week of pregnancy counting from the first day of the last menstrual period. That is, at this point all the organs and tissues are made but all of them have still to grow until the fetus is mature enough to be born. This period of growth occupies approximately the last 28 weeks of pregnancy.

THE GROWTH OF THE FETUS

A newborn baby is not just a miniature adult. It has to go through the neonatal, child, puberty and adolescent phases before becoming an adult. (See Chapter I, Fig. 1.6.) Moreover growth in different organs and limbs proceeds at different speeds *in utero* so that as compared with the adult some organs and parts are in advance of others. This is the principle of allometric growth and is to be distinguished from isometric growth in which the growth proceeds at the same rate in all parts. During the growing phase of post-natal development (Chapter I) it was possible to distinguish skeletal, neural, genital and lymphatic (Fig. 1.7) types of growth which were all proceeding at different rates. So it is in the fetus. It has been noted before that the newborn baby has a peculiar shape when compared with the adult. The head is large, the trunk is cylindrical with the abdomen rather protuberant, whilst the limbs are rather small and the ratio between the arms and legs, as regards length, is about unity, whilst in the adult the ratio is much less than unity as a result of the postnatal growth in length of the legs. This relatively peculiar shape is a biological adaptation to the development of the human being, who is born in a state of helplessness but with immense capacity for post-natal growth and learning. Thus the locomotor apparatus can be left to be developed at a later date, but the nervous system must be developed sufficiently at birth so that the child can begin to learn new responses at once. In addition the baby obviously has to survive, so that it must have a respiratory, cardio-vascular, alimentary and secretory apparatus ready to go into action as soon as it is born. Until birth the respiratory, excretory and alimentary functions of the fetus are all carried out by the placenta. With the discarding of this mode of existence these systems must begin to function at once. This is a remarkable example of pre-adaptation brought about by evolutionary mechanisms, for the lungs and chest, for instance, must be ready to perform their respiratory function at birth even though the fetus has never before been in a gaseous environment. It is perhaps not surprising that death at birth is not too uncommon when the immense changes required of the baby are recognized. Especially is death likely if the baby is born prematurely, that is before about 36 or 37 weeks of gestation, or if it is born weighing less than about 5½ lb. (2·5 kg.). In fact the care of premature and small babies has so advanced of recent years that many of them will survive, but even so there is a very high death rate of babies born before the 34th week of pregnancy and of those weighing less than 3½ lb. (1·6 kg.). It would seem that a certain amount of time and of growth *in utero* is needed for physiological mechanisms to mature so that the baby will be able to survive in the extra-uterine environment. Comparatively little is known of the differential growths of different parts of the fetus so no more will be said of it here. Most is known of the growth of the fetus as a whole as measured especially by weight, though there are also some data on lengths of various kinds such as crown–rump length and crown–heel length as well as foot length.

THE WEIGHT OF THE FETUS

It might seem to be easy to construct a curve of the growth of the fetus, but it can in fact only be done by cross-sectional methods in which large numbers of fetuses are obtained at varying lengths of gestation and the results averaged. But it must not be forgotten that many of the babies born or delivered prematurely are so delivered because of some disease either in the mother or the baby so that investigation may not be of the normal. However, attempts have been made to exclude all these extraneous factors which could affect the weight of the fetus and representative curves have been constructed. Figure 19.1 is an example. In absolute terms the weight of the fetus is slow at first and then enters an accelerated phase from about 20 to 36 weeks and then tails off slightly.

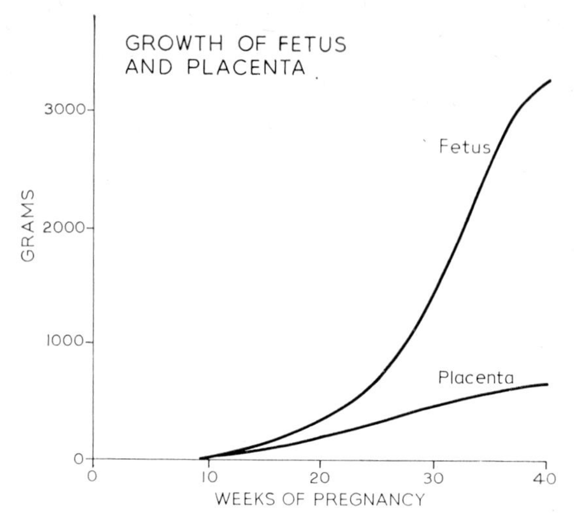

FIG. 19.1.

Hytten and Leitch (1964) give the following average figures, which are a useful guide:

Weeks	Weight of fetus in grams.	Weight of placenta in grams.
10	5	20
20	300	170
30	1500	430
40	3300	650

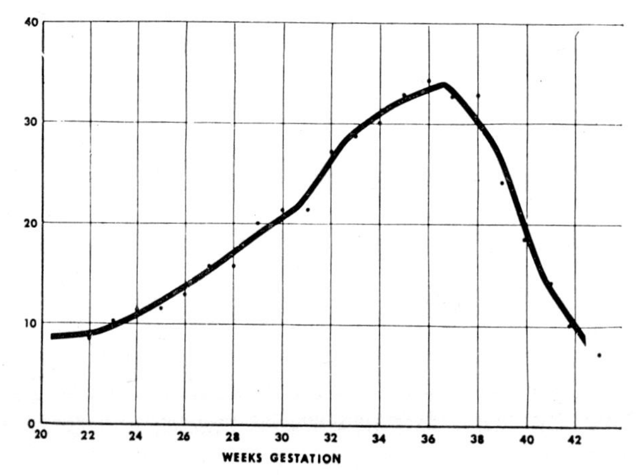

FIG. 19.2.

From *Textbook of Obstetrics and Gynaecology*, by David N. Danforth (1966). Hoeber Medical Division.

From these figures and the graphs it will be seen that the products of conception are mainly chorionic at first, but by about 17 to 18 weeks of pregnancy the fetus has largely outstripped its placenta. Nevertheless, both fetus and placenta continue to grow until birth, though the rate of growth tails off as term approaches. This is especially well shown in the graph of mean daily growth (Fig. 19.2). This demonstrates that the increments in weight fall off greatly from the 36th week. The ratio of fetal to placental weight tails off, although the ratio rises from about 3 to about 5·5 between the 24th and 40th weeks of pregnancy (Fig. 19.3). The ratio shows by how much the fetus is outstripping its placenta during gestation, for at the 10th week of gestation the ratio of fetus to placenta is

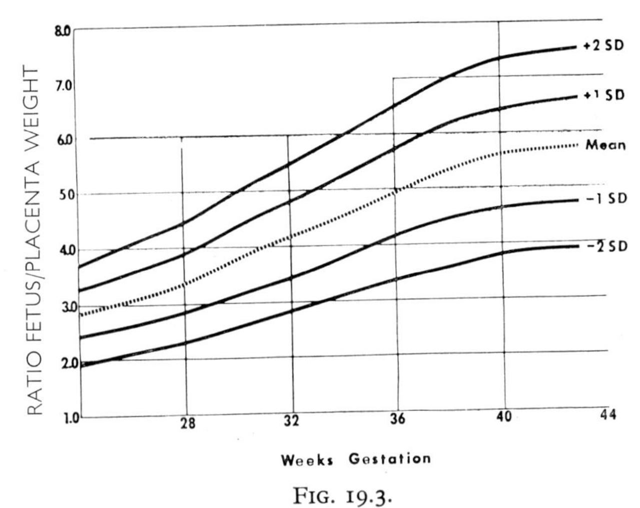

FIG. 19.3.

From *Textbook of Obstetrics and Gynaecology*, by David N. Danforth (1966). Hoeber Medical Division.

1/4. It is surprising that the relatively slow growing placenta can support the life of the much faster growing fetus. The reasons are probably that the placenta has an enormous reserve of function and that for most of pregnancy its reserve is far in excess of the demands which will be made upon it; and also it is probable that placental function increases as pregnancy advances. Such improving function depends upon (1) increasing uterine maternal blood flow, (2) abler transfer of materials to the fetus by the functioning cells of the placenta, and (3) an increasing fetal blood flow to the placenta. Estimates of uterine blood flow and of fetal blood flow have been made in the human at various stages of pregnancy and Fig. 19.4 shows the result. It will be realized that the uterine blood flow is not exactly the same as the maternal placental blood flow, though the two probably run parallel. The graph shows that some of the increasing placental efficiency is due to the maternal and fetal blood flows which follow each other fairly closely.

THE BODY COMPOSITION OF THE FETUS

Estimates have been made of the make-up of fetuses at various times in pregnancy, especially as regards fat and total body water. Fat is not laid down in any great quantity until about the 28th week of pregnancy but thereafter increases fairly rapidly forming about 12 per cent of the total body weight at birth. It is well known that premature babies look

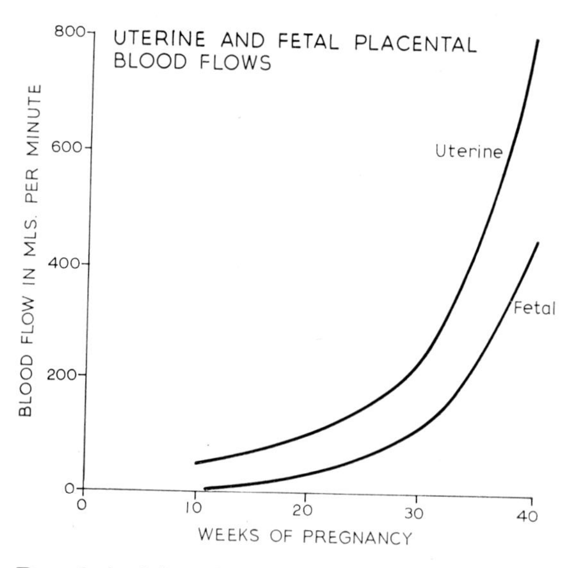

FIG. 19.4. Data derived from Assali, N. S., Rauramo, L. and Peltonen, T. (1960) *Amer. J. Obstet. Gynec.* **79,** 86, and Hytten, F. E. and Leitch, I. (1964) *The Physiology of Human Pregnancy*. Blackwell.

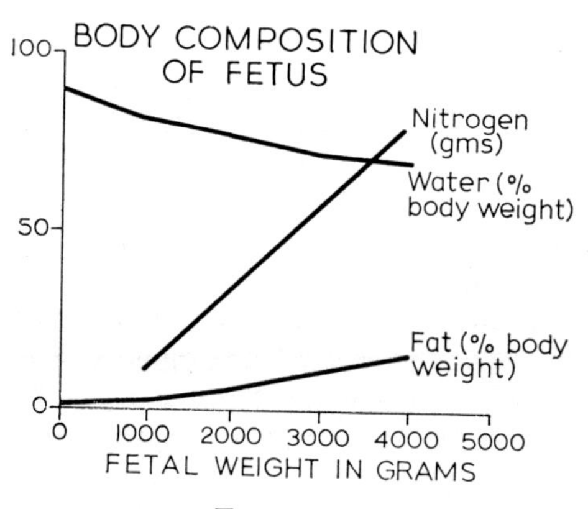

FIG. 19.5a.

wrinkled because of their relative small amounts of subcutaneous fat. A young lean adult male may contain about 10 to 15 per cent of his weight as fat, whilst a young woman may contain about 20 to 25 per cent of fat.

The percentage of the body weight which is water is about 90 at a weight of about 100 grams (say 15 to 16 weeks) but at a weight of about 3000 grams (nearly term) the total body water, abbreviated to T.B.W., has fallen to about 75 per cent (Fig. 19.5a). Absolute values are not easy to come by, but the graphs show the general trends. The total body weight

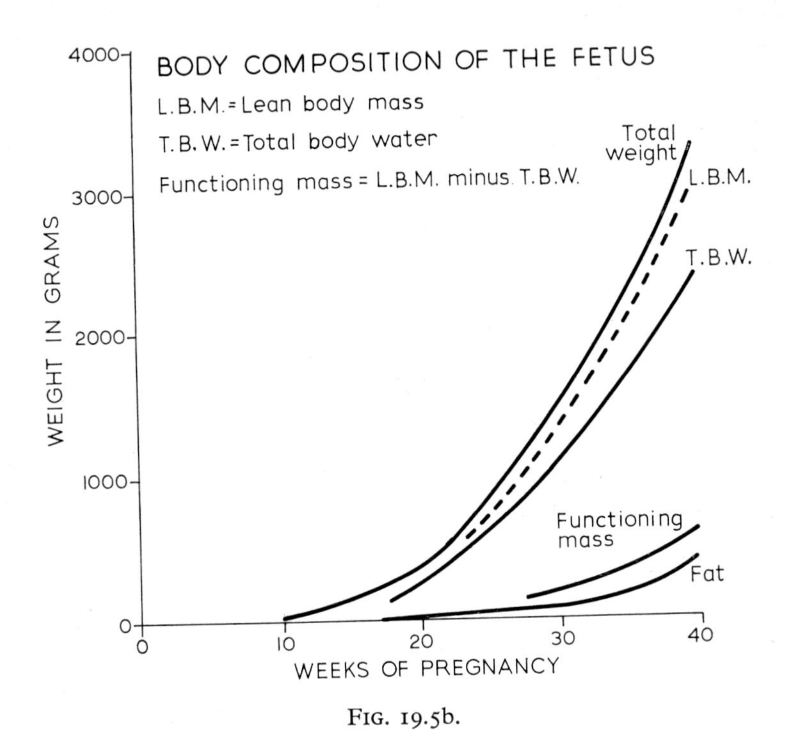

FIG. 19.5b.

is as in previous graphs. Total body weight is made up of fat plus a remainder which is called the lean body mass (L.B.M.). The L.B.M. is the functioning metabolizing tissue made up of cells. The major component of these cells is water.

Figure 19.5b shows that as pregnancy proceeds the fat increases. As it does so the L.B.M. inevitably makes up a smaller proportion of the total body weight. The curve for L.B.M. can be seen falling away from that for the total body weight as term approaches. The total body water (T.B.W.) forms a declining proportion of the L.B.M., and this is shown by the percentages previously mentioned. The curve for T.B.W., therefore, falls away from that for L.B.M. If the L.B.M. is increasing whilst the

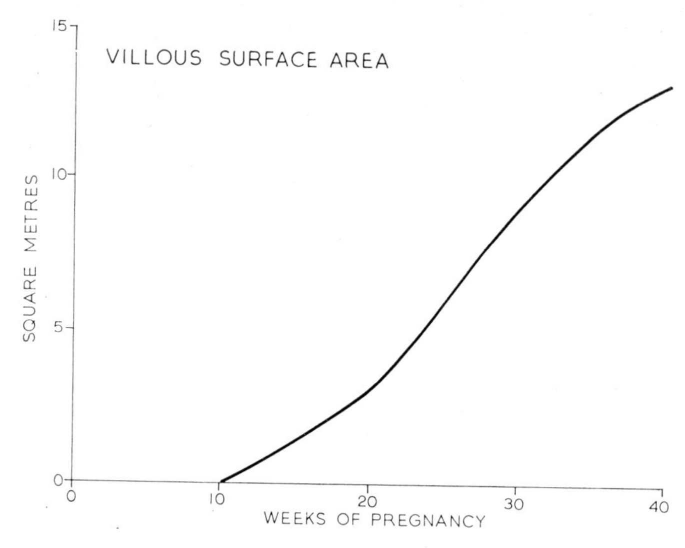

FIG. 19.6.
(Derived from Aherne and Dunnill (1966).

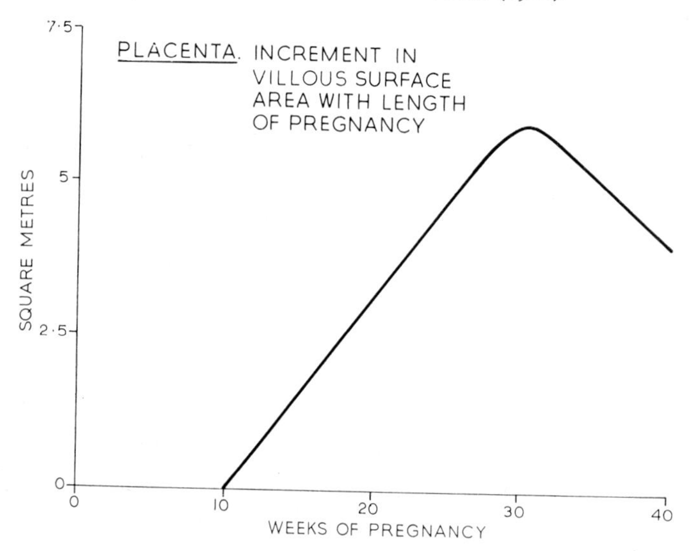

FIG. 19.7.
Recalculated from Aherne and Dunnill (1966).

proportion of water within it is declining then the functioning mass of the cells is increasing at the expense of the water.

The general trend of growth *in utero* is therefore to lay down a very well hydrated lean body mass. At first there is no large amount of fat except for that needed by all cells in their metabolism. By about the 30th week of pregnancy fat begins to be laid down in some quantity, cells increase their functioning tissue whilst the amount of water in them diminishes. All these changes are, of course, dependent on the placental changes, especially in blood flows, which have been described.

The villous surface area of the placenta has been estimated and it increases throughout pregnancy as Fig. 19.6 shows, though like other features of the placenta the rate of growth declines towards the end of pregnancy (Fig. 19.7).

The following table gives some idea of the magnitude of some of these changes.

Weeks	10	20	30	40
Fetal weight	5	300	1500	3300
Fat	—	2	80	408
L.B.M.		298	1420	2892
T.B.W.		270	1136	2394
Functioning mass		18	284	598
Fetal blood flow (mls./min.)		40	90	450
Uterine blood flow (mls./min.)		100	220	800

These are all derived figures simply to show the sorts of results to be expected and are not to be used other than as the roughest of guides, for much work still remains to be done on these problems. Results so far are tentative.

THE FAILING PLACENTA

Despite its prodigious feats in maintaining the growth of the fetus, there are some signs that the placental function begins to fail some time before the end of pregnancy. This is especially well demonstrated when the growth in weight of the fetus per week is calculated. It will be seen from Fig. 19.8 that the peak increase in weight is attained at 34 weeks and thereafter the rate of weight increase falls. Even more interesting is that the weight increase of the neonate picks up rapidly after birth and this shows that it has the capacity for rapid growth if only it can be supplied with the right environment. It seems that the placenta cannot supply this environment in the last six or so weeks of pregnancy.

ALLOMETRIC GROWTH

The shape of the newborn has already been commented upon (page 204). The general shape of the fetus can be inferred by a comparison of weight with length. Fig. 19.9 shows that at first the growth of the fetus is mainly in length. Thereafter it begins to fill out laterally. The maximal rate of increase in length occurs at about the 4th month of pregnancy and

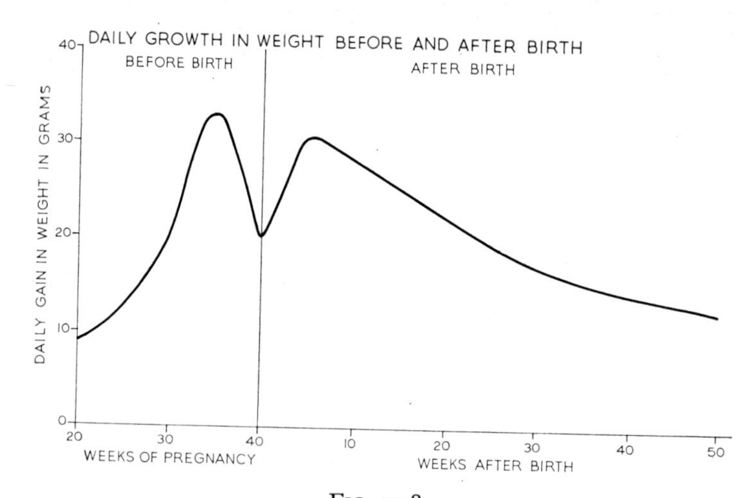

FIG. 19.8.

Based on Hendricks, C. H. (1964), *Obstetrics and Gynaecology*, **24**, 357 and Documenta Geigy Scientific Tables (1956).

subsequently falls to term. The rate of increase in length picks up a little immediately after birth (Fig. 19.10). Figure 19.11 shows varying growth in head, trunk, lower limbs and sitting height.

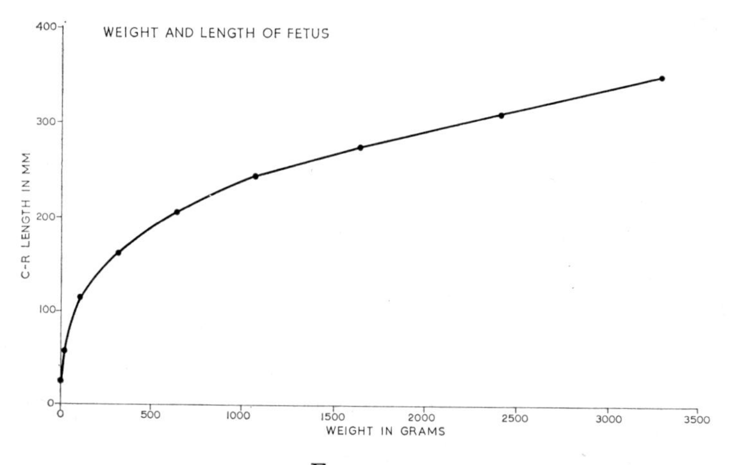

FIG. 19.9.

FACTORS AFFECTING THE GROWTH OF THE FETUS

Fetal growth must obviously depend upon genetic propensities for growth and upon environment. The environment means its supply and use of nutrients such as oxygen and all foodstuffs as well as other factors. In the final analysis all these depend upon the functional efficiency of the

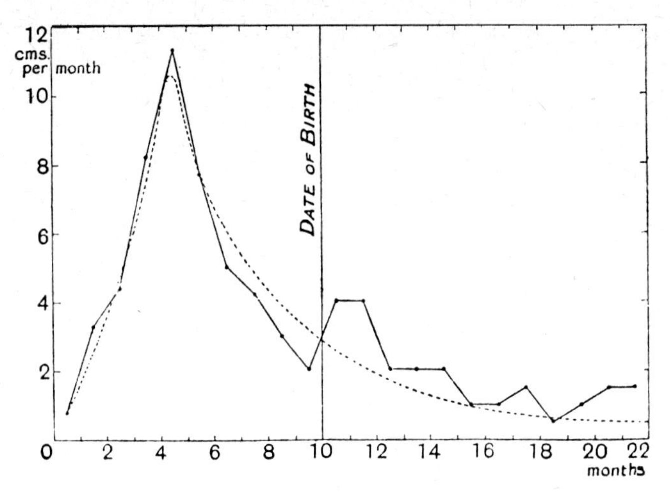

FIG. 19.10. Body length increase each month. Re-drawn from *Human Biology*. Harrison, Weiner, Tanner and Barnicot, Oxford. 1964.

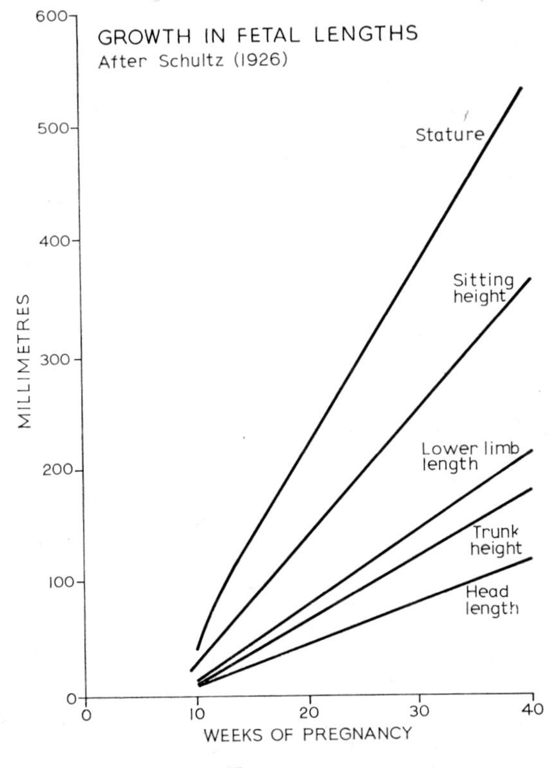

FIG. 19.11.

placenta as well as upon the inherent capacity of the fetus to employ the materials that are brought to it. Therefore anything which affects the function of the placenta will affect the growth of the fetus. In Chapter XVIII it was emphasized that placental function depends upon (a) the maternal circulation in the placental lake, (b) the parenchyma of the placenta, i.e. its villi, and (c) the fetal circulation. It will be realized that fetal growth is the outcome of many factors and so it has become usual to talk of the feto-placental unit whose functions are integrated. In general, however, impairment of growth of the fetus will be related to placental function.

The following are factors which are known to affect fetal growth:

1. *Maternal stature.* In general smaller women will have smaller babies, though this is not an absolute rule. Stature is usually simply measured by maternal height, which is a complex variable dependent upon genetic endowment together with nutrition during prenatal and postnatal life but especially during childhood and adolescence. In general, the smaller the baby at birth, the less are its chances of survival and normal development, so this observation of the correlation of birth weight with the height of the mother underlines the important fact that good reproductive performance depends on the whole genetic endowment and nutrition of females from conception up to the time of their own childbearing.

2. *Social class.* Statistically the higher the social class the bigger will the babies be at birth. The reason for this is that girls of the upper social classes receive better nutrition during their growing period than do the girls of the lower social classes. The girls of the upper social classes are generally taller than those of the lower social classes and it has already been noted that the birth weight of the baby tends to be correlated with maternal height. How maternal nutrition affects placental function so that babies are better nourished *in utero* if the mother is taller is not known, but it seems that the effect is environmental and not genetic.

3. *Nationality.* There is some evidence that West Indian babies are about 4 ounces lighter at birth than European babies even though the pregnancy may have lasted the same length of time in each group. Moreover, there is evidence too, that gestation length may be about three days shorter in the West Indian than in the European. Here is a hint of genetic factors acting, but it is difficult to exclude nutritional factors since the West Indians who have been investigated in London tend to come from the lower social groups. However they are on average a little taller than similar European social groups. (Barron and Vessey, 1966, *B.M.J.* **I,** 1189.)

4. *Numbers of previous children.* Statistically as a woman has more children so does the birth weight tend to rise. That is the birth weight tends to rise with parity. Parity is the expression used to denote the number of babies which a woman has had previously. The lower social classes tend to have more children per family than the upper social classes, perhaps

because the lower social classes lack the requisite knowledge of contraception or the desire to apply such knowledge. This is an interesting observation since it tends to run counter to the observation that birth weight is less in the lower social classes, but a little thought will show that this is not so for the birth weight of the sixth baby of a woman from the upper social classes will be bigger than that of the sixth baby of a woman

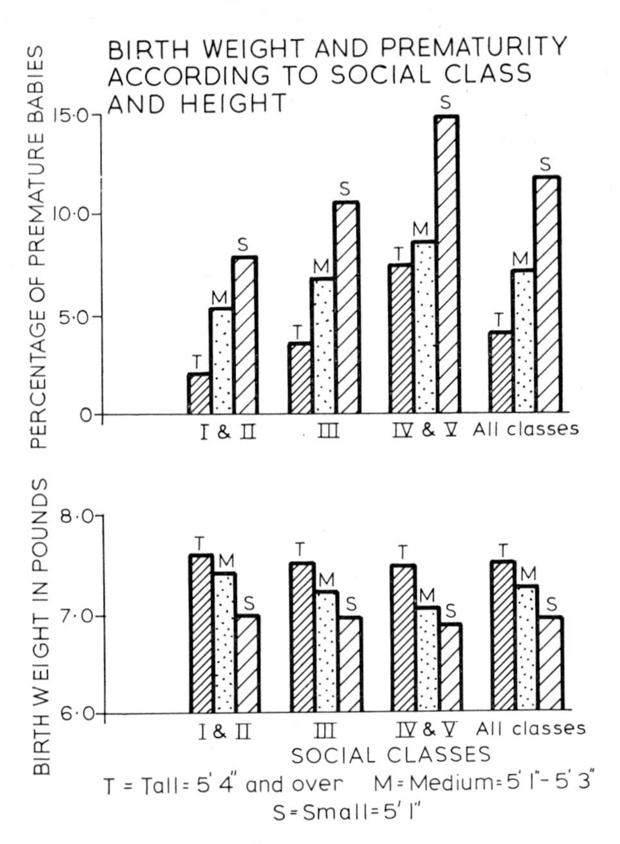

FIG. 19.12.

Based on Baird, D. (1962). *Combined Textbook of Obstetrics and Gynaecology.*

from the lower social classes. It would seem that the intra-uterine environment of the fetus improves with each succeeding pregnancy, but what this means in physiological terms is not yet known.

5. *Multiple pregnancy.* The birth weights of twins tend to be less than those of single babies, and that of triplets tends to be less than that of twins. The placentas in twins and triplets are larger than in single pregnancy, but it would seem that there is competition between the occupants of the uterus for available nourishment, and that there seems to be a

physiological limit to what the mother can supply in the way of an intra-uterine environment. A single normal baby at birth will weigh in the region of 7 lb. (3·3 kg.) and twins will, in fact, grow as fast as the single baby until the combined weight of the twins reaches about this level. Thereafter the growth of both babies slows down. Twins are also born before single babies and the gestation length is of the order of 36 to 38 weeks in twins as compared with the 40 weeks of singletons. This suggests that the uterus is only able to accommodate a certain maximum volume of contents and when this maximum is reached it expels its contents in labour, for the combined weight of twins may be of the order of 10 to 14 lb. (4·5–6·0 kg.).

6. *Hypertension.* Gross hypertension is usually associated either with infertility or fetal death. But hypertension of the order of 160/100 mm. Hg. throughout pregnancy may be associated with very small babies, e.g. 2 to 3 lb. (1–1·5 kg.) at term. The reason is thought to be a diminution of the maternal circulation through the placental lake. It may be decreased to about 200 ml. per minute from a more nearly normal value of about 600 ml. per minute. The observations on which this belief is based have been much criticized of late.

7. *Work in pregnancy.* If a woman undertakes fairly heavy work as in a laundry during the last twelve weeks or so of pregnancy then her baby will probably be lighter than if she had not worked. This may be because if blood is required for the heavy muscular work then the uterus gets relatively less and the placental circulation may be impaired. There is some direct evidence to suggest that this may be so but it too has recently been criticized. It should be noted that only poorer women tend to work late in pregnancy and these are already inclined to have smaller babies. This fact of having smaller babies when work is carried on into late pregnancy is officially taken note of in that maternity allowances of cash are paid out by government agencies in the United Kingdom from the 29th week of pregnancy till term, that is for the last eleven weeks of pregnancy.

8. *Smoking.* The smoking of cigarettes in pregnancy, say about seven per day, leads to the production of smaller babies, which may weigh up to a pound less than the babies of non-smokers. The mechanism is obscure but may be an effect of one or many of the constituents of tobacco on the maternal blood vessels leading to the placenta, or it may be that those who smoke are nervous and tense and may have labile vascular responses. It should be noted that in general it is the women of the lower social classes who nowadays smoke the most.

9. *Starvation.* It is astonishing that even quite severe starvation may lead to no general fall in birth weights, though in India where starvation is rife the birth weights of the babies are less than those in European countries. But the starvation has to be very severe. The fetus is a parasite who will see that its needs are met even at the expense of its host, the mother.

10. *Diabetes mellitus.* The infants of diabetic mothers are generally larger than those of normal mothers. The increase may be of the order of 2 to 3 lb. (1 to 1·5 kg.) and sometimes more. This has been attributed to the circulation of growth hormone but this seems to be unlikely, and it has also been attributed to the high blood sugar overnourishing, as it were, the fetus. But in many cases the blood sugar of the mother is not unduly raised, and moreover the high birth weight of the fetuses is often seen in the condition of pre-diabetes where the blood sugar is not necessarily raised though there may be changes in carbohydrate metabolism. There is also the possibility of some change in steroid production affecting anabolism.

Chapter XX

FETAL ADAPTATIONS TO INTRA-UTERINE LIFE

CIRCULATION

The anatomical adaptations of the fetal circulatory system *in utero* are related to the fact that respiration occurs in the placenta instead of the lungs. The physiological adaptations are related to the low arterial oxygen tensions which this imposes, the ever-changing proportions of the fetal organs and placenta in relation to each other during development, and to the development of cardiovascular control.

Anatomical adaptations. Figure 20.1 illustrates the probable course of the mammalian fetal circulation once the major channels have developed. During fetal life the most oxygenated blood circulates from the placenta in the umbilical vein; the liver is perfused by the greater part of this flow which joins the inferior vena cava through the hepatic vein, only a small portion by-passing the liver through the ductus venosus. Inferior caval blood in the fetus is divided as it enters the heart by the crista dividens of the foramen ovale; the greater proportion of the flow is directed into the left auricle where it mixes with the small volume of pulmonary venous blood and leaves the heart by the left ventricle for the head and upper extremities. The smaller stream of inferior caval blood is diverted to the right auricle, mixing with venous blood from the superior vena cava and the heart; the greater part of the flow leaving the right ventricle short circuits the lungs and flows via the ductus arteriosus into the descending aorta to supply the body of the fetus or become oxygenated in the placenta.

The relative distribution of blood between the fetus and fetal placental circulation changes during growth. When the embryo is young the greater part of the blood volume circulates in the placenta, but about half way through the gestation period when the placental size is almost reaching its maximum, the position is reversed. The circulating blood volume in the fetus increases, while that in the placenta remains constant. The static aspects of this relationship are shown in Fig. 20.2 where the fetal-placental weight ratios have been plotted against gestation length. The anatomical limit having been set, increased fetal requirements are met by

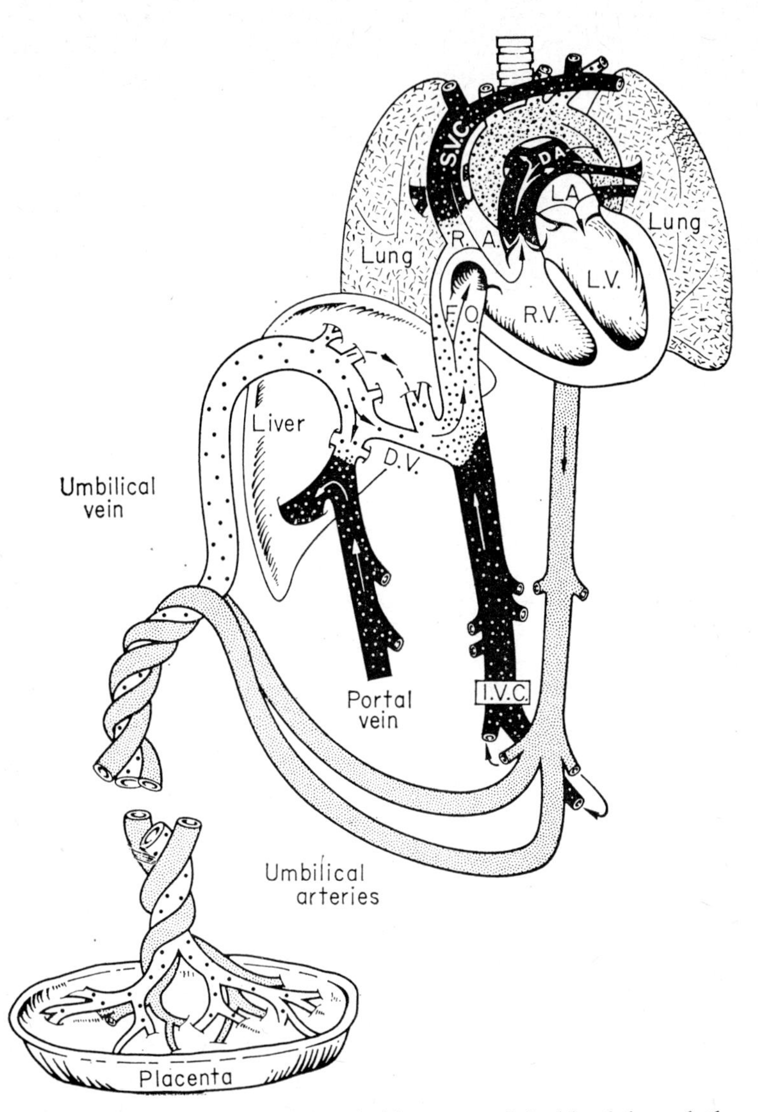

FIG. 20.1. Fetal circulation and probable course of the blood through the fetal heart. DV, ductus venosus; DA, ductus arteriosus; FO, foramen ovale. From Dawes, G.S. (1964) *Textbook of Physiology and Biochemistry*, ed. Bell, Davidson & Scarbourgh. Edinburgh: Livingstone.

an increase in turnover rate of the blood in the placenta. The increasing umbilical blood flow depends upon a reduction in placental resistance and a rise in fetal arterial pressure. The latter is effected by an increase in cardiac output and vasomotor tone. The umbilical arterial blood flow is probably 60 per cent of the cardiac output at term. The liver blood flow from the umbilical vein is high *in utero* and equivalent to the placental flow. The cerebral flow must also be relatively high and the pulmonary blood flow is known to be low.

Physiological adaptations. The character of the circulation in the fetus has to be extrapolated from measurements made in the newborn infant:

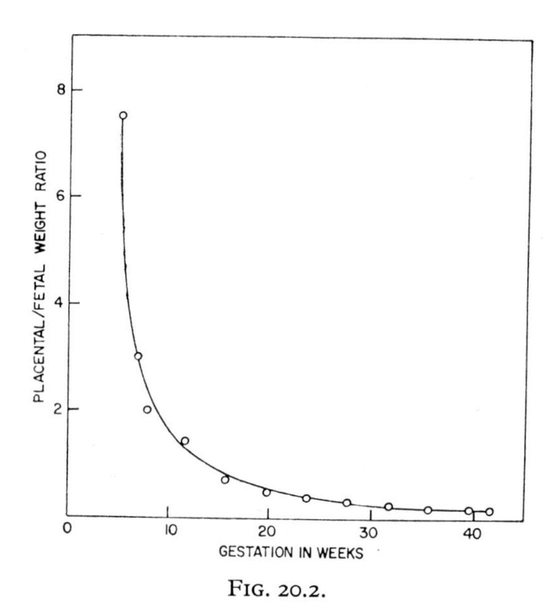

FIG. 20.2.

these are described fully on page 194. In general the heart rate is high and resting vagal restraint is small: the mean perfusion pressure is low but the cardiac output and peripheral blood flow are high, suggesting a low peripheral resistance. The fast circulation rate is similar to the circulatory adaptations to the low arterial oxygen tensions, which are found at high altitudes in the adult.

OXYGEN CONSUMPTION AND ENVIRONMENT

The oxygen consumption of the human fetus has been calculated from A-V oxygen differences and either uterine or umbilical blood flow measurements. Both have given values of 4–6 ml./kg./min., which is very similar to those of the resting adult.

The anatomical relationships (page 191) of the two circulations in the placenta make it impossible for the oxygen tension in the fetal blood to be as high as that in the maternal: moreover, the placental oxygen consumption in series between the two circulations must further reduce the oxygen tension in the fetal blood. Although the arterial oxygen tension is no greater than 30 mm. Hg., the fetal arterial blood in the umbilical vein may be 80 per cent saturated: in the carotid artery, supplying the brain, the saturation will be slightly less due to admixture of blood from the inferior vena cava. The adaptations of the fetal oxygen dissociation curve which enable uptake of oxygen from the placenta and release to the fetal tissues, over the narrow oxygen tension range of 20–40 mm. Hg. have been described (page 198). In spite of this low oxygen tension the oxygen consumption is the same as that in the adult. The blood lactate levels are normal, 1 m. Eq./l., and there is no evidence of anaerobic metabolism.

THE LIVER

Though the liver is large in relation to the total body mass, the detailed morphological characteristics of the hepatic cell are recognizable only late in gestation. During intra-uterine life the liver is perfused by practically all the blood from the placenta, the umbilical vein being the functional portal vessel. The regulatory and excretory activity of the placenta leaves some hepatic functions relatively untried at birth.

Carbohydrate stores are high in all fetal tissues; in most species the concentrations rise first in the lungs and heart, then the placenta and later in the skeletal muscle and liver. Cardiac muscle and liver glycogen is mobilized during asphyxia, providing for both the maintenance of a circulation and a supply of glucose to the brain. Synthesis of liver glycogen has been shown to be under the control of the pituitary, through the adrenal cortex, in the fetus. Though the concentrations of hexokinase and phosphorylase are low, synthesis is very active when glucose, galactose and fructose are used as substrates, but not with pyruvate. Glucose-6-phosphatase, concerned with glycogen breakdown to glucose, is also low in fetal liver.

The liver accumulates amino-acids early in intra-uterine life and nucleic acid metabolism and protein synthesis are very active in both the fetus and newborn: part of the hepatic activity may be related to its haemopoietic function which is usually relinquished by the end of the gestation period. With the exception of the γ globulins, plasma proteins do not cross the placental membrane, neither are they synthesized by the placental tissue. Albumin synthesis has been demonstrated in early human embryonic liver slices. At term the plasma protein level is 1 g. per cent below that of normal adults. Plasma fibrinogen levels are as in the adult, but prothrombin is low and synthesis may be stimulated by vitamin K. The enzymes associated with amino-acid breakdown and the urea cycle are very low at term.

Lipid synthesis occurs very actively from pyruvate in fetal liver, particularly under anaerobic conditions. The liver contains large stores of triglyceride fat at term but the plasma levels are low.

GASTRO-INTESTINAL TRACT AND KIDNEYS

The gut is motile and capable of secretory activity about half way through the gestation period. Activity, from the oesophagus through to the bowel, has been demonstrated *in utero* by the intra-amniotic injection of radio-opaque substances. The swallowing observed is thought to contribute to the turnover of the amniotic fluid pool. The peristaltic movements increase during gestation and are enhanced by hypoxia, when meconium, the normal content of the fetal bowel, may be passed.

Mucosal glands start to appear at four to five months of intra-uterine life and at term the structures are nearly as deep as in the adult. The gastric contents at birth are practically neutral.

Urine may be found in the bladder by three months of age and may contribute to the amniotic fluid volume. Renal excretion probably plays no part in the regulation of the "internal milieu", the placenta maintaining homoestasis *in utero*. Infants born with no kidneys are normal as regards their blood chemistry.

ENDOCRINE CONTROL

Early growth. Early intra-uterine growth is slow and determined predominantly by genetic factors controlling cell differentiation, together with an endometrial response dependent upon progesterone and chorionic gonadotrophin, the mechanisms of whose action are unknown. Vitamin requirements are high and fetal to maternal ratios above unity are maintained in the tissue fluids. Vitamin deficiencies may possibly cause congenital malformation. Later in gestation, as the rate of growth increases, maternal nourishment becomes quantitatively important. Larger amounts of both oestrogen and progesterone are also necessary for the uterus to be able to accommodate the conceptus. The oestrogen is required both for the enlargement of the uterine arteries, to ensure blood flow and nutrient supply to the developing placenta and fetus, and for the growth of the uterus which houses them. Progesterone maintains the muscle in a relatively quiescent state. Endometrial and myometrial accommodation to the conceptus improves with each succeeding pregnancy in the young mother.

Pituitary and thyroid glands. The regulation of growth *in utero* is apparently not dependent upon the growth hormone nor thyroxine. Anencephalic infants with no pituitaries and fetal rabbits, decapitated *in utero* by tying the neck below the thyroid glands, both have a normal birth weight. The hypophysis stores the active hormone early in development, the placenta also contains a similar hormone, and the plasma levels of growth hormone are high at birth. All the protein containing hormones cross the placenta in small amounts, but neither maternal acromegaly nor insulin given to the diabetic mother influences the fetal birth weight. The fetal pituitary gland does not control the early development of any of the endocrine organs, nor the gonads; but later, both it and the thyroid gland determine the size of the adrenal cortex and the deposition of glycogen in the fetal liver.

Labelled iodine uptake studies show that secretory activity of the developing thyroid is dependent upon TSH late in gestation. In the human neither iodine nor thyroxine administration are contra-indicated during pregnancy, but fetuses of hypothyroidic mothers may be hypothyroidic due to TSH stimulation and neonatal thyroids tend to be large in goitrous areas. Athyroidic infants grow normally *in utero*. Normally both the protein bound and butanol extractable iodines are similar in fetal and maternal bloods though the latter is higher than the non-pregnant level. The thyroid is very active just before birth.

Adrenal cortex. The rapid growth of the fetal adrenal during the last two months of pregnancy is dependent upon its own pituitary, though not apparently upon its secretion of ACTH. The enlargement is due to a layer of cells adjacent to the medulla known as the X zone. This zone may form about 85 per cent of the whole gland at birth and it contains weak androgens, some aldosterone and hydrocortisone. The adrenal is small in anencephalic infants, which of course do not develop a pituitary gland.

Adrenal medulla. The development and catechol-amine content of the adrenal medulla vary considerably among the species at the end of the gestation period. Most of our information on the control of the release of both noradrenaline and adrenaline has been obtained in the lamb and calf, and it has been observed that the amine concentrations of the gland are often inversely proportional to its capacity to release them. Early in development the cells containing noradrenaline respond to asphyxia, a reduction to 5 mm. Hg. arterial pO_2 being necessary to do this, irrespective of the pCO_2 and pH. As the gland becomes innervated by the splanchnic nerve, adrenaline is also released and, as development proceeds, the relative proportion of adrenaline to noradrenaline secretion increases. The splanchnic response is triggered at an arterial pO_2 of 10–15 mm. Hg. Under normal conditions *in utero* the arterial oxygen tension is well above that which would stimulate the cells directly or through the splanchnic nerve. During labour very low oxygen tensions may occur and both catechol-amines would probably be released, though noradrenaline is likely to predominate causing peripheral vasoconstriction and, therefore, maintaining a circulation to the head; it will also be less likely to exhaust the metabolic reserves.

NERVOUS SYSTEM: NEURO-MUSCULAR DEVELOPMENT

Central nervous system. The growth rate of the human infant's brain increases during the second and third trimester, reaching a maximum just before birth (Fig. 20.3), when it is conspicuously large, a quarter of the weight of the adult brain. Convolutions of the cerebral hemisphere appear at 24 weeks gestation and the adult pattern of sulci and gyri is complete soon before term: fissuring of the cerebellar hemispheres occurs earlier, between 12 and 24 weeks gestational age. The chronological

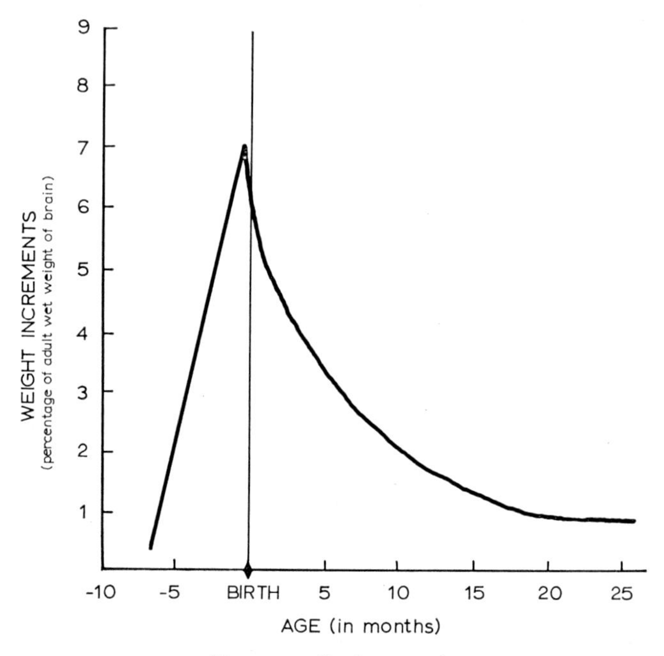

FIG. 20.3. Brain growth.

development of neuronal centres is still uncertain, but it is known that functional activity can occur before myelination of the axons. All neurones are present seven and a half months from conception, and no mitotic figures are detectable after this age. The subsequent growth of the brain is due to an increase in vascularity, elaboration of dendritic processes, and to the deposition of myelin. The imprecise histological studies of myelinization have been supplemented by myelin extraction and measurement in developing animals. Malnutrition has been shown to retard the deposition of myelin, due to the slow production of the necessary enzymes and to an insufficient supply of precursors. As this vulnerable period of development coincides with prematurity and labour in the human infant, it is very probable that the stresses associated with this period can cause permanant impairment of structure and function. Once formed the myelin sheath is metabolically stable. The growth of sympathetic ganglia is stimulated by a specific "nerve growth factor".

Cerebral electrical activity is detectable early in fetal life and develops at a constant rate irrespective of conceptual age at birth: the normal adult pattern is not established until 13 years of postnatal age.

Neuromuscular development. The spinal motor neurones differentiate first and determine the characteristics of the myoblasts they innervate. The final maturation of the muscle fibres which appear in the latter part of intra-uterine life and in the first few weeks after birth is not dependent upon a continued neuronal connection. However, the development of the subneural apparatus, and limitation of the acetyl choline sensitivity of the muscle fibre to this region, is dependent on continuous motor innervation to the motor end plate. The cholinesterase which also appears at this site is thought to be induced by the acetyl choline liberated at the nerve ending. End plate potentials are low in frequency at birth. d-Tubocurarine abolishes neuromuscular transmission in fetal phrenic nerve diaphragm preparations and both spontaneous and induced activity in skeletal muscle. The development of intrafusal fibres from the myotubes and muscle spindle differentiation is also dependent upon sensory innervation.

Little is known of the mechanical behaviour of isolated developing skeletal muscle but the responses of all the muscles to motor stimulation are slower than that of "slow" adult muscle, both active state and force-velocity characteristics being involved. The differentiation into the "slow" and "fast" muscle maintained through adult life, occurs rapidly after birth and is determined by its connections with the central nervous system.

Behaviour. The area supplied by the trigeminal nerve is the first to respond to stimulation *in utero*. Contralateral neck muscle responses occur as early as 5–6 weeks and by two months the motor reponse is widespread with flexion of the trunk and neck. The palmar and solar responses of the hand and foot also appear at this time. Pupillary responses occur at 28 weeks, blinking at 30 weeks and the traction response at 32 weeks: their development is related to postconceptional age, and is not interrupted by birth, and are therefore a useful guide to maturity.

IMMUNITY

At term the lymphoid system is still immature and lacking the antibody producing plasma cells, or small lymphocytes. Immunoglobulins are, however, transferred from the maternal to the fetal bloodstream. New-born plasma contains the immunoglobulin IgG, in the greatest amount with traces of IgM and IgA. The specific antibodies of the IgG class present are those for poliomyelitis and measles, and the incomplete Rh antibody. These protect the offspring against infection but may also interfere with subsequent active synthesis of that particular antibody by the infant. Weak antibodies to syphilis may be actively acquired *in utero*, since treponemata are present in great numbers in the placenta.

Why is the fetus not destroyed by the immunological response of its mother ? Three reasons have been suggested. Firstly, anatomical isolation from the mother. However, the chorionic buds which may be swept off into the maternal circulation and embed in her tissues do not induce an effective immune response. Secondly, lack of antigenicity of the placenta. Circulating placental antibody has been demonstrated in the maternal plasma post partum. If this is produced during gestation it may be absorbed by the large trophoblast. Thirdly, immunological inertia of the pregnant woman. The operation of some kind of tolerance is suggested by the fact that antibody production is more active for the Rh factor than for the ABO system. The placenta itself contains immunologically competent cells which makes the success of the placenta as a homograft even more remarkable.

Chapter XXI

THE LIQUOR AMNII AND THE UMBILICAL CORD

After fertilization the ovum begins to divide in the Fallopian tube and soon forms a mass of cells called the morula. This quickly cavitates to give rise to the inner cell mass, which will ultimately give rise to the embryo, amniotic vesicle and yolk sac, and also to the trophoblast which will become lined by chorion to form first the chorionic villi and later the placenta. The amniotic vesicle is present by $7\frac{1}{2}$ days after ovulation and is visible microscopically in all ova which have embedded in the decidua. This fact is of some interest when the origin of the liquor amnii is considered. The amniotic vesicle grows very quickly during early pregnancy and soon comes to surround the embryo entirely. Figure 21.1 illustrates how the amniotic cavity develops so that it comes to have the anatomical relationships seen at term or indeed at any time from about the 16th week of pregnancy when the placenta has formed as a disc.

THE VOLUME OF THE LIQUOR AMNII

In the very early days of pregnancy the volume of the liquor exceeds that of the embryo and in this is like the chorion. Figure 21.2 shows that the fetus "overtakes" the placenta at about the 16th week of pregnancy and the liquor at about the 20th week. In the early weeks of pregnancy the volume of the liquor is reasonably accurately known since it can be measured directly on intact products of conception removed at therapeutic termination of pregnancy. In the last few weeks of pregnancy the volume has been measured by injecting a dye into the amniotic cavity through the abdominal wall and then removing some liquor a little later. By estimating the amount of dilution of the dye it is possible to calculate the volume into which it has been injected. It should be realized that the range of values at all stages of pregnancy is extremely wide under quite physiological circumstances and the range at term may be of the order of 500 to 2000 ml. and at 15 weeks 140 to 210 ml. There are almost no exact figures for the liquor volume in the middle three months of pregnancy nor even up to 34 weeks of pregnancy for there are no clinical indications

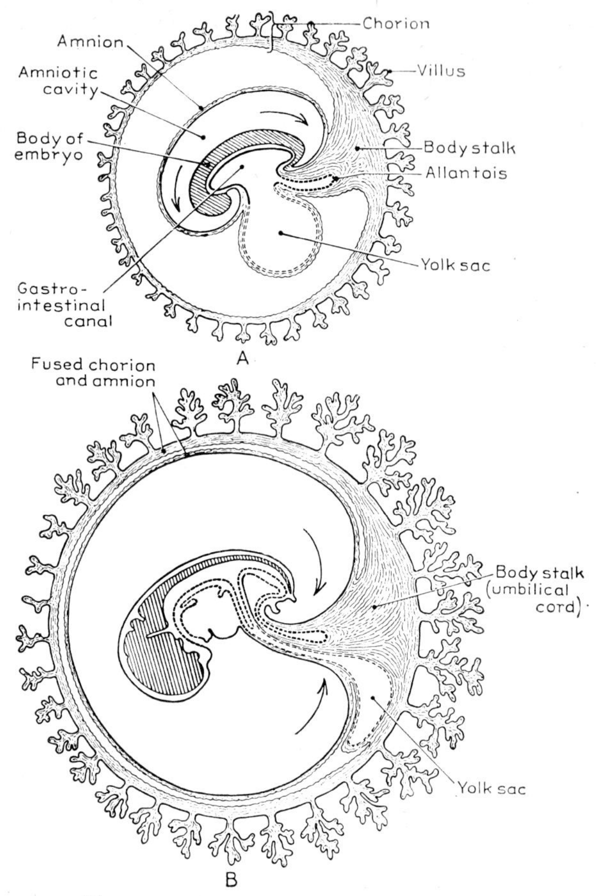

FIG. 21.1. Growth of the amniotic cavity.
Reprinted by permission from Greenhill's *Obstetrics*, 13th edition (1965).

for experiment upon the normal person. With these very severe reservations it is possible to give rough average values for the volume of the liquor at different stages of pregnancy and to compare these with the averages for the placenta and fetus.

Weeks of pregnancy	Fetus (gm.)	Placenta (gm.)	Liquor (ml.)
10	5	20	30
13	15	—	100
15	100	—	150
20	300	170	300
30	1500	430	600
40	3300	650	800

It is by no means certain that the liquor accumulates in the amniotic cavity at an even rate, but since the volume of 30 ml. at 10 weeks is fairly

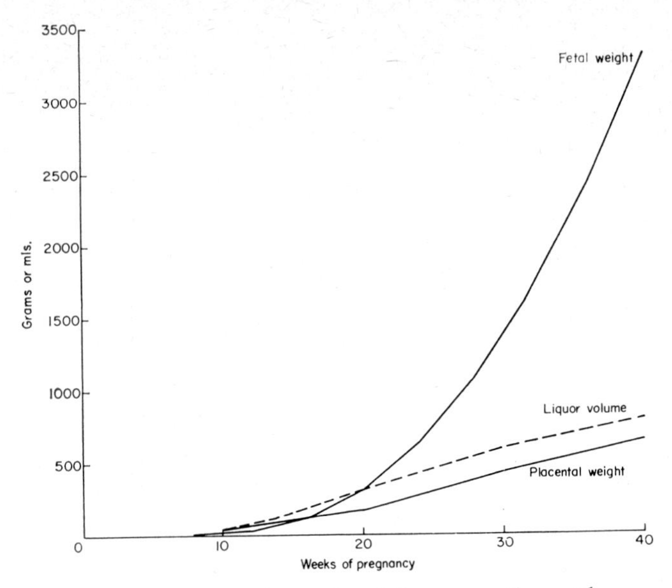

FIG. 21.2. Fetal, placental and liquor amnii growth.

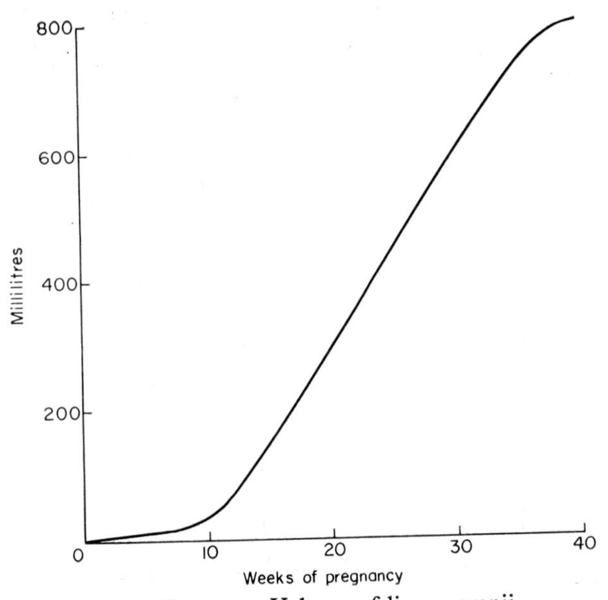

FIG. 21.3. Volume of liquor amnii.

accurate and so is that of 800 ml. at term, by calculation it would seem that any physiological processes must account for the increase in liquor amnii to be of the order of 25 ml. per week (i.e. $30 \times 25 = 750 + 30$ ml.).

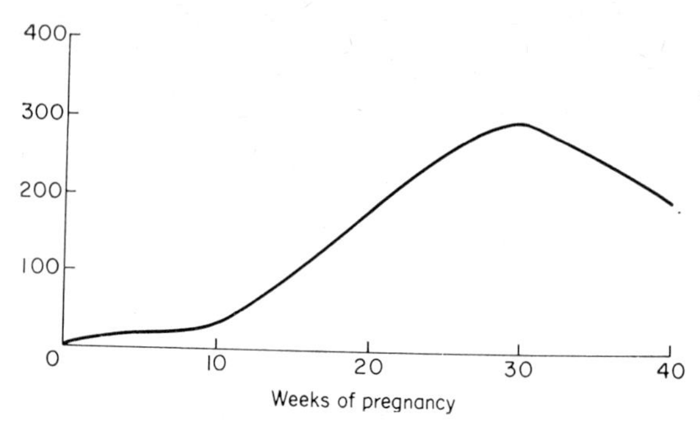

FIG. 21.4. The graph of increments in liquor amnii volumes, derived from the above table shows that the peak retention of fluid, the resultant of greater inflow than outflow, probably occurs at about 30 weeks of pregnancy. Before and after this time, retention is relatively less.

THE ORIGIN AND CIRCULATION OF THE LIQUOR

For a long time it was rather tacitly assumed that the liquor was a stagnant pond. Now, however, it is realized that the liquor circulates. Clinical observations first drew attention to this circulation. In the condition of hydramnios there may be a great excess of the liquor amnii up to as much as 5 or 6 litres and this is often, though not always, associated with abnormalities of the fetus which prevent it from swallowing. It is well known that the fetus does swallow liquor since radio-opaque materials introduced into the amniotic cavity can always be seen in the fetal stomach and intestine. When there is complete atresia of the oesophagus hydramnios results, or if the mechanism of swallowing is absent as when the fetus is anencephalic then hydramnios also results. At the other extreme there may be no liquor at all, or very little, in the condition called oligohydramnios, and often this is associated with absence of the kidneys when no urine is formed *in utero* or when there is blockage of the urethra so that no urine can be passed into the amniotic sac. These observations suggest therefore that liquor consists at least in part of fetal urine and that the liquor is at least in part removed from the amniotic cavity by swallowing. From the gut of the fetus it is probable that the liquor is absorbed into the bloodstream and from there its water is probably passed through the placenta into the maternal bloodstream. Observations on the composition of the liquor bear out the at least partial origin from fetal urine, the fetal kidney therefore being involved. However, it is odd that oligohydramnios does not occur when the fetal kidneys are absent until about the 35th week of pregnancy, and this suggests that before this time the liquor comes from some other source.

It has already been noted that the amniotic vesicle is formed at $7\frac{1}{2}$ days after ovulation. It is probable that the fetal heart does not begin to beat till 22 days after ovulation and the mesonephros, which may not in fact function as a kidney, is not present till 26 days after ovulation, so at least in the early days the liquor could not come from fetal urine.

Other sites of origin of the liquor have been suggested such as the amnion itself, either where it lies over the placental site or where it lines the chorion which is next to the uterine wall. The amniotic membrane has a very complex structure of five layers (Fig. 21.6): epithelium, basement membrane, compact layer, fibroblast layer and spongy layer, but there is no blood supply to any of these. The thickness of the amnion is very variable in different parts of the membrane and ranges from 0·02 to 0·5 mm. However, electron microscopy shows the presence of large numbers of microvilli on the amniotic epithelium and these elsewhere are

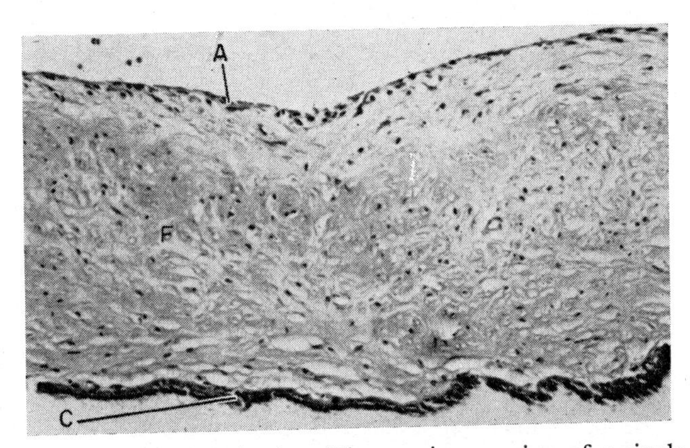

FIG. 21.5. Amnion and chorion. The amnion consists of a single layer of flattened cells resting on connective tissue formed by the fusion of the amniotic and chorionic stroma. The cells of the chorionic plate form the lower surface of the membrane. A = Amnion. C = Chorion. F = Fusion of amniotic and chorionic fibrous tissue.

usually associated either with absorption or secretion. The chorion lying between the amnion and the uterine wall, except in the area of the placenta, is composed of four layers which are cellular, reticular, pseudobasement membrane and trophoblast. It has a thickness varying from 0·02 to 0·2 mm. but it does not have a capillary blood supply. Within the amnion electron microscopy shows that there are innumerable canaliculi between and within the cells and these might have an absorptive function. However, whether the amnion has a role in the production or absorption of liquor, is still an open question.

The umbilical cord is usually covered by a thin layer of flattened cells over the Wharton's jelly which contains the umbilical vessels, but in some areas the epithelium is infolded and then is more columnar. These areas have been suggested as a source of production or absorption of the liquor but there is no proof one way or the other for this. Similarly columnar epithelium of the lungs and air passages have been assigned some part in the exchanges going on in the liquor. If the trachea of the fetus is tied in experimental animals, the lungs blow up with fluid. Normally this fluid presumably escapes into the liquor through the mouth and nose. Fetal skin too might also be a source of liquor but certainly late in pregnancy it is covered by a fatty material, the vernix caseosa, which would seem to preclude it from playing a role in liquor exchanges. Indeed it would seem that the vernix is expressly designed to prevent the skin becoming soggy through prolonged immersion in a warm water bath.

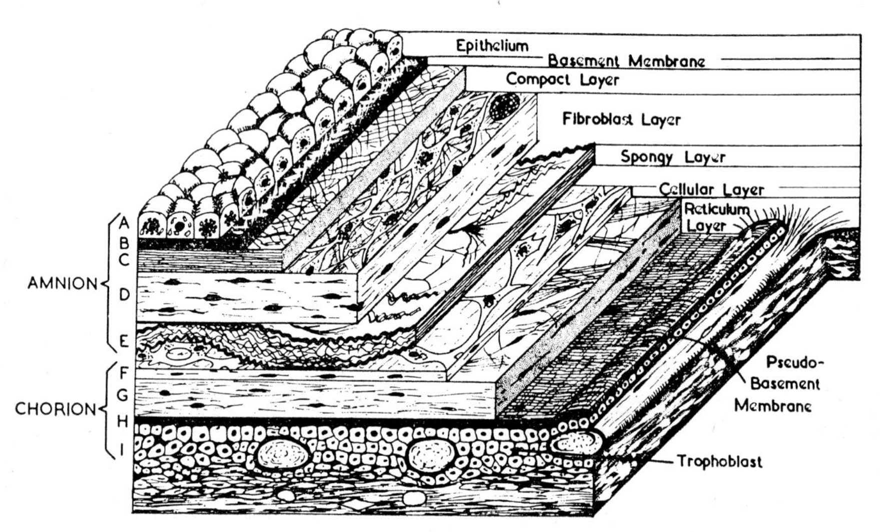

FIG. 21.6. Diagram of amnion taken from Fig. 1, p. 5, in *The Human Amnion and Chorion* (1962), by G. L. Bourne. Lloyd-Luke (Medical Books) Ltd.

Radio-isotope studies have been done on the rate of water exchange in the liquor and they suggest a very rapid turnover of water, perhaps of the order of 8 litres per day, but this is hard to square with clinical evidence. This last derives from women in whom the amnion ruptures long before labour and yet the escape of liquor from the cervix is only of the order of a few ounces per day and it is often very little indeed. If there were a vast flow of liquor through the amniotic cavity one would expect much more of it to escape, though it might be that when the amniotic cavity is breached the production of liquor virtually ceases. However, exchange across the amniotic membrane may be great without there being an actual flow through the cavity.

With so little known of the origin and disposal of the liquor the control of the volume is not known at all. The only thing that is certain is that the rate of production must slightly exceed that of disposal till a late stage in pregnancy. After term the amount of the liquor diminishes slightly when disposal rates must presumably outweigh production, which is another example apparently of failing physiological "efficiency" at the end of pregnancy. It has been suggested that the rate of increase of the liquor volume might be of the order of 3 to 4 ml. per day. If this were doubled the amount of liquor at term would be increased from 800 ml. to 1600 ml. or thereabouts, and this gives some insight into the nicety of the control.

THE FUNCTIONS OF THE LIQUOR AMNII

Fluid at physiological pressures is incompressible and therefore any external trauma to the maternal abdomen will be diffused through the liquor so that the fetus is protected. The liquor also allows some freedom of movement to the fetus and when there is a very small amount of liquor there may be limb deformities probably caused by the pressure of the uterus on the fetus. Also when there is too little liquor there may be malpresentations of the fetus making the process of labour and delivery more difficult with a higher mortality than usual for the babies. But perhaps the most important function of the liquor is to act as a water bath which is kept at a virtually constant temperature by the mother. This means that the fetus does not have to use its metabolic processes to maintain its temperature above that of its surroundings, which is what the animal in air has to do. Indeed the major part of a terrestrial animal's metabolism (measured by the basal metabolic rate) is devoted to just this. In an adult with an output of 2500 calories per day about 1800 is devoted mainly to heat production. Since the fetus is spared this kind of demand upon its metabolic processes most of the processes can be devoted to growth.

Early in pregnancy the volume of the liquor is much greater than the volume of the fetus and it may be that the liquor volume is one of the factors in making the uterus grow to accommodate the products of conception. But this is not the sole cause of uterine growth for some seems to be induced by progesterone and oestrogen.

Some time just before labour or during the first stage of labour the fetal membranes rupture and some liquor escapes. It has been suggested that this is a flushing mechanism to clear the lower birth canal, and some have thought that the liquor in the amnion in front of the presenting part of the fetus forms a wedge which during uterine contractions helps to dilate the cervix in labour. Neither of these functions of the liquor is now thought to be of much importance.

THE COMPOSITION OF THE LIQUOR AMNII

The main constituent of the liquor is water but in addition there is some particulate matter and many solutes. The particulate matter consists mainly of epithelial cells desquamated from the fetal skin and also there is vernix caseosa, the fatty material secreted by the skin to form some protection against the long immersion in water.

Comparatively little has been done on the chemical composition of the liquor despite its easy accessibility, but the following is a guide to what is known:

Water	Over 99%
Protein	100–500 mgm./100 ml.
Glucose	40–50 mgm./100 ml.
Sodium	75–165 mEq./litre
Potassium	4·0 mEq./litre.
Calcium	9–10 mgm./100 ml.
pH	7·0–7·5
Specific gravity	1010

The turnover of the water of the liquor shown isotopically is of the order of 500–600 ml. per hour, but that of sodium and potassium is relatively much slower, being about 13 mEq. and 0·6 mEq. per hour respectively.

An interesting point is that the maternal blood antigens of the ABO system can be found in the liquor if the mother is a "secretor" of these antigens, but they are not to be found in the fetal serum. This is evidence that these antigens reach the amniotic fluid not through the fetus but perhaps through the amnion directly.

The concentration of glucose in the liquor seems to follow the level of glucose in the fetal serum rather than that of maternal serum. It may be raised in diabetes, however, and during glucose infusions of the mother.

In addition to the above substances there are also small quantities of urea and creatinine.

The study of the liquor amnii will probably be much intensified in the near future because it is becoming recognized that examination of it during pregnancy may give indications of fetal health or otherwise. It is now not uncommon to perform amniocentesis in which a needle is passed through the abdominal and uterine walls in order to withdraw some liquor. It is especially used in the diagnosis and management of erythroblastosis, when there is Rhesus incompatibility between the mother and her fetus. In this disorder bilirubin may be found in the liquor and its quantity can be measured. With the removal of a few fetal cells from the liquor it is possible to determine the sex of the fetus before birth by examining the cells for sex chromatin, though this is not often done because the mother may have set her heart on having a baby of the opposite sex to the one she is carrying, and such knowledge before birth may be upsetting, yet when she sees her baby any disappointment she might have felt will disappear. Also congenital adrenal hyperplasia can be diagnosed ante-natally from the presence of pregnanetriol in the liquor.

It is probable that in time many more fetal abnormalities may be diagnosed by the investigation of the liquor amnii.

THE UMBILICAL CORD

The umbilical cord is the conduit containing two arteries and a single vein running from the fetus to the placenta. The vessels are encased in a loose gelatinous matrix called Wharton's jelly. Surrounding this is a single layer of flattened cells of amnion, which in places are columnar. The average length of the cord is about 55 cm. (22 inches) but the variability

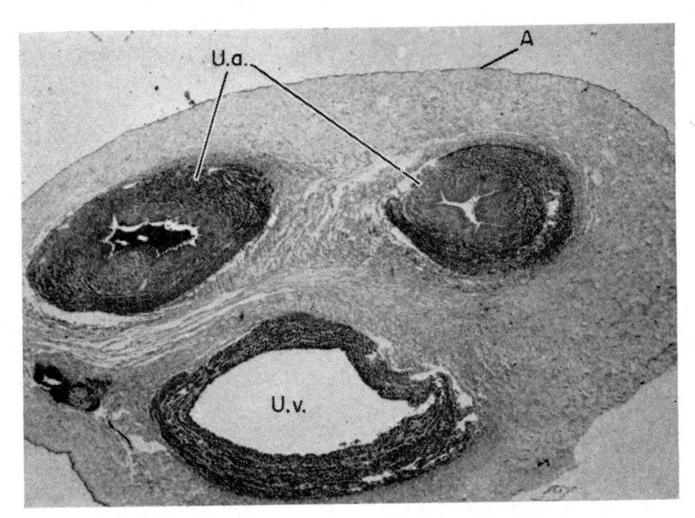

A = Amnion. U.v. = Umbilical vein.
U.a. = Umbilical arteries.

FIG. 21.7. Umbilical cord. The cord contains three large vessels, two arteries and one vein embedded in mucous connective tissue called Wharton's jelly. The cord is covered by a single layer of amnion.

is very great, from a few inches to 4 or 5 feet though these extremes are rare. The cord grows during pregnancy and roughly parallels the total length of the fetus. The control of the growth is unknown. The length is probably needed to allow the fetus some freedom of movement without snarling of its lifeline. In perhaps about 1 per cent of cases there is only one artery in the cord and when this happens there is an increased chance that the baby has some congenital abnormality.

Chapter XXII

MYOMETRIAL ACTIVITY

Labour is the process whereby the products of conception, that is the fetus, umbilical cord, liquor amnii, placenta and fetal membranes are expelled outside the mother's body by the contractions of the uterus and of the abdominal muscles. The cervix is closed throughout pregnancy but it has to be fully opened before the fetus can be pushed out of the uterus and into the vagina on its way to the outside world. Labour begins with the onset of painful uterine contractions. The contractions first are directed to dilatation of the cervix and from the onset of labour to the cervix being fully dilated is designated the first stage of labour. This takes about 18 hours in a woman having her first baby and about 12 hours in a woman having later babies, though the times are very variable, and may range from an hour or two to 36 or even 48 hours, though the longer times are usually deemed to be abnormal.

Once the cervix is fully dilated the fetus can be pushed down the birth canal and this is done by the contractions of the uterine muscle aided by contractions of the voluntary muscles of the abdominal walls which raise the abdominal pressure. These contractions are called "bearing-down efforts". When the uterus contracts the mother takes a breath, holds it and then presses down with her diaphragm on to the fetus so forcing it towards the introitus of the vagina. At the same time the flank muscles of the abdominal wall are also contracted. From full dilatation of the cervix to the birth of the baby is called the second stage of labour. It may last up to about 2 hours in the woman having her first baby and is about an hour long in a woman having later babies, but the range may be from about 5 minutes up to 3 hours.

When the baby has been born the umbilical cord is cut between ligatures and the baby is then removed to a warmed cot. The third stage begins after the baby is born and is devoted to the expulsion of the placenta and membranes, that is the amnion and chorion. The liquor escapes from the genital tract at varying times during the labour. As the fetus is expelled from the vagina the uterus contracts down on the placenta which is sheared from the uterine wall and further contractions push it down into the lower part of the uterus or into the upper part of the vagina. The mother than "bears down" again with her abdominal muscles and the placenta is expelled from the introitus. Since the membranes are attached to the placenta they are peeled off the uterine wall as the placenta is delivered. The separation of the placenta from the uterine wall always involves some loss of blood from the maternal circulation but normally this should not amount to more than about 250 ml. and may well be less, though occasionally the loss can be of two or three pints, i.e. 1 to 1·5 litres. The prevention of serious blood loss is by the firm contractions of the uterus which compress the maternal blood vessels as they run through the uterine wall. Later, clotting occurs in the open mouths of these vessels.

MYOMETRIAL CONTRACTIONS

It will be seen that the process of labour depends upon the contractions of the uterine muscle or myometrium. It has long been known that the uterus contracts throughout the whole of pregnancy, but only comparatively recently has it been recognized that the uterus contracts throughout the whole of reproductive life, so that labour must now be seen as only a special variety of uterine muscle activity.

STRUCTURE OF THE UTERUS

In the non-pregnant state the uterus is made up of three ill-defined layers of smooth muscle. In the region of the internal os of the cervix they are mainly circular though some circular fibres surround the whole of the uterine cavity. There is also a thin layer of obliquely directed fibres, and longitudinal fibres take origin mainly from the region of the internal os,

CLINICAL ENLARGEMENT OF UTERUS

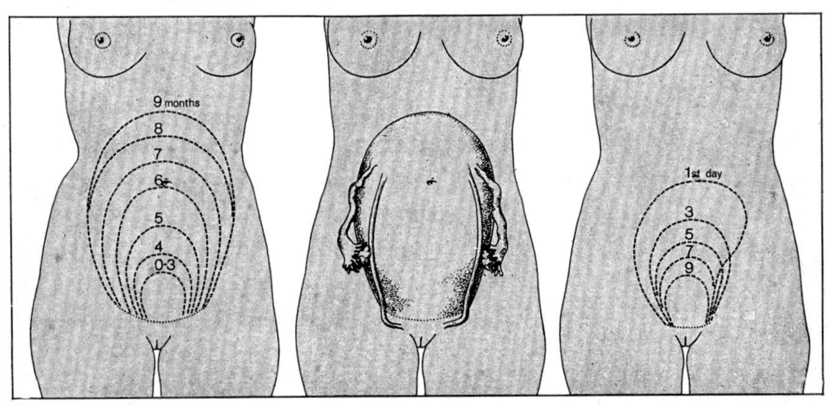

Enlargement during pregnancy | Size at full term | Contraction after labour

FIG. 22.1. Clinical enlargement of uterus. Reprinted from *Expecting a Baby*. B.B.C. Publications.

pass up over the fundus of the uterus and down the back again to the region of the internal os. To the upper part of the cervix are attached the ligaments, namely the pubo-cervical, cardinal and utero-sacral. One role of the cervix during pregnancy is to remain closed and keep the products of conception within. The body of the uterus has to expand to contain the growing products of conception. In labour the body of the uterus has to expel the products of conception after the cervix has been forced to dilate. Whilst the body of the uterus is made up of muscle, the cervix is mainly fibrous tissue, estimated as 80 per cent of its structure whilst the rest is muscle. The structure of the various parts of the uterus therefore seems to be adapted to these main functions, and it will be seen that the supporting ligaments can only be attached to the cervix, since if they extended on to the body they would interfere with the expansion of the uterus during pregnancy.

The growth of the uterus in pregnancy seems to depend in large measure on the sex steroids and particularly progesterone. At least in early pregnancy the uterus grows in advance of its contents. After about the 12th week the products of conception fill the endometrial cavity and might therefore by their growth force the uterus to distend, but this seems to be unlikely and the muscle may grow independently of its contents though the growth of them both is co-ordinated. Progestagens given therapeutically will cause the uterus to increase in size in the non-pregnant patient.

It is thought that the number of muscle cells in the uterus does not increase during pregnancy and that growth is by increase in size of the muscle cells that are already present when pregnancy begins. The growth is prodigious since the non-pregnant uterus weighs about 2 ounces (60 gm.) and the pregnant uterus at term is about 16 times heavier (960 gm. or more).

THE CHEMISTRY OF UTERINE CONTRACTIONS

The chemistry of uterine muscle seems to be of the same basic type as that of muscle elsewhere. Actomyosin molecules expand and contract as they do in smooth muscle and this may be an electro-magnetic process. The recovery phase after contraction is dependent upon an adequate supply of oxygen and the presence of creatine phosphate and adenosine tri-phosphate. Energy is supplied by glucose and there are probably anaerobic and aerobic phases of recovery. The chemistry may be affected by the sex steroids and the types of contraction differ when the uterus is under the influence of oestrogen from when it is under the influence of oestrogen and progesterone together. How these steroids may bring about their effects is not known.

ACTIVITY DURING THE MENSTRUAL CYCLE

In the human, the activity of the uterus is measured by pressure recording devices introduced into the cavity of the uterus through the cervix. In the first half of the cycle when oestrogen predominates, the contractions come at the rate of about 120 per hour and the pressure developed is about 0·2 mm. Hg. Each contraction lasts about 30 seconds. Near ovulation when oestrogen secretion from the ovary is at its height the frequency of contractions doubles to 240 per hour. These small frequent contractions are designated A waves. In the second half of the cycle when the ovary is secreting both oestrogen and progesterone the frequency of contractions drops to 30 per hour and the pressure increases to 2 mm. Hg., which is a ten-fold increase. These are called B waves and they carry on even through menstruation and slowly die out as they give place to the A waves again.

ACTIVITY DURING PREGNANCY

Although the A type waves continue in the myometrium, B waves become increasingly prominent and during the first months of pregnancy they come about every hour and the pressure inside the uterus rises to 10 or 15 mm. Hg. Towards the end of pregnancy the waves come every 15 to 20 minutes and the pressure within the uterine cavity is of the order of 40 mm. Hg. The diagram attempts to summarize the main features of the contractions.

Thus there is always a certain basal level of activity in the uterine muscle. Amplitude increases and frequency decreases during the luteal

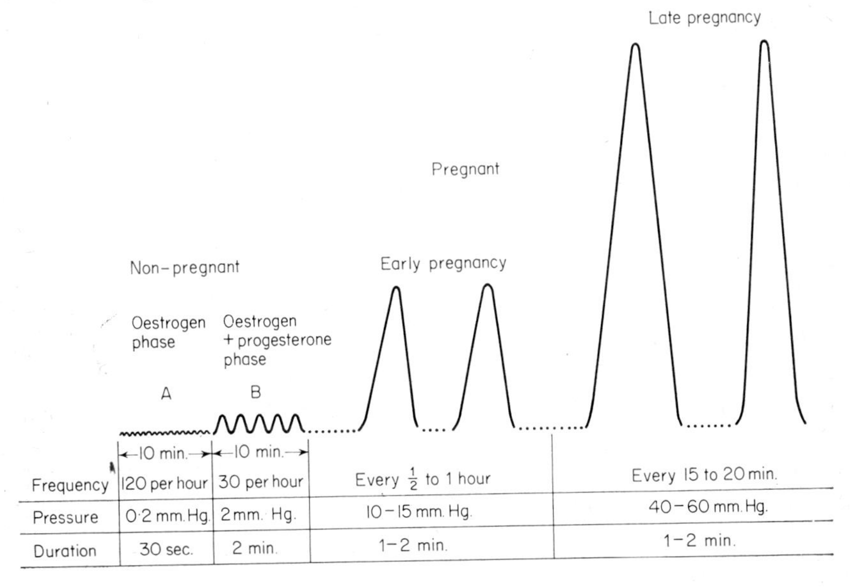

	Oestrogen phase	Oestrogen + progesterone phase	Early pregnancy	Late pregnancy
Frequency	120 per hour	30 per hour	Every $\frac{1}{2}$ to 1 hour	Every 15 to 20 min.
Pressure	0·2 mm. Hg.	2 mm. Hg.	10–15 mm. Hg.	40–60 mm. Hg.
Duration	30 sec.	2 min.	1–2 min.	1–2 min.

FIG. 22.2.

phase of the cycle. As pregnancy proceeds the B waves come to overwhelm the A waves even though the A waves are still present. As term approaches both the frequency and the amplitude of the contractions increase. The contractions during pregnancy have long been known, and have been named after the man who first described them, Braxton Hicks. The pregnant woman and the doctor with a hand on the abdomen in late pregnancy can feel these contractions as a hardening of the uterus and the patient feels no discomfort.

ACTIVITY IN THE FIRST STAGE OF LABOUR

Just as there is no qualitative difference between the contractions of the luteal phase of the cycle and the early part of pregnancy, so there is no qualitative difference apparently between the contractions of the end of pregnancy and those at the onset of labour. There is no sharp change, only a gradual transition. In fact there is no obvious point at which the clinician can say for sure that labour has begun, although he can do so in retrospect. Clinically, labour begins in one of three ways, (i) by the membranes rupturing with escape of a little liquor (ii) by a "show" of blood and mucus which has been expelled from the cervical canal (iii) by regular painful contractions of the uterus. The outstanding feature of the first stage is the dilatation of the cervix. The dilatation is shown by the breaking of the membranes which cannot withstand the intra-amniotic pressure without the support of the internal os of the cervix. The same is true of the expulsion of the plug of mucus. As far as the contractions are concerned the feature is that they are painful whereas the contractions of pregnancy are painless.

The cause of the pain of labour is still not fully decided but it is probable that it is mainly due to the dilatation of the cervix. The pain is felt in the lower part of the abdomen and over the sacral region, both of which areas are sites of pain referred from the cervix in other conditions. It used to be thought that the uterine contractions caused muscle ischaemia but in fact the intra-amniotic pressure in labour is usually only very little greater and often no greater than the pressure developed at the end of pregnancy. At least in the early part of labour the myometrial contractions seem to be exactly the same as they are at the end of pregnancy. The only apparent difference between these two times is the dilatation of the cervix. The pressure developed within the uterine muscle as distinct from that within the cavity can be measured, but this does not rise much above 50 mm. Hg. and so it too stays below the diastolic pressure so making ischaemia as a cause of pain probably untenable.

POLARITY OF THE UTERUS

If the myometrial contractions of pregnancy are apparently so little different from those of labour it needs to be explained how those of labour dilate the cervix. The answer seems to lie in the concept of polarity.

At the end of pregnancy when the uterus is an abdominal organ the pressures within it may be measured by inserting instruments into it through the abdominal wall. Recording heads can be passed either into the amniotic cavity or into the actual muscle of the uterus. Also it is possible to measure the amniotic pressure indirectly by the guard-ring tokodynamometer. (Tokos is the Greek for labour.) All these methods have been employed. When recorders are passed into the uterine muscle to measure the intramyometrial pressure it has been found that during pregnancy the contractions are not propagated in an orderly fashion throughout the uterus. But during labour the intra-myometrial pressure rises first in the fundus of the uterus and then a wave of pressure passes down the uterus towards the cervix. That is, there is polarity from the top to the bottom of the uterus. It is this change which constitutes the great difference between the end of pregnancy contractions and those of the onset of labour.

Further evidence of polarity comes from the technique of electro-hysterography, which is a method for recording the propagation of electrical impulses through the uterine walls. This shows that the electrical wave that precedes the contraction wave starts in the region of the right utero-tubal junction at the fundus of the uterus. From this uterine "pacemaker" the wave spreads in all directions and ultimately down to the cervix. The electrical and physical properties of the uterine muscle are essentially the same as those of any other smooth muscle. Differences seem mainly to be in the differing time relationships of different kinds of muscle.

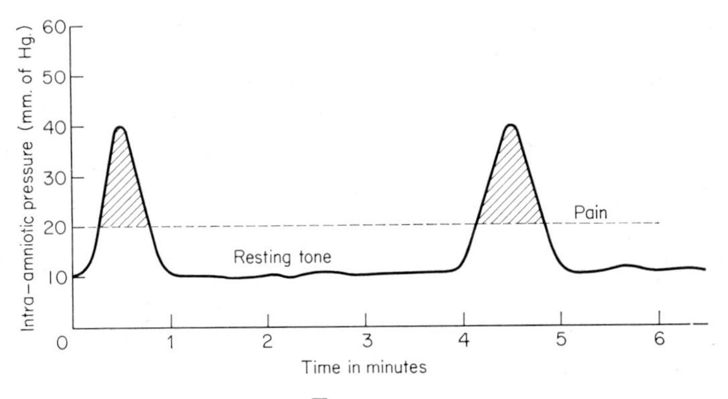

FIG. 22.3.

In the early part of the first stage of labour contractions come about every 15 to 20 minutes. As the cervix becomes almost fully dilated the contractions occur about every 3 minutes. Despite the increase in frequency the pressures developed in the amniotic cavity do not rise above 40 to 60 mm. Hg. As a general rule the resting tone in the cavity is about 10 mm. Hg. As the contraction gains strength pain is felt at about 20 mm. Hg. The pain increases as the contraction reaches its peak and the pain goes as the pressure reduces below 20 mm. Hg. again. A hand on the patient's abdomen can follow these changes in the uterine contractions even though the exact pressures are not being measured, and Fig. 22.3 illustrates this. R. Caldeyro-Barcia of Montevideo, who has pioneered much of the study of myometrial activity, attempts to measure the work done by the uterus by Montevideo units. This is simply the multiplication of the number of contractions in 10 minutes by the average pressure developed in the uterus. Thus, if there are 3 contractions in 10 minutes at an average pressure of 40 mm. Hg. this amounts to 120 Montevideo units. As the frequency of contractions increases and as there is a slight rise from about 40 to 60 mm. Hg. in pressure as labour progresses in the first stage, there is a gradual increase in the work as measured by these units. The following are representative:

Time	*Number of Montevideo Units*
Pregnancy (early)	10
Pregnancy (late)	20–50
Labour (early first stage)	100–150
Labour (late first stage)	150–200
Labour (second stage)	200–250

The number of contractions required to dilate the cervix fully is rather variable and seems not to depend upon the intra-amniotic pressures developed. Dilating the cervix from about 3 cm. diameter, which is the earliest point at which one can be certain that a woman is in labour takes between 25 and 150 contractions, the greater number being found in general in those having the earlier babies.

The intra-uterine pressure is less when the patient is on her back and greater when she is on her side during labour. Turning her on to her back may reduce the pressure from about 40 mm. Hg. to about 20 mm. Hg., though this is somewhat offset by a slight increase in frequency, but the overall activity in Montevideo units is reduced by about one-third. As a matter of clinical observation very few women lie on their backs during labour since they feel most comfortable lying on their sides.

The cause of the onset of labour is not known and especially why polarity is established. It may be due to changing concentrations of circulating sex steroids, there being a fall off in the production of progesterone whilst oestrogen production tends to carry on rising. It has been suggested that progesterone as it were keeps a brake on the inherent power of the uterus to contract and expel its contents. Also it has been thought that in the immediate area of the placenta the local concentration of progesterone may be high and therefore the position of the placenta within the uterus may affect the time of onset of labour. Those who hold this view feel that if the placenta is in the region of the "pacemaker" at the right of the fundus, then labour may be delayed and pregnancy prolonged. The theory is not yet proven.

Oxytocin, the hormone of the posterior lobe of the pituitary gland, undoubtedly can make the uterus contract and is used therapeutically. It would be expected that it would be secreted in labour and would control uterine contractions, but it is difficult to assay in small amounts and it has not been demonstrated satisfactorily either in blood or in urine in labour so far. Better methods of detection might change this observation.

Although nervous activity might affect the way in which the uterus contracts it is certain that the uterus does not have to be connected to the central nervous system for labour to begin. Women who are paraplegic or animals in whom the spinal cord is removed go into labour quite normally. Women in whom the presacral nerves (autonomic) have been removed surgically also go into labour normally.

THE DILATATION OF THE CERVIX

Once labour is established the cervix dilates. It does not do this evenly. Progress from 0 to 2 cm. dilatation is apt to be slow in the latent phase.

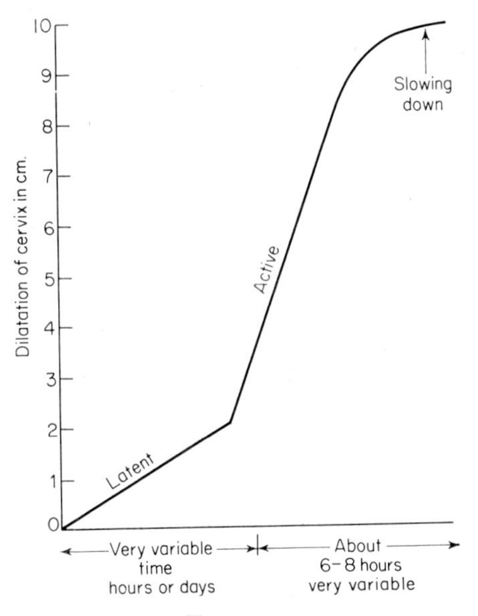

FIG. 22.4.

In women who have had babies before, the cervix may be 2 cm. dilated before labour begins, and the same may be true even of those having first babies. This is another expression of the fact that the boundary between pregnancy and labour is not clear-cut. After 2 cm. dilatation, progress is much more rapid in established labour and this carries on till 9 to 10 cm. dilatation, though the last centimetre takes rather a long time and progress slows down. Figure 22.4 illustrates this.

With each uterine contraction the cervix is forced open a little and as the contraction passes off it closes down slightly though not usually to the same position that it was before the contraction. At the height of a contraction the diameter of the cervix may be increased by say 0·5 cm. but after the contraction the increased dilatation attained may only be 0·2 cm. or less.

The way in which the cervix is made to dilate as a result of the contractions of the uterine body is not clear. Normally the head of the fetus fits firmly into the lower uterine segment above the cervix and it is partly the pressure of the head against the cervix that forces it open. But even when there is no part of the fetus in the lower segment the cervix opens partially. It would seem that cervical dilatation is partly due to the pressure of the fetus and partly due to muscular action lifting the cervix over the head. X-ray studies of the fetus in labour show that as the uterus contracts the fetal body which is normally curved a little is straightened out so that the fundus of the uterus rises slightly in the abdomen. The straightening of the fetal body must be brought about by the contraction of the circular muscle fibres of the uterus. With the fetus held firmly the long fibres arching over the fundus of the uterus contract to push the head against the cervix. Since the head cannot descend whilst the cervix is in the way the powerful contractions of the muscle of the body of the uterus can only pull upwards on the fibres of the lower uterine segment and cervix and this upward pull dilates the external os. During the first stage of labour the external os of the cervix does not rise up in relation to the bony pelvis and the head descends scarcely at all. The pressure of the head against the cervix at the height of a contraction has been measured as about 200 mm. Hg.

ACTIVITY DURING THE SECOND STAGE OF LABOUR

With the cervix dilated, one obstruction to birth is overcome and the fetus now has to negotiate the vagina and the introitus of the vagina, surrounded by muscles of the levator ani group mainly. The uterus contracts as before and the frequency of contractions and the pressure developed are similar to those at the end of the first stage. They are still painful and in passing it is worth noting that the maximum pain felt during labour is usually at the end of the first stage when the cervix is maximally stretched and this would accord with the view that the pain of labour is due to the forcible dilatation of the cervix. The frequency of contractions is about one every 3 minutes and the intra-amniotic pressure developed by the uterus is about 60 mm. Hg. However, in addition to the uterine activity, the woman "bears down" with her abdominal muscles in time with the uterine contractions. These abdominal muscles add a pressure of about 40 mm. Hg. to that of the uterus so that the peak pressures in the second stage are about 100 mm. Hg. Since the woman usually "bears down" two or three times during the uterine contraction the peak pressures oscillate from about 60 to about 100 mm. Hg. fairly rapidly.

As the fetus moves down the vagina and as its head and then body emerge from the perineum the uterus contracts down upon an ever diminishing volume of its contents. Much has been made of this by clinicians and they have felt that muscle fibres have shortened during a contraction and then although the contracted length has been held the uterus has entered a short phase of rest before gathering strength for a further contraction. To this process of being apparently relaxed in the

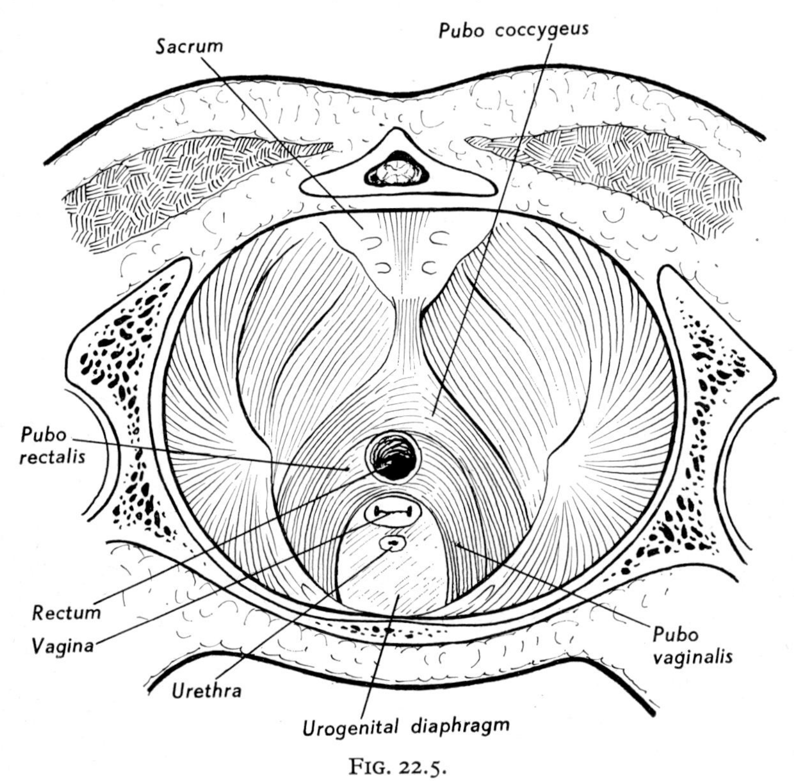

FIG. 22.5.

From Smout, *Introduction to Midwifery*. Edward Arnold.

contracted condition the name of retraction has been given. It is doubtful if this is a real entity. The muscle fibres certainly shorten but the resting tone in the uterus remains as before since the uterine volume diminishes. Something similar must occur in the urinary bladder when voiding, the only difference between the uterus and bladder being that voiding is a continuous act whilst the contractions of the uterus make progress in step-wise fashion. Retraction would seem to be another word for the maintenance of tone. This is a difficult enough concept in itself but need not be made more difficult by the introduction of a superfluous word.

ACTIVITY DURING THE THIRD STAGE OF LABOUR

At the beginning of labour the fundus of the uterus is at about the level of the xiphisternum. As the fetus is delivered the uterus contracts until its fundus is just below the level of the umbilicus (see Fig. 22.1), a difference of about 4 or 5 inches. If the weight of the fetus can be taken as a measure of its volume then the volume of the uterine cavity diminishes from about 4 litres to perhaps about ½ a litre, that is from containing the fetus, liquor and placenta the uterus comes to contain only the placenta. The intra-uterine pressure during this stage of labour is of the order of 140 mm. Hg. and the frequency of contractions is once every 5 to 10 minutes. This high pressure prevents undue loss of blood from the blood vessels of the placental site which have been opened up by the separation and delivery of the placenta. It is of interest that despite this high pressure the contractions are quite pain free, and this again is evidence that it is the forcible dilatation of the cervix which is responsible for the painful contractions of labour, and not ischaemia of the uterine muscle.

ACTIVITY DURING THE PUERPERIUM

The high pressures recorded just after delivery of the placenta gradually die away over the course of the next few days and the frequency of these powerful contractions diminishes. Slowly there is a return to the basal activity of the oestrogen phase of the ovarian cycle and a return to the A waves. Ovulation may occur at about 6 to 12 weeks after the baby has been born and then the B waves of the luteal phase return.

If the woman breast feeds her baby contractions of the uterus may be very powerful during the process of suckling. The expulsion of milk from the breast is mediated by oxytocin secreted by the posterior lobe of the pituitary gland. It is probably this hormone which as a side-effect causes the uterus to contract. The woman herself notices this since the loss of lochia (blood and débris from the uterus) increases whilst she is feeding her baby.

Figure 22.6 attempts to summarize the myometrial activity during the end of pregnancy and the stages of labour as well as in the puerperium.

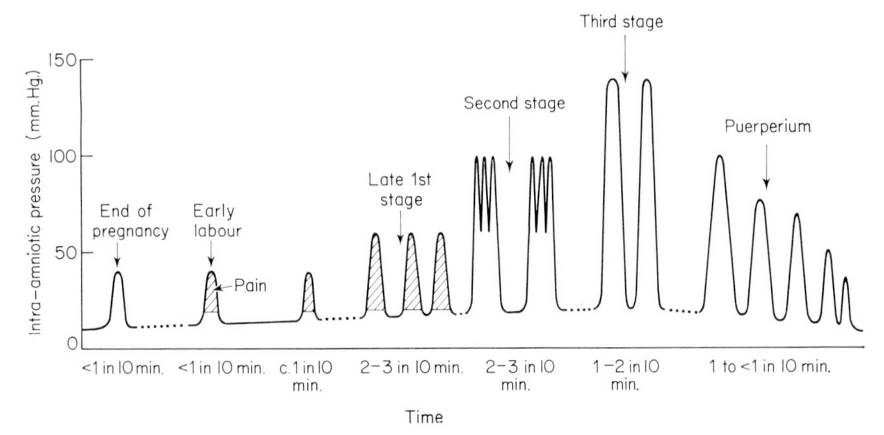

Phases of myometrial activity. Note the rise and fall of the resting tone.

FIG. 22.6.

The blood flow through the maternal side of the placenta is probably affected by uterine contractions. It is believed that as the intra-myometrial pressure rises it cuts off the venous outflow from the placental lake and therefore blood is held temporarily in the placenta. As the contraction wears off the venous flow increases again. The pressure in the intervillous space, i.e. the lake, rises during a contraction due to the rise in pressure in the amniotic cavity being transmitted through the placenta. The intervillous pressure must by its rise somewhat reduce the rate of inflow of blood from the maternal arterioles, but the input of blood on the maternal side continues until the pressure in the intervillous space equals that in the incoming arterioles.

Chapter XXIII

THE PHYSIOLOGY OF LABOUR

Labour is the process of expulsion of the products of conception and the role of the uterus in this has just been considered in the last chapter. The repeated contractions of the uterus which may continue for 12 to 18 hours or thereabouts involve muscular work. As at other times such work involves physiological adaptations and these will be outlined as they occur in the mother. In addition the fetus is subjected to physiological stresses which must also be dealt with.

Uterine contractions show a gradually increasing tempo during labour as shown by the increasing frequency of contractions and by the increasing intra-amniotic pressures developed. Therefore it is to be expected that the greatest physiological changes will occur as labour nears its end, whilst at the beginning of labour there will be relatively little difference from the physiology of the end of pregnancy. Although this is not the place to consider psychological changes it should be realized that these are an inevitable concomitant both of pregnancy and of labour and indeed of the whole childbearing process. Such psychological changes may have physiological effects especially when fear and anxiety dominate the scene, and these may show themselves in increased pulse rate, increased respiration rate, frequency of micturition, indigestion, headaches and so on. Labour is painful and pain too may have physiological effects as well as psychological ones, so that labour can be seen to be a complex phenomenon affecting the whole person. Physiological changes must be looked at in this light.

THE MOTHER

The first stage of labour. As labour proceeds the pain tends to increase and is first appreciable enough to need the administration of analgesics when the cervix is about 2–3 cm. dilated. The pain is usually most severe at about the time when the cervix is almost fully dilated, i.e. about 10 cm. diameter. This is liable to be associated with intense anxiety and fear. It will be noted that the pain is most severe during the acceleration phase of dilatation of the cervix. (See page 244.)

The circulatory system. During labour there is a tendency for the pulse rate to rise steadily by about 10 beats per minute. In addition the blood pressure rises by about 5 to 10 mm. Hg. in the diastolic. The rise in the systolic may be a little more and both pulse rate and blood pressure are raised further still if the patient is distressed, anxious or in great pain. The pulse rate and blood pressure also rise during each uterine contraction and the cardiac output during a contraction may rise by 20 to 30 per cent over the basal level of about 5 litres per minute. The increase of cardiac output is brought about mainly by a rise in pulse rate rather than by an increase in stroke volume. It has been thought that during a contraction blood is forced from the placental site into the venous system, the quantity being estimated at about 500 ml. If this were so then the increased venous return might be responsible for the change in heart rate. However, it is likely that the contractions of the uterus prevent the escape of venous blood from the intervillous space and so the venous return would be perhaps slightly diminished, suddenly to be increased as the contraction wears off, so that any increase in pulse rate would be seen after each uterine contraction. It is probable, therefore, that the rise in pulse rate and in blood pressure are the results of the pain of the contraction.

It has been generally accepted that the effective perfusion pressure in the maternal placenta is the uterine arterial blood pressure less the pressure within the uterine cavity. During pregnancy, Alvarez and Caldeyro found the intra-uterine pressure to be 5 mm. Hg., and at term about 10 mm. Hg.; when the uterus contracted the pressure rose, from which it was assumed that the perfusion pressure fell and that blood might be actually squeezed out of the intervillous space with each contraction. Simultaneous recordings of the intra-amniotic and intervillous pressures have also been made and observed to be about equal both when the uterus is relaxed and during contraction; however, the increase in intervillous pressure lags behind the rise in intra-amniotic pressure during the contraction. Following a strong contraction which increases the intra-amniotic pressure to 50 mm. Hg. there is also an increase in maternal arterial pressure and cardiac output and it is suggested that the intervillous space volume is slightly reduced during the early phase of contraction; once the intra-amniotic pressure exceeds that in the uterine vein venous drainage will cease and the intervillous volume become expanded as the arterial inflow continues. The oxygen supply to the fetus, though slowed during uterine contraction, may continue for some time and the spongy structure of the placenta and large venous sinuses allow local pressure differences to be distributed and prevent retroplacental haemorrhage. The increased pressure in the intervillous space will be transmitted to the fetal vessels causing a reduction in umbilical blood flow and a further reduction in oxygen supply to the fetus.

Much more remains to be done on the physiology of the cardiovascular system in labour and especially in the time relationships of cardio-vascular responses with uterine contractions. Although the above account has emphasized the role of pain it is at least possible that the uterine contractions squeeze blood out of the venous system of the uterus in the non-placental areas and this could appreciably increase the venous return with the usual physiological effects. Measurements of the femoral venous pressures show that they rise during a contraction and this is probably due to an increase in venous return through the internal iliac veins so that the venous return from the external iliac veins is slowed. The rise in venous pressure is accentuated if the patient is on her back when it is probable that there is some occlusion of the inferior vena cava. Turning the woman on to her side rapidly reduces the femoral venous pressure.

The respiratory system. The respiration rate may rise a little during the progress of the first stage but the change is not marked and may be affected by the psychological state. Virtually nothing has been done on the consumption of oxygen during labour nor of carbon dioxide output. Lactic acid levels in the blood rise as they do in any other muscular work. Little is known of this either.

Nervous system. It has been pointed out that nervous connexions to the uterus are not essential to the successful conclusion of labour. But of interest clinically is whether pain with its psychological concomitants can affect the contractile behaviour of the uterus and perhaps cause labour to be prolonged. The nervous connexions to the uterus are through the sympathetic system via the presacral nerves and the parasympathetic system via the sacral nerves with central connexions at S 2, 3 and 4. In animals neither removal of the nerve supply nor stimulation of it seems to change the course of labour but it may not be possible to extrapolate these findings to women. The evidence is inconclusive either way but it would seem at the moment that problems of labour as they occur in the abdomen and pelvis are more likely to be the cause of pain and psychological upsets than that the psychology of the woman seriously affects the capacity of the uterus to contract.

Alimentary system. There is good clinical evidence that the motility of the stomach is largely in abeyance at least during the later part of the first stage of labour. If food is administered it is likely to be returned undigested many hours later if the patient vomits. Indeed vomiting is a hazard of labour in that food may find its way into the lungs, so that when labour is fully established it is well not to give food though small frequent drinks are in order.

Urinary system. As the fetal head moves down the pelvis there may be interference with the process of micturition and catheters may have to be passed. The cause of the difficulty may be elongation of the urethra since the bladder is pushed up into the abdomen in labour. Otherwise the urethra may be pressed on directly by the fetal head since the head almost fills the bony pelvic cavity.

The urine becomes more concentrated as labour progresses. This is an expression of the relative dehydration since the intake of fluid by mouth is reduced because with the recurrent pain the woman has little inclination for food or fluids. At the same time the insensible loss through lungs and skin probably rises above the level of about 1 litre per day. The insensible loss rises because of the increased work being done and the vasodilatation in the skin which is one of the physiological changes of labour. Since the insensible loss is the first call on reserves of water the kidney compensates by reducing its output, sometimes below 500 ml. in the 24 hours. Such reduction in the excretion of water may raise the specific gravity of the urine to 1018 or thereabouts. Also it is probable that the RBF is decreased as it is with all hard muscular work.

If labour is at all prolonged it is common to find ketone bodies in the urine. The mechanism of the production of ketonuria is far from clear. It seems to be associated with the relative lack of intake of carbohydrate during labour and might be an expression of depletion of liver glycogen. In clinical practice the appearance of ketonuria is the signal for the giving of glucose and water intravenously, when the ketonuria disappears.

In a large proportion of labouring women, perhaps up to a half, proteinuria is present. The mechanism of this is obscure too. Proteinuria can appear in healthy young adults after vigorous exercise and the proteinuria of labour may be similar in origin.

Skin. The temperature is apt to rise a little during labour and it is clinically certain that the skin blood flow increases in response to the exertions. Women may often sweat and they refuse all but the lightest of bed coverings. The extent of the increase in blood flow has not been measured.

Endocrine system. It would be surprising if there were not great changes in endocrine secretions during labour but they have been little investigated. From the ketonuria so often seen it seems probable that there may be changes in insulin production. The thyroid might be implicated in the metabolic changes. The adrenal medulla might be expected to be active especially in the nervous patients but there have been no measurements of catechol-amines in the urine even. The cortisol output is almost certainly raised and so probably is the aldosterone. If a labour is at all prolonged, say up to 24 hours, there is a relative rise of potassium in the serum and a slight fall in the sodium.

As far as the pituitary is concerned it would be expected that there would be some oxytocin in the blood but evidence on the point is conflicting, but this may be due to the difficulties of assay.

The second stage of labour. This is the phase of expulsion of the fetus and involves the work of the abdominal muscles added to that of the uterine muscle. With each contraction the breath is held and the woman bears down. The pattern of breathing is therefore changed. The rise in the intrathoracic pressure causes a tremendous rise in venous pressure and the woman goes blue in the face and her neck veins stand out like cords. This is the equivalent of a violent Valsalva experiment. The effect is to diminish the venous return to the heart the rate of which is therefore slowed. As the contraction wears off the venous return increases and the rate suddenly increases with a slight rise of blood pressure. Reflexly these two changes call into play the mechanisms to counteract the change and therefore the heart now slows whilst the blood pressure falls a little only to be gradually restored and the same cycle of events is then repeated with the next contraction.

Virtually no other measurements have been made of physiological changes during the second stage and presumably they are similar in kind to those described for the first stage of labour.

The third stage of labour. This is the phase of expulsion of the placenta and membranes. The events in the uterus have been described in the previous chapter. For the woman there is a sense of relief after the baby is born and the pain of the distension of the birth canal has gone. When the placenta has been separated from the uterine wall by muscular contraction it is then expelled from the birth canal by the patient bearing down in similar fashion to the second stage. There are few physiological changes of note during this short time except venous congestion and there are no cardio-vascular changes unless haemorrhage from the torn maternal placental vessels is great, when the phenomena of "shock" are seen, i.e. rising pulse, falling blood pressure, pallor and skin coldness.

It is thought that the sudden reduction in volume of the uterus during the third stage must force quite a large volume of blood into the venous system but the extent of this increase has only been guessed at as of the order of say 500 ml. However, this may be offset by the loss from the uterus to the exterior. No great changes in pulse rate or blood pressure occur, though it is felt that this stage may constitute a threat of cardiac failure in those with heart disease. Nowadays the issue is somewhat clouded by the fact that it is usual to give ergometrine to most patients and this drug whose main effect is to cause contraction of uterine muscle may have secondary effects on the great vessels, and if the large veins constrict an appreciable amount of blood may be returned rather rapidly to the right side of the heart. Certainly it is not uncommon to see a rise of blood pressure soon after the end of the third stage, but the full mechanisms involved have not been elucidated.

An odd phenomenon often seen after the third stage is uncontrollable shivering. No satisfactory explanation of this has been found, though it might be due to the foreign protein of the fetus, blood and liquor amnii, finding its way into the maternal blood stream through torn placental sinuses.

Rhesus sensitization probably usually occurs, if at all, at the time of separation of the placenta. As the barrier between the two blood streams is broken fetal red cells escape into the mother. They can be demonstrated on blood films of the mother by differential staining (Kleihauer technique). If the cells of the fetus are Rh +ve and the mother is Rh −ve then she may become sensitized and produce Rh antibodies. If she becomes pregnant again with a Rh +ve fetus the antibodies may cross into the fetus and cause haemolysis with varying degrees of anaemia and bilirubinaemia. In some cases anaemia may be very mild and in others severe enough to cause death *in utero* probably by cardiac failure. A gross degree of bilirubinaemia ($>$ 20 mgm. per cent) may damage the basal ganglia of the brain (kernicterus) after birth, but before birth the placenta is able to keep the bilirubin in the blood at a low level however great the haemolysis.

It has now been shown that the maternal immunological response to the Rhesus positive cells of her fetus can be suppressed by the intramuscular injection of anti-D globulin obtained either from women who have developed antibodies or by the injection of Rhesus positive cells into Rhesus negative male volunteers. This has now become routine clinical practice, and Rhesus negative women shown to have borne a Rhesus positive fetus are now injected with anti-D globulin within 36 hours of delivery. This passive immunization destroys the Rhesus positive cells which have reached her circulation from the fetus in the third stage of labour. With the destruction of these antigenic cells she is then not stimulated to produce antibodies and so her subsequent Rhesus positive fetuses are protected from having their red cells haemolysed.

THE FETUS

The journey of babies to the outside world is a potentially dangerous one and about one to two per cent of them will die in the process or soon after. The cause of death is usually anoxia and this emphasizes the respiratory role of the placenta. For the relatively short time of labour the other placental functions viz. nutrition, excretion and endocrine production and possibly the formation of liquor amnii, are of virtually no importance. Adequate respiratory exchange is, on the other hand, literally vital.

In Chapter XX the respiratory exchanges of the fetus at term have been detailed. At the beginning of labour there will be little change but as the myometrial contractions increase in frequency and to some extent in power, as measured by intra-amniotic pressure, the flow of blood through the placenta may become impaired. It has been suggested that with each contraction the venous sinuses on the maternal side of the placenta are occluded first whilst blood is still being pumped in under the arterial pressure. Soon the inflow will cease as the intervillous pressure comes to equal that of the arterial input. At the height of the contraction the blood in the intervillous space is probably temporarily stagnant. Provided, however, that the fetal circulation remains intact oxygen can still be extracted from the maternal blood since the blood leaving the placental site is never fully de-oxygenated. However, it may be that the fetal circulation is interfered with too as contraction follows contraction with increasing frequency. At birth the newborn blood pressure is of the order of 70/45 mm. Hg. and an intra-amniotic pressure of 60 mm. Hg. might well seriously interfere with flow along the umbilical vessels and especially in the smaller vessels of the placenta itself, where the pressure is inevitably lower than that of the diastolic pressure. It can therefore be seen that if contractions follow one another with great frequency the fetus may be at the extremity of its oxygen deprivation tolerance and might die of anoxia.

From the clinical point of view the fetal heart is virtually the only measurement of value that can be made to assess the state of the fetus *in utero*. Therefore the study of the heart assumes great importance. Recently it has been found possible to pass an instrument through the cervix and take small samples of blood from the fetal scalp. On this blood, estimates of the partial pressures of oxygen and carbon dioxide and pH can be made and these may give useful information about the state of the baby. Attempts have also been made to use aspirated liquor amnii for measuring gas tensions but these have not yet been found clinically useful.

In addition to the effects of uterine contractions on the placental circulations they also have an effect on the intra-cranial pressure of the fetus. With each contraction the fetal head is pressed firmly against the lower uterine segment and the cervix, and the pressure may be up to 200 mm. Hg. When the cervix is fully dilated the head is pushed down the vagina and out through the introitus. All these parts of the birth canal have to have their resistance overcome. The fetal head is soft and compressible since the sutures are open and have to be so to allow of growth after birth. That the head is compressed is well-known clinically since the bones can be felt to overlap in the process called moulding. Moreover, it is usual to see some oedema of the scalp immediately after birth, the oedema having been caused by the pressure of the birth canal upon the scalp veins.

The fetal heart. When the human fetal heart first begins to beat its rate may be measured in hundreds per minute. By the time that the heart becomes audible with a stethoscope through the maternal abdominal wall at about 28 weeks of pregnancy the heart rate is about 160 to 180 beats per minute. At term the average rate has dropped to about 132 per minute. The fall in heart rate is partly due to the increasing size of the fetus but also it is due to the control of the heart by the developing brain through the vagus nerves. The nervous system does not mature all at once but like all other parts has to mature and become operational. The maturing process continues even after birth. With a fetus at term the basal heart rate is about 132 per minute with a normal range from 120 to 160 beats per minute. If the rate goes beyond these ranges persistently during labour the fetus is said to be distressed and is usually deemed to be in danger of its life.

With each contraction of the uterus the fetal heart slows to about 100 per minute and occasionally to 80 per minute, but if the fetus is not distressed it restores to the basal level within about one minute of the end of the contraction. If the fetus becomes seriously anoxic as shown either by scalp vein sampling or by the clinical condition of the baby at birth, the fetal heart slows to about 80 to 100 per minute or even less and often it becomes irregular. A further sign of anoxia is that the fetal bowel may contract and expel meconium (the normal fetal bowel content) into the liquor amnii from where it escapes to the exterior and is recognisable by its greenish or brownish colour. Once the fetal heart is permanently slowed and fails to recover during the intervals between contractions of the uterus then rapid delivery of the baby is needed to save it.

A somewhat earlier sign of distress of the fetus is a permanently raised heart rate to the region of 155–160.

Until recently this was about all that was known of the fetal heart rate in labour. Now Caldeyro-Barcia and his school at Montevideo have clarified what happens. They have done this by recording the fetal electro-cardiogram with one electrode on the scalp introduced through the cervix and the other electrode passed through the maternal abdominal wall into the fetal buttock. The R waves of the QRS complex were then fed into a tachometer so that there was a continuous record of the changes in fetal heart rate. The following summarises the main findings:

1. The basal heart rate was about 132 to 140 beats per minute between fluctuations. When the heart rate was raised above about 155 per minute basally the fetus was likely to be distressed by anoxia as measured by the Apgar score (see page 277) at birth.
2. There are rapid fluctuations in rate of 1 to 8 beats per minute occurring 3 to 10 times per minute. These seem to be without serious significance and are absent with anencephalic fetuses.
3. There are transient ascents in the rate of about 21 beats per minute and they last about 26 seconds. There is very slight correlation of these ascents with uterine contractions but they are not indicative of fetal distress.
4. There may be *spikes* which are rapid brief falls in the fetal heart rate of 20 to 90 beats per minute which last about 8 seconds. They bear no relation to fetal distress.

5. There may be *dips* in the rate. These are transient falls in rate due to a uterine contraction. The amplitude of the fall is the difference between the basal and minimal rates at the bottom of the dip. The lag-time is the interval in seconds from the peak of the contraction to the bottom of the dip. On the basis of the lag time Caldeyro-Barcia distinguishes between Type I and Type II dips. The Type II dips are of serious significance and the fetus may die of anoxia. The lag-time is of the order of 30 to 50 seconds whilst in the Type I dips it is very much shorter, and with Type I dips the fetus is not distressed. The Type II dips have been correlated with the blood changes and have been found to occur when the pO_2 falls to 20 mm. Hg. or below, which is a 30 per cent oxygen saturation at pH 7·2.

The control of the fetal heart seems to be by both the sympathetic system and by the vagi. During the experiments cited above atropine was injected into the fetal buttock and this abolished the rapid fluctuations, the transient ascents, the spikes and the Type I dips. Therefore all these variations in rate would seem to be due to modifications in vagal activity either increasing or decreasing. The atropine also increases the basal heart rate so that the heart is under the influence of the vagus all the time. Atropine diminishes the amplitude of Type II dips, therefore this response of the fetal heart to anoxia is not entirely due to vagal activity though this plays a part. When the fetal heart rate rises in fetal distress as it may do, atropine causes a further rise in rate, meaning that something is overriding the vagal action. It is thought that this overriding factor is increase in sympathetic activity, and it may be mimicked by the injection of laevo-epinephrine.

The teleological "explanation" of these changes in heart rate is suggested to be the increase of feto-maternal exchanges by the increase in heart rate, and this is clinically the first response of the heart to anoxia. If anoxia is more severe then the heart slows and this may be because of the necessity to conserve the energy stores (? glycogen) in the heart muscle itself. Apparently with diving mammals such as seals and whales the heart rate slows during the dive just as it does in the human fetus which is short of oxygen, and this too is thought to be a mechanism of conservation of energy stores in the heart.

From the clinical point of view these researches are of enormous importance. The diagnosis of fetal distress in labour is apt to rest upon a comparatively short time of counting the fetal heart rate. If these times are too short, rapid fluctuations, transient ascents and spikes may suggest gross irregularity of the heart and this is deemed to be a bad prognostic sign. Moreover it may be difficult to differentiate Type I from Type II dips over a short interval. Therefore when the state of the fetus is in doubt the heart rate should be counted for 20 consecutive periods of

15 seconds each period, taking a 5 second interval between each period of 15 seconds. The results should be charted and the chart must also show the time of the peak of each contraction. When this is done it can usually be easily seen whether the fetus is really distressed or not.

SCALP VEIN SAMPLING

Within the last few years it has been found possible to take blood from the scalp of the fetus during labour and especially to measure the pH of the fetal blood. The method and the inferences to be drawn from it are still in their infancy, but even the crude measure of pH seems to show that when it is below 7·25 it is highly probable that the fetus is at risk from anoxia. Above this level the fetus is likely to be safe. More work is in progress on pO_2 and pCO_2 values but these are more difficult to determine than pH and since they are time-consuming they are not of help yet in the clinical situation where time is always pressing when the fetus is distressed. Fully to understand the total situation also requires measurements on maternal blood and decisions about whether a given acidosis (acidaemia) is of the respiratory or metabolic types and whether maternal acidosis alone has any significance for the fetus.

OXYGEN ADMINISTRATION DURING LABOUR

During fetal "distress" the administration of 100 per cent oxygen to the mother cannot raise the oxygen content of the umbilical vein blood appreciably, for three reasons. First, the cause of the fetal distress is probably a reduction in maternal placental blood flow and, therefore, the maternal-fetal exchange is also reduced. The exchange is unlikely to be improved until the blood flow can be increased. Second, it is not possible appreciably to increase the oxygen content of the blood leaving the normal maternal lungs. While breathing 100 per cent oxygen, the tension and amount of oxygen in the maternal alveoli may increase five-fold. Her blood O_2 tension may also increase five-fold but the oxygen content can only be raised by the amount physically dissolved in the plasma, 1 vol per cent. Lastly, the placenta lies in series between the maternal and fetal circulations and, since it will also be in need of oxygen when the maternal circulation has been impaired, its oxygen debt is likely to be satisfied first. The placenta has a high oxygen consumption, 10 ml./kg./min., double that of the fetus and, under normal circumstances, at term, it captures one third of the oxygen leaving the maternal placental blood. Respiratory gas gradient and diffusing capacity measurements, comparable with those in the lung are, therefore, impossible for the placenta, but may be calculated using an inert gas. Finally, there is some evidence that high arterial oxygen tensions in the mother may cause a further reduction in maternal placental blood flow, and enhance the oxygen debt, acidosis and bradycardia in the fetus.

OXYGEN DEBT

The oxygen debt of the infant during a normal labour is small: at birth the plasma lactate is slightly raised, it doubles during the first minutes of life as lactate is washed out of the tissues, but falls to normal levels within an hour. During difficult labours considerable oxygen debt may occur, causing a metabolic acidosis with raised plasma lactates at birth which do not return to normal levels for many hours or even days.

Chapter XXIV

ADAPTATION IN THE NEWBORN

At birth the fetus exchanges its intra-uterine existence with its dependence upon the placenta for a separate existence in a totally new environment. The conditions in which the new life is lived are quite different and yet much of the intra-uterine life has been devoted to "pre-adaptation" of the organs and physiology of the fetus so that it may be ready to cope with the hazards of extra-uterine life. These hazards may be great at and shortly after birth and this can be a very dangerous time of life. Some adaptations have to be sudden, as in respiration, others can be slower, as in nutrition and temperature control.

Usually the baby takes its first breath before the umbilical cord has been severed, but for purposes of description it is easier to consider the time of ligation of the cord.

THE TIME OF LIGATING THE UMBILICAL CORD

It has long been known that no change in the infant's arterial pressure occurs when the cord is tied in man following a vaginal delivery. In this respect the observations in sheep fetuses are misleading. A rise in umbilical arterial pressure can also be observed in the human infant *in utero* when the whole cord is compressed, temporarily removing the low resistance of the placental circulation.

The problem of whether cord clamping should be immediate or delayed after delivery has never been satisfactorily settled from the clinical point of view. This is chiefly due to the unsatisfactory nature of the experimental observations. Few clinical trials to settle the question have been accompanied by gravimetric evidence, which is the simplest to obtain, for the amount of blood gained per unit weight of infant. Cord stripping with the fingers has been shown to be the most effective method of ensuring the transfer of 50–120 ml. blood from the cord and placenta to the infant. Holding the infant below the level of the uterus does not necessarily mean that blood will be transferred by gravity because the umbilical vessels may be constricted and incapable of allowing flow.

The rapid increase in blood volume, of 25–50 per cent, which may occur following a uterine contraction or cord stripping does not appear to embarrass the newborn circulation which has remarkable elasticity. Transiently higher arterial pressures are observed in such infants, but seldom reflex bradycardia. Their glomerular filtration rate is high and marked diuresis is observed. The umbilical venous pressure and atrial pressures are also transiently raised. The transverse diameter of the heart is not increased unless the musculature is weakened by more than moderate hypoxia during labour.

The newborn blood haemoglobin concentration is normally high at birth. The so called "physiological" jaundice observed in the newborn period may be more marked with higher blood volumes and the risk of kernicterus (deposition of bile pigment in the basal ganglia of the brain) greater.

In general it would seem to be best to allow the baby to receive as much blood as possible from its placenta. This will entail a delay of a few minutes before the cord is tied. The blood can be transferred from the placenta by holding the baby below the level of the placenta and allowing some blood to drain out of the umbilical vein. This transfer may be assisted by "stripping" or "milking" the cord towards the baby and perhaps the pressure of the contracting uterus in the third stage of labour may force some blood along the umbilical vessels. There is little to be said for waiting for the pulsations in the cord to cease before tying it since pulsation does not necessarily mean that blood is flowing along the arteries. Moreover the beating may continue for up to 40 minutes or more and this would be an unnecessarily long time to wait and there would seem to be no concomitant benefits.

Although there would at least seem to be no harm in attaining the maximal blood volume in healthy babies it is possible that a large volume might be an embarrassment to a premature baby in whom the effects might be to increase the work of the heart and the amount of bilirubin as haemolysis sets in. The premature baby is already more liable to kernicterus from hyperbilirubinaemia than the mature one and it might therefore be helpful to minimize this situation. The same is probably true of the baby which might be affected by Rhesus incompatibility.

In animals the umbilical vessels close as the mother bites across the cord. Rolling the human cord between the fingers can also stop pulsation and cutting across the cord then is not attended by bleeding. However the vessels may open up again when the vessels are warmed and therefore they should always be tied or clamped so that the baby may not suffer from severe haemorrhage. The physiological closure of the vessels is perhaps mainly brought about by mechanical stimulation but it is also probable that the vessels walls may be sensitive to rising oxygen tensions in the blood stream, just as the ductus arteriosus is. In animals, if the newborn are kept hypoxic, bleeding from the cut cord goes on for longer than when they are allowed their full quota of oxygen.

THE FIRST BREATH

The purpose of the first breath is to establish the respiratory mechanisms which maintain the correct ventilation-perfusion relationship in the lungs, so that the arterial oxygen tension is about 90–100 mm. Hg. The capacity of the newborn to sustain this adult arterial oxygen tension by its own respiratory efforts is viability in its narrowest sense. Normally, respiration takes off to a "flying start" with the first good cry and the arterial oxygen tension rises within minutes to 70–80 mm. Hg. The haemoglobin is, therefore, 80–90 per cent saturated with oxygen shortly after birth and reaches adult levels within hours or days.

Stimulus for the First Breath. Complete failure to start breathing is rare, but it is important that the whole mechanism should be established within a few minutes of birth. The stimuli for this breath are multiple and difficult to quantitate; their effectiveness depends both upon maturity, and upon the internal environment of the infant at birth. The neural regulatory pathways are capable of initiating and sustaining respiratory activity in the human infant, born as early as 24 weeks gestational age. Respiratory efforts have been demonstrated *in utero* by the entry of radio-opaque substances into the lungs from the amniotic fluid, though such movements may be a response to anoxia.

Rhythmic activity of the respiratory centre in adult animals is dependent upon sensory input and, at birth, the medulla is submitted to a flood of new sensory impulses which probably initiate this rhythmic activity. Foremost amongst these stimuli are those of cold from the skin, and proprioceptive impulses from the muscles, as the infant ceases to be "weightless". The fetal respiratory acidosis which developed during labour may also enhance the excitability of the respiratory centre to these stimuli for the newborn respiratory centre is known to be particularly sensitive to carbon dioxide, once respiration is established. The recent observation that carotid chemoreceptor activity is low until respiration is established suggests that these receptors play no part in initiating the first breath when the oxygen tension falls during labour. The sensitivity of the pulmonary afferent endings is also anomalous in early life and the Hering Breuer reflex does not operate until 24–48 hours after birth. Gentle inflation of the newborn lungs causes Head's paradoxical reflex, that is an increase in depth of respiration, before expiration occurs. This may make assistance of respiration by positive pressure ventilation more effective.

Mechanical changes in the lungs. Serial radiography of an infant's chest during the initiation of respiration shows that the diaphragm plays a larger part than the other respiratory muscles in changing the volume of the thoracic cage and in developing the potential negative pressure in the pleural space. The success of the subsequent rhythmic breathing depends upon the expansion of the alveoli and the development of a functional residual capacity. Intrathoracic pressures, measured in the oesophagus, fall 40–60 cm. H_2O below that of the atmosphere for about a second during the first few breaths, and positive pleural pressures of 20–30 cm. H_2O may also be developed during the first expirations. This relatively large inspiratory pressure is used to overcome three sets of resistance before air can enter the lungs.

The first resistance offered to entry of air into the newborn lung, is that of the viscous fluid which fills the upper respiratory tract and alveoli at birth. Only a small proportion of this is amniotic fluid inhaled during labour. Most is produced by the fetal alveoli during intra-uterine life.

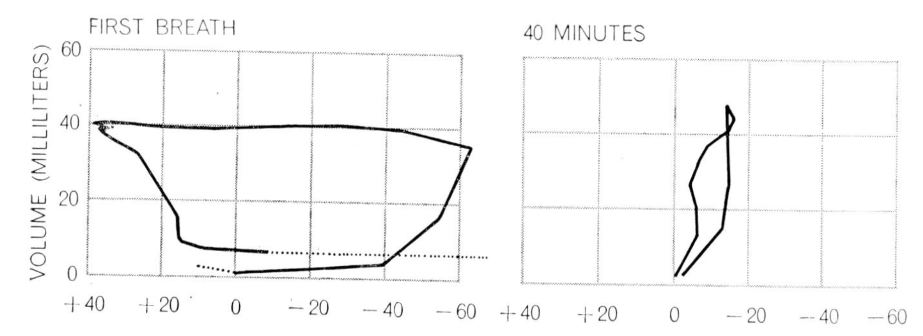

FIG. 24.1. Inspiratory effort is traced by plotting negative pressure required to open the lungs (*horizontal axis*) against the volume of air intake (*vertical axis*). Pressure is given in centimetres of water. The base line represents the pressure of one atmosphere. The infant whose respiration is traced here breathed 33 seconds after birth and had to exert −60 centimetres of pressure to achieve an intake just below 40 millilitres.

The second loop, made 40 minutes after birth, shows that by then −15 centimetres of pressure could cause a 50-millilitre intake. From Smith, C. A., *Scientific American*, **209**, 4 (1963).

They have a glandular appearance, their secretory capacity is considerable and can cause a rise in intra-tracheal pressure, if the trachea is tied in experimental animals. A striking feature of this alveolar fluid is its acidity, pH 6·40; the protein and bicarbonate concentrations are low; and the chloride concentration high. It may contribute to the amniotic fluid volume. Alveolar fluid assists the initial inflation of the lungs (Fig. 24.1) and disappears quickly as respiration is established, a large proportion being accounted for by the increase in pulmonary lymph flow. Some may be expressed through the mouth by the high intra-uterine pressure transmitted to the fetal thorax during labour.

The greatest resistance to initial inflation of the lungs lies at the air–liquid interface in the terminal bronchioles where the radii of curvature are small. Figure 24.2 shows the pressure volume curves of an excised lung. In the first the lungs contain fluid, and complete expansion occurs with a 1 cm. H_2O pressure change, because the compliance of lung tissues is high and offers little resistance to distension. The second graph shows

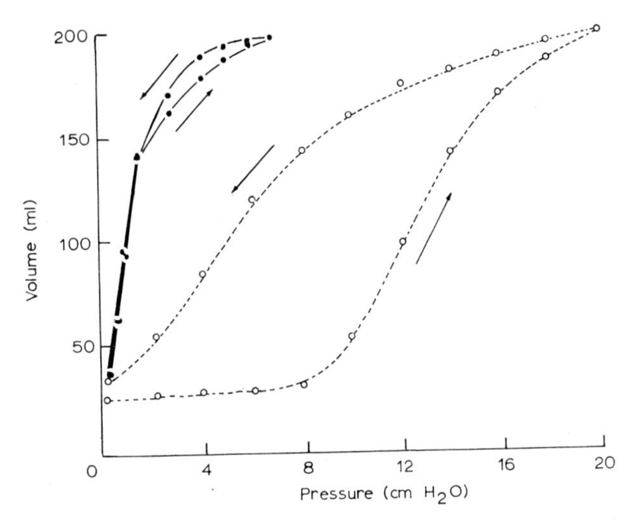

FIG. 24.2. Pressure volume curves of an excised cat lung. ●, the low inflation pressure required with saline demonstrates the low resistance of the lung tissue: ○, the larger inflation pressure required with air demonstrates the high resistance at the air-lung interface; the hysteresis, see text, demonstrates instability of the alveoli. From Young, M., *A Companion to Medical Studies*, Passmore, R. and Robson J. F., Eds. Blackwell Scientific, (1968).

the lung containing air: the opening pressure is higher, 8 cm. H_2O, and complete expansion requires a further increase in pressure. The surface tension at the interface between liquid and air, in the alveoli particularly, decreases the compliance and increases the resistance. By contrast, in the fluid-filled lung the volume changes do not follow the same curve when the pressure is reduced. This is known as hysteresis and suggests inherent instability of the system and the need for a stabilizing material. A surface wetting agent, present in the lung fluid at birth, can be extracted from lungs and is thought to line the alveoli and stabilize them so that those of unequal size may remain connected to each other. According to La Place's law the surface tension at an air–liquid interface is greater when the radius of curvature is small, and the smaller alveoli would, therefore, empty into the larger. The surface active substance is a lipoprotein, known as surfactant. It is thought to disappear slowly when pulmonary oxygenation is poor, the resulting lack of stability of the alveolar lining causing their collapse and atelectasis, particularly in the newborn period.

CHANGES IN THE PULMONARY CIRCULATION AND LARGE FETAL CHANNELS AT BIRTH

The first breath initiates the changes in the course of the blood streams in the heart. Expansion of the lungs decreases the resistance in the pulmonary arterioles and the resulting increase in pulmonary blood flow raises the left atrial pressure, above that in the inferior vena cava, closing the foramen ovale functionally; this closure is assisted by the temporary fall in inferior vena caval pressure, following occlusion of the umbilical vessels. The whole volume of inferior caval blood now joins the superior caval blood in the right atrium to maintain the high pulmonary blood flow. As a result of the reduced pulmonary vascular resistance the pulmonary arterial pressure falls below the systemic level and blood flow through the ductus arteriosus is diminished.

The mechanism for the initiation of the reduction in pulmonary vascular resistance with the first breath depends upon inflation of the lungs with a gas of the correct composition. The alveolar and lung parenchymal tissue pO_2 must be raised and the pCO_2 lowered from the fetal levels for, even when denervated, the newborn pulmonary arterioles are readily constricted by hypoxia and hypercapnia. In the fetal lamb the pulmonary blood flow increases about ten-fold with lung inflation; the pulmonary arterial pressure is reduced by about a half, to approximately 35 mm. Hg., during the immediate postnatal period in the human infant. The mechanism of closure of the ductus arteriosus is not yet clear. The muscle is sphincter-like and poorly innervated: closure occurs when the arterial pO_2 is raised, and dilatation when the oxygen tension is low; the reduction in pressure, below the critical opening pressure of the ductus, as the flow falls with the first breath, may also play a part in its physiological closure.

The changes in the small pulmonary vessels and in the large cardiac channels all start abruptly, but take some weeks to complete. The unstriped muscle walls of the pulmonary arterioles gradually become reduced in thickness, and anatomical closure as well as physiological closure of the foramen ovale, ductus arteriosus and ductus venosus take place. A mid-systolic murmur may be heard, when listening to the heart sounds of normal infants during the first two weeks of life, probably indicating intermittent patency of the ductus arteriosus. The direction of flow will depend upon the pressure differences between the aorta and pulmonary artery, and should be from left to right, a reversal of the fetal condition. These major changes in the course of the circulation at birth do not vary amongst the species.

In the human umbilical cord, both arteries and the umbilical vein have thick muscular walls which are apparently not innervated. The musculature is normally very responsive to cooling and mechanical stimuli, but the effectiveness of the contractile mechanism is impaired by asphyxia. The ductus venosus is still patent in the human infant at term, but must close after birth otherwise an Eck fistula will result, with the portal blood short-circuiting the liver and passing through the portal sinus and straight into the vena cava. Patency of the ductus venosus is necessary clinically, for cardiac catheterization through the umbilical vein.

PULMONARY VENTILATION

Chemical control. Chemical control of respiration, both central and peripheral is similar in the newborn and in the adult. The carbon dioxide sensitivity curve is steeper in the newborn than in the adult, but the

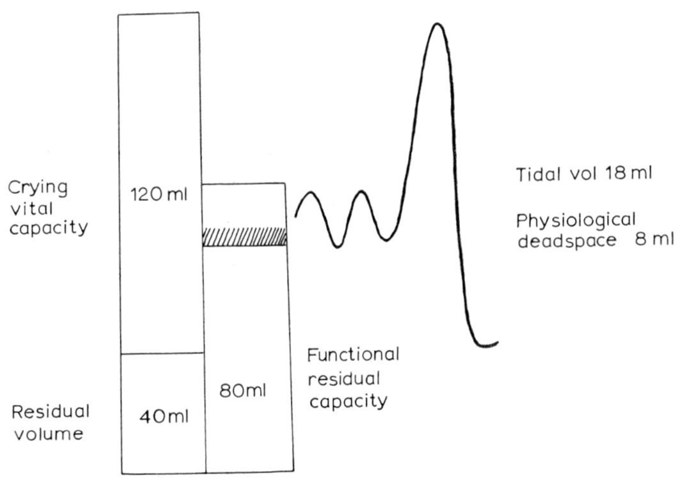

FIG. 24.3. Subdivisions of lung volume in the newborn human infant. From Young, M., *A Companion to Medical Studies*, Passmore, R. and Robson, J. F., Eds. Blackwell Scientific, (1968).

adult and newborn slopes are identical when the increase in ventilation for a given increment in alveolar pCO_2 is related to body weight. The reason for the low resting arterial pCO_2, 32 mm. Hg., is not at present known. There is some evidence to suggest that the responses to both metabolic acidosis and alkalosis are similar to those in the adult. The administration of 15 per cent oxygen causes stimulation of respiration which is not sustained and 100 per cent oxygen causes a temporary reduction in ventilation. These observations demonstrate both sensitivity of the peripheral chemoreceptor mechanisms and some resting tonic activity.

Lung mechanics. The important ventilatory values and subdivisions of lung volume in the newborn are summarized in Fig. 24.3. It is difficult to collect the expirate in the infant and many of these figures have been obtained using the body plethysmograph (Fig. 24.4), otherwise the methods which have been used are, in general, similar to those employed

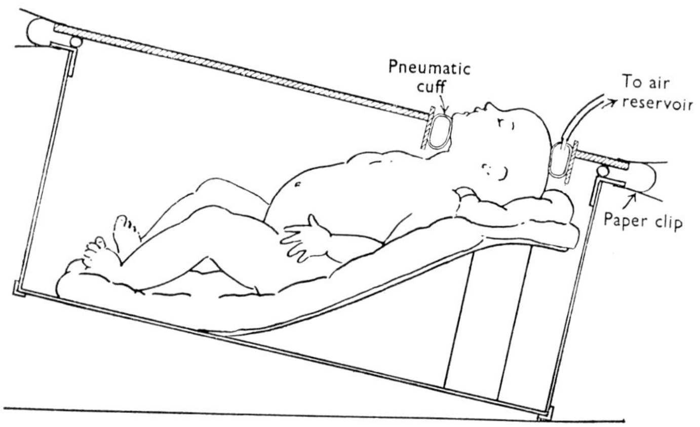

FIG. 24.4. Diagrammatic longitudinal section through body plethysmograph for respiratory studies in infants. The head seal is shown and the outlet for pressure volume recording. The baby is clothed and covered with blankets. From Cross, K. W., *J. Physiol.*, **109**, 460 (1949).

in the physiology of lung mechanics in the adult. Advantage is taken of crying to measure the vital capacity. The respiratory rate is usually periodic and calculation shows that the work of respiration is minimum at a rate of 37/min. The physiological dead space is relatively high and alveolar ventilation, per unit of lung volume, is also high but in proportion to the oxygen consumption. Lung compliance is apparently very low but comparable with the adult when expressed in relation to the functional residual capacity. Airway resistance is high in the newborn, due to the smaller diameter of the bronchioles, and the intrathoracic pressure swings are greater than in the adult.

RESPONSES OF THE SYSTEMIC CIRCULATION

The character and responses of the newborn circulation differ conspicuously from the adult subject. In the human infant the mean aortic pressure, measured with a strain gauge manometer, is 70/45 mm. Hg. and the mean perfusion pressure is therefore half the adult value. The cardiac output measured by the dye dilution technique, is about 200 ml./kg./min. (average birth weight 3·3 kg.) and the blood flow in the extremities, measured by the plethysmographic method, is 3–6 ml./100 g./min. Both are at least twice the adult values on a weight for weight basis. The peripheral blood flow is probably chiefly through the skin. The blood volume is variable but relatively high, 85 ml./kg., and increases by about 20 per cent in the newborn period. The haematocrit is usually above 50 per cent and blood viscosity must also be elevated.

The nervous pathways for the cardiovascular reflexes are laid down early *in utero*, but the experimental evidence relating to the development of activity of these reflexes at birth is conflicting and suggests that one cannot assume that the newborn circulation behaves in the same way as in the adult. In the experimental animal, changes in pressure in the intact carotid sinus cause the appropriate alterations in arterial pressure at birth. The change in peak frequency discharge of the baroceptor mechanism, for a given percentage change in arterial pressure, is the same in the newborn, with a low mean arterial pressure, as it is in the mature

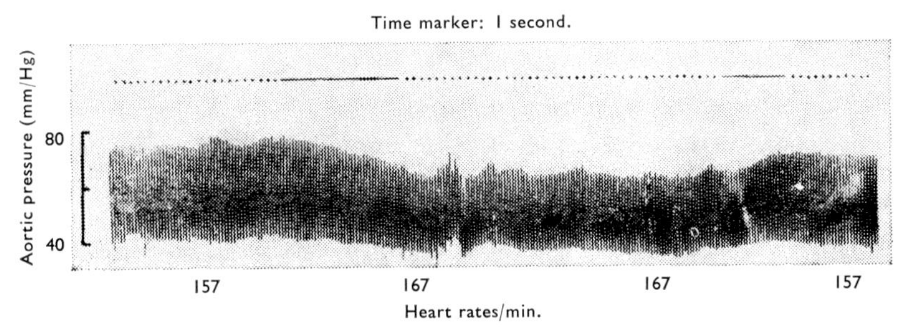

FIG. 24.5. The removal of 20 ml. blood at the first signal caused a reduction in both pulse and diastolic pressures, measured in the aorta. Tachycardia also occured. These changes persisted until the blood volume was restored 40 seconds later. From Young, M. & Cottom, D. 'An Investigation of Baroreceptor Responses in the Newborn Infant' from "The Heart and Circulation in the Newborn and Infant". Edited by D. E. Cassels. Grune and Stratton, New York (1966).

animal. However, baroceptor responses in the intact newborn animal, following a reduction in pulse pressure, induced by a fall in venous filling pressure during haemorrhage and passive tipping into the head up position, suggest increasing cardiovascular reflex activity during postnatal life. The responses to chemoreceptor stimulation, during asphyxia, also suggest an increasing cardiovascular reflex activity with age. There is also some evidence that the high cardiac output cannot be further increased during hypoxia in the newborn animal.

In the human infant the arterial blood pressure is not maintained as well as in the adult during a reduction in blood volume or with asphyxia, suggesting poor reflex vasoconstrictor activity, but there is some evidence for reflex control of the heart rate. The capacitance of the venous system must be large and it is remarkable that the infant's circulating blood volume can be increased swiftly, by about 30 per cent, during a uterine contraction, with only a small increase in right auricular pressure. Newborn skin blood vessels constrict readily in a cold environment but reflex vasodilatation in response to a rise in body temperature is not regularly observed until three to four days of life.

If both reflex and chemical regulation of the cardiovascular system were as active in the young as in the adult, there are two reasons why the final pattern of response may be different. First, there are differences in relative body proportion and, probably, of the cardiovascular bed under reflex control. For instance, the human infant's head is about four times the size of the adult's in relation to its body volume, the trunk proportions are the same but the limb volumes less than one half the size. Second, the intermittent patency of the fetal channels may influence the effectiveness of changes in vasomotor tone and of cardiac output.

BLOOD FORMATION

Before haemopoietic centres are developed in the bone marrow blood formation occurs *in utero* in islets in the liver. At birth the haemoglobin content of umbilical venous blood is 15–20 gm./100 ml. and the haematocrit over 50 per cent. These high values are chiefly due to the large mean corpuscular volume, 113 μ^3, and mean corpuscular haemoglobin of 37 μg. The red cell count is only slightly elevated. Reticulated cells form 1 per cent of the red cell count during the last four months of intra-uterine life, then numbers decline in the newborn period as the circulating haemopoietin falls, following the rise in arterial oxygen tension. Both the R.B.C. count and haemoglobin concentration rise during the first hours of life but the reduction in bone marrow activity causes a subsequent fall. This physiological anaemia is maximum at about 2–3 months of age, when the R.B.C. count is between 3–4 million/mm.^{3}: at this time the M.C.V. and M.C.H. are nearly equal to the adult values.

The white cell count is elevated at birth by the relatively high proportion of polymorphs. This declines during the first week of postnatal life as the lymphocyte number increases. Platelet counts are normal at birth, but the platelets are small and agglutinate more readily than in the adult.

TEMPERATURE REGULATION

As the newborn leaves its tropical environment *in utero* the changes in environmental temperature provide the second major stress, after asphyxia, to which it has to respond. Inequality of heat production and heat loss, which is reflected in the instability of the body temperature of the newborn mammal is partly due to inadequacy of the heat producing mechanisms and, to a greater extent, the structural disadvantage of the small animal, which has a large surface area in relation to its body mass.

Metabolic rate. True basal metabolic rates are impossible to measure in the newborn but the minimal oxygen consumption related to surface area is low during the first days of life. In the human infant the minimum oxygen consumption in a neutral thermal environment is 8 ml./kg./min. at 14 days of age and about double the value at birth. The critical environmental temperature which stimulates heat production falls during development. The neutral temperature range, in which there is no change in heat production, heat loss being regulated by vasomotor and sudomotor responses, expands with age as the maximum potential heat production increases.

Heat production in response to cold is seldom brought about by shivering in the newborn, and has been shown to take place in brown fat. This is a specialized form of adipose tissue resembling the adrenal cortex, with rich nervous and blood supplies, a high mitochondrial and cytochrome content and an oxygen consumption *in vitro* equivalent to that of cardiac muscle. In the human infant it is chiefly found on the floor of the posterior triangle of the neck, surrounding the subclavian and carotid vessels and in the perirenal areas. Brown fat is also present in considerable amounts in hibernating and cold-adapted animals. The heat production is brought about by the hydrolysis of triglycerides, which is known to be a highly exothermic reaction. The free fatty acids are oxidized *in situ* and glycerol is liberated into the blood stream: a rise in glycerol is observed following exposure to cold. These reactions are probably initiated and maintained by the local release of noradrenaline at the sympathetic nerve endings. Stimulation of the sympathetic nerves to brown fat increases local heat production and removal of the brown fat abolishes the thermal response to cold, without altering the basal metabolic rate. Heat production in response to cold is reduced by hypoxia and by starvation which reduces the supply of substrate for the resynthesis of triglyceride.

Physical factors. Just as basal metabolic rate in the mammal is related to surface area, so is the metabolic response to cold, and both vary according to the size of the newborn. Environmental thermal demand depends upon air temperature, air movement and radiant temperature, and to changes in heat loss due to alteration in insulation brought about by vasomotor and postural changes. In spite of the considerable increase in heat production of which the newborn is capable, the battle with its own large surface area is frequently lost. Thermal loss is enhanced by the high conductance of the skin, for the subcutaneous tissue layer is thin. In the immediate postnatal period the evaporation of surface liquor must cause considerable heat loss in those young, such as lambs, born in the open. This is limited to some extent by skin vasoconstriction in the human infant, but the baby seal is unable to reduce its heat loss, and may drown in the snow which melts in a pool around it.

Premature infants are at a disadvantage when they are not kept in a warm environment because, although they are able to increase their heat production, this is insufficient to keep up with the loss from the relatively large surface area and also their brown adipose tissue is not rich in lipid. Body temperatures may fall as low as 25°C in newborn infants, insufficiently clothed and exposed to wintery conditions, and the "cold injury" syndrome may develop, with lethargy and hypoglycaemia. The metabolic processes are first exhausted in an effort to maintain the body temperature, but are then slowed down by the low body temperature.

LIVER

Carbohydrate metabolism. Liver glycogen stores fall rapidly during and after birth and are not replenished for several days (Fig. 24.6). The R.Q. is 1·0 in the early neonatal period.

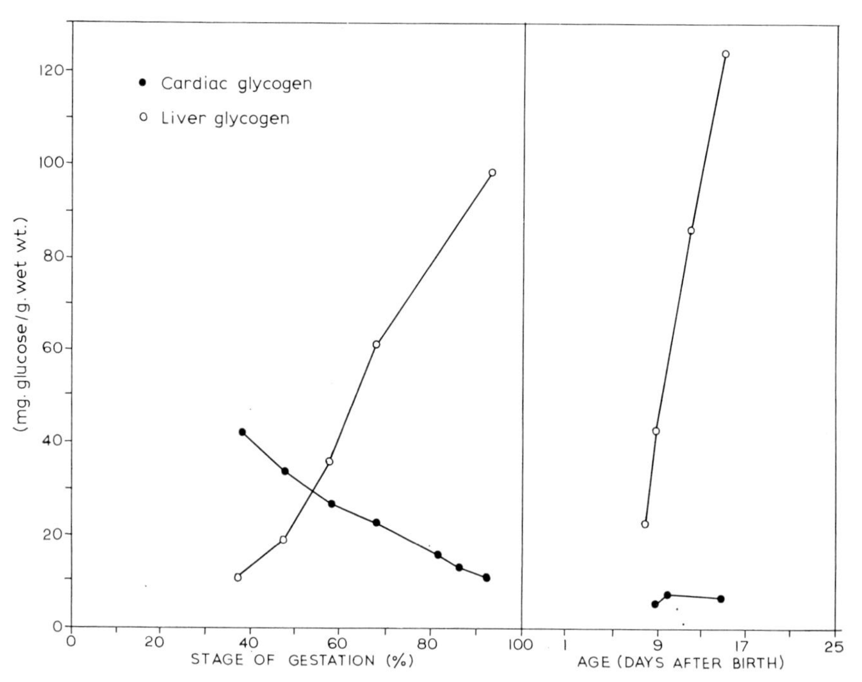

FIG. 24.6. Adapted from Shelley, H. J., *Brit. Med. Bull.*, **17**, 138–140 (1961).

During labour the blood sugar usually falls to levels which would cause neurological disturbance and coma in the adult; it does not reach adult levels for about 2 days after birth. Apnoea and cyanosis occur when the blood sugar is less than 20 mg. per cent for long periods and may be relieved by the administration of glucose. Glucose loads in the normal infant do not cause the expected rise in blood sugar and only a small, delayed release of insulin, in spite of the large pancreatic islet store. Neither are adrenaline and glucagon effective in raising the blood sugar until phosphorylase and glucose-6-phosphatase concentrations rise in the liver. Gluconeogenesis must play an important part in the regulation of blood glucose in the newborn period because milk contains relatively small amounts of carbohydrate.

The liver contains large stores of triglyceride fat at term but the plasma levels are low. Soon after birth the liver stores decline and plasma triglyceride and free fatty acid rise, the R.Q. falling with the increasing utilization of fat.

Excretion of Bile Pigments. The concentration of the liver enzyme systems, conjugating bilirubin and glucuronic acid, is low in the fetus at term, but increases rapidly after birth and the water soluble glucuronide starts to be excreted into the bile ducts and gut. Physiological jaundice is very common for short periods in the neonatal period, starting about 2 hours after birth and reaching a maximum at 4–6 days. Bilirubin is toxic to the nervous system and, because the blood-brain barrier is very permeable in the newborn period, damage to the basal ganglia, kernicterus, occurs more readily than in the adult. Extreme jaundice, with kernicterus is seen most commonly with haemolytic disease of the newborn, due to rhesus incompatibility in which the liver enzymes are not developed quickly enough to conjugate the quantity of bilirubin formed. Exchange transfusions to reduce the concentration of bile pigments and treat anaemia are carried out when the plasma bilirubin is about 20 mg. per cent. Jaundice may also occur in the hypoxic premature infant because the usual rise in liver enzyme concentration is depressed.

The increases in enzyme concentration which occur at delivery may be regarded as metabolic adaptations to the change in environment at birth. The exact stimulus is not known, but the changes are delayed until after delivery in post-term infants and take place more slowly after premature births. This synthesis of new enzyme protein, is inhibited by the restriction of food intake and inhibitors of protein synthesis. However, neither induction by substrate accumulation nor increased adrenocortical activity, both of which stimulate enzyme production in the adult and might be expected to occur at birth, are effective in the newborn.

GASTRO-INTESTINAL TRACT

Motility. In the newborn the whole gut is very distensible, and considerable variability of size, topography and activity is found. The muscle layers are relatively thin, in comparison with the adult, and in the stomach the longitudinal muscle fibres are particularly deficient over the greater curvature. The stomach capacity is 30–60 ml. and barium studies have shown that the emptying time is slower in the newborn than at any time of life, though the major portion of a feed has left by 3–4 hours. Large amounts of milk in the stomach may limit the respiratory movements of the diaphragm in unfit infants. The emptying time of the stomach is not influenced by the large quantity of air which may be swallowed during crying. This air passes rapidly from the stomach into the intestine where the muscle tone is low, and the segmentation and "puddling" observed may be abolished by the parasympathetic stimulant, mecholyl. The colonic musculature is relatively well developed and the activity of the large bowel is, in contrast with the adult, greater than in the small intestine.

The gastric contents of the newborn stomach contain all the expected enzymes and hydrochloric acid and though their concentrations are less than in the adult, milk is readily digested. The secretory mechanism also responds to gastrin. Intestinal, pancreatic and bile secretions, though weaker than in the adult, are also adequate for the function they have to perform. All the enzymes are present, with the exception of pancreatic amylase which does not appear for several months. Ptyalin is present in saliva during the last half of gestation. The anti-anaemic factor and secretin are only present in small amounts in the gastric and intestinal mucosae.

The first stool may be passed 12 hours to 48 hours after birth and the meconial characteristics disappear by about the fourth postnatal day. Meconium was thought to consist only of the shed mucosal cells of the upper respiratory tract, together with epidermal cells and fat from the vernix caseosa. Recently, it has been shown that the major component is mucopolysaccharide, representing the residue of the mucous secretions of the entire alimentary canal, immune from the action of the proteolytic enzymes. This mucoprotein has a high blood group specific activity, particularly in secretors. Bilirubin, and biliverdin, provide the characteristic dark colour.

Absorption. Less than 10 per cent of the nitrogen, calcium, phosphorus and fat ingested in milk is recoverable from the faeces and therefore absorption of the feed, following digestion, must normally be adequate. The composition and energy equivalent of milk are found on page 289. Cow's milk is rich in protein and electrolytes and is usually diluted for consumption by the human infant. Nitrogen excretion is low during growth, due to the preponderance of anabolism over catabolism, and during starvation, because the protein sparing capacity of the newborn is good. The fetal content of both calcium and iron is largely related to birth weight. Any deficiency in either declares itself some time after birth and may be prevented by adequate ingestion in the neonatal period. Calcium retention is temporarily interrupted at birth and returns to the fetal rate after a few months, depending upon the intake. Marginal stores of iron are present at birth. Milk has a low iron content and it is not known whether the newborn can store iron. The high vitamin requirements of growth are supplied by milk, though supplementation of both fat and water soluble vitamins with cod-liver oil and orange juice is valuable. The plasma levels are high *in utero* but no appreciable storage occurs.

ENDOCRINE REGULATION

Regulation of growth in the neonatal period was thought to be independent of the growth hormone because hypopituitary dwarfism is usually not apparent until two to three years of age. However, recent evidence has shown that growth may be retarded earlier.

The thyroid is very active just before birth but is temporarily depressed in the neonatal period in the human infant. Infants born of hyperparathyroid mothers may be hypoparathyroid either due to suppression by the maternal hormone or to the high fetal plasma calcium. Hypofunction may occur as the plasma calcium falls in the neonatal period. Hypertrophy of the glands is also common, possibly due to the high phosphate content of the milk.

The adrenocortical response to stress in the newborn period is weak and plasma corticoid levels lower than *in utero*. Adrenocortical insufficiency may be partly responsible for the neonatal fall in blood glucose, and premature infants treated with ACTH frequently have high blood sugars, and some sodium loss which may be prevented by cortisone. The adrenal cortical hormones begin to be excreted a few days after birth as the true cortex develops whilst involution of the X zone occurs. In the newborn the secretion of noradrenaline plays a part in the initiation of the metabolic response to cold by brown fat.

At birth the newborn sex organs are usually well developed, and thereafter regress until puberty. The high levels of placental hormones in the fetal plasma stimulate the mammary glands of both male and female newborn, and "Witches' milk" may be expressed at birth.

BEHAVIOUR

The responses which can be elicited from the newborn are legion and largely dependent on the spinal and lower brain centres. Withdrawal from painful stimuli occurs, turning of the head to free the nostrils, crying when hungry, rooting for the breast, milking the breast and swallowing. The grasp reflexes and the Moro embrace reflex may be phylogenetic hangovers, but the latter is particularly useful clinically for it is readily abolished by cerebral disturbance in the newborn. It is elicited by placing the baby on a firm surface and then slapping the table loudly and firmly, when the normal baby will flex its legs and the arms will be brought together as in putting them round the neck. Because the infant lies curled up it is thought to exhibit "flexural tone". In the supine position the limb movements are gross and appear purposeless, but in the cuddling or prone positions these random movements are replaced by more organised climbing or crawling features.

Some cerebral functions must also be present. The eyes are open, and the infant appears awake, responding to light and sound in a variety of ways, changing the respiratory pattern, startled, blinking and sometimes turning its eyes and head. Slow occipital rhythms are not observed on the EEG until about 4 months postnatal age. Smiling may also be observed.

The newborn sleeps for about three hours at a time with shorter periods of wakefulness. The first phase of sleep, as in the adult, is irregular and continuous movements of eyelids and eyeballs are readily observed with small movements of other parts of the body. The respiratory rhythm and heart rate are variable. In the second phase eye movements cease, only occasional body movements occur, and the respiratory rhythm and heart are slower and more regular: oxygen consumption is reduced. Sleep rhythms first occur in the EEG at 34 weeks of gestational age.

RENAL EXCRETION

At birth, the glomerular capsule of the nephron is still covered by a tall columnar epithelium but the renal tubule appears well developed. The function of neither structure is comparable with that of the adult until one month of age. The newborn kidney is unable to eliminate water, salt or acid loads quickly, nor conserve electrolytes. The maximum concentration of the urine is rarely more than 0·5 osmole/litre, less than twice that of plasma, in comparison with 1·3 osmole/litre in the adult. Because of these slow responses to their internal environment newborn infants readily become hydropenic, when water intake is inadequate. They may become oedematous, if the water load is only moderate, premature infants being at a particular disadvantage. Following vomiting and diarrhoea the electrolytes must be restored quickly, with half osmolar saline in isotonic glucose.

The dynamic relationship between the glomerular capillary pressure, the oncotic pressure and intrarenal pressures differs from the adult pattern in the newborn period. Both the plasma protein concentration and the mean arterial pressure are low. p-Amino hippurate clearance is also low in the newborn suggesting a low blood flow, but little is known of the secretory capacity of the tubules for hippurate at this age. The glomerular filtration rate, measured by inulin clearance, is low in the newborn, 35 ml./min., when expressed in relation to body weight or surface area. However, better agreement with the adult figure is obtained when the total body water is used as a basis of comparison. The value of 80 ml./min./42 litres body water found at birth increases to 120 ml./min./42 litres by one month of age.

The posterior pituitary gland contains pitressin early in fetal life and stores of the hormone at birth. The antidiuretic hormone can probably be released in the neonatal period for it is recovered in the urine during water deprivation. The urinary osmolarity is, however, not increased, suggesting insensitivity of the renal tubule and collecting ducts to the hormone at this age. The human infant's kidney can respond to exogenous pitressin and to a water load by six days of age. The diuresis following the water load is, however, caused by an increase in GFR with little change in osmolarity, in contrast with the adult in which there is no change in GFR but decreased tubular absorption. There is evidence in the human infant for activity of the renin-hypertensinogen mechanism during intra-uterine life.

SKIN

At birth the skin is thin with little cornification. Over the body it is frequently covered with vernix which is a secretion of the sebaceous glands containing some epithelial cells. Vernix is said to have bacteriocidal properties. It does not persist in the newborn.

The insulating properties of the skin, in particular the part played by the response of the skin blood vessels to changes in environmental temperature, have been discussed briefly under circulatory responses and control of body temperature.

IMMUNITY

The immunoglobulins which are present in the fetal blood stream at birth are passively acquired by transfer from the maternal blood stream across the placental membrane. This fetal immunoglobulin consists chiefly of the IgG type, and, because the lymphoid system is still immature, the concentration falls during the first 3–10 weeks of life. All the immunoglobulins start to appear in the plasma in significant quantities during the second and third month of life (Fig. 24.7). The immunoglobulin IgM quickly reaches adult levels but the concentrations of IgG and IgA do not reach adult values until between 1–4 years of life. This active immunity is acquired when the production of circulating plasma cells, or small lymphocytes, is induced by exposure to infection. This occurs first from the thymus, which controls the subsequent development in other lymphoid tissue. Premature infants can gain full immunological competence at the same rate as mature infants. Germ free animals develop no immunity. Immunological competence towards all antigens is not affected in the same time interval. Delayed sensitivity reactions, such as that to BCG vaccinations against tuberculosis, also occur in infants. Some homograft immunity may belong to this type of response.

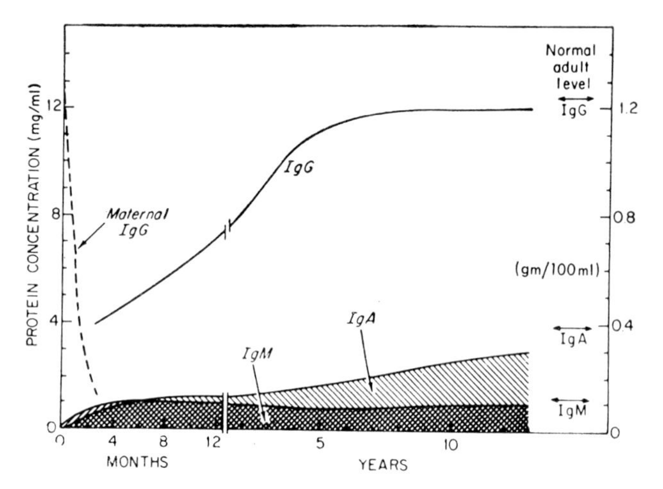

FIG. 24.7. Plasma immunoglobulin changes during the first year of life and to puberty; a comparison with adult values. Note the initial fall of IgG acquired from the mother. From Fahey, J. L., *J.A.M.A.* **194**, 255 (1965).

SURVIVAL DURING ASPHYXIA

Asphyxia causes general excitability, hyperpnoea and marked bradycardia of vagal origin in newborn animals. These changes are followed by primary apnoea and are succeeded, within a few minutes, by strong slow gasping, due to stimulation of arterial chemoreceptors, while the arterial pressure and heart rate decline further. Secondary or terminal apnoea follows, in which the heart continues to beat slowly and feebly. This is in contrast to the adult animal, in which the heart stops before the terminal apnoea. Resuscitation after the last gasp is possible but it has been shown at autopsy in the monkey that there is already damage to the subcortical centres, the mid-brain and brainstem. Postmortem caesarian section suggests that the human infant may withstand up to 25 minutes of fetal anoxia, which is a little longer than the time found for the rhesus monkey.

The mechanisms concerned with the great ability of the fetus and newborn to survive asphyxia are not fully understood. In young animals, the survival time to the last gasp, in an atmosphere of nitrogen, is directly related to the carbohydrate stores of the heart and liver. The energy requirements of the heart are provided by the anaerobic conversion of glycogen to lactic acid and survival time is shortened if this is inhibited by iodoacetate. The energy equivalent of anaerobic glycolysis is only one sixth of that provided by aerobic glycolysis. So long as a circulation is maintained, the liver glycogen stores can be mobilized to supply glucose to the brain.

It is not known whether irreversible damage to the tissue is mainly due to the absence of oxygen and therefore, the supply of energy, or to the fall in pH as the lactic acid accumulates. The process is self-limiting because lactate production is reduced as the tissue pH falls. Survival time is prolonged if the acidaemia is corrected with alkali; glucose enhances this action but is ineffective alone. The potassium content of anoxic fetal kidney slices is maintained longer than in adult tissue and, if this is true for the fetal heart and brain, it may explain the maintenance of excitability for long periods during asphyxia.

SUMMARY

When the maternal adaptation to her conceptus has been good, the respiratory and circulatory adaptations of the baby, necessary for immediate survival, are normally accomplished within a minute of birth. The stores of glycogen and brown fat, present at birth, need supplementing without delay for current energy sources as well as for growth. The functions of the gastro-intestinal tract are quite adequate for the necessary intake, digestion and absorption of milk.

A metabolic response to cold occurs in the immediate postnatal period, but fights a tough battle with the heat loss from the large surface area of the small infant, unless it is properly clothed. Liver enzyme concentrations are low at birth and the detoxicating mechanisms, particularly, are inadequate, but develop swiftly in the neonatal period. The endocrine organs are not very active in the immediate postnatal period, and endocrine regulation may take up to a year to become fully established. Renal regulation of the body constituents is marginally effective in the neonatal period, but improves swiftly within weeks of birth. Passive immunity, acquired from the mother is weak and transient and replaced by active immunity, in response to the environment by two to three months of age.

THE APGAR SCORE

Clinically it is desirable to make an estimate of the state of the baby at birth. In many centres this is done by the Apgar score, which is named after the doctor who first described the system. Assessments are made of pulse rate, respiratory effort, muscle tone, response to stimuli, and skin colour. According to what is found so a score of one or two is given to each physical sign. If all is well the score for any one factor is two and since there are five factors the maximum score is 10. A dead baby will score 0 and there are all ranges between. The following table summarizes the method. The score is made 60 seconds after the full birth of the baby and then can be made at intervals afterwards if this is thought to be necessary.

Apgar scoring

Sign	0	1	2
Heart rate	Absent	Slow (below 100)	Over 100
Respiratory effort	Absent	Slow Irregular	Good Crying
Muscle tone	Limp	Some flexion of the limbs	Active movement
Response to catheter in nostril	No response	Grimace	Cough or sneeze
Colour	Blue Pale	Body pink Extremities blue	Completely pink

It will be seen that this is a method of determining the state of the cardio-vascular, respiratory and central nervous systems, and their reflex responses.

Chapter XXV

BODY COMPOSITION IN PREGNANCY

Of recent years interest has turned to the concept of constitution. Constitution is the sum of all that goes to make up an individual, and therefore is an expression of the genetic inheritance with which he was born together with all that has happened to that inheritance. All individuals differ from one another because they have different genes and all are acted upon by different environments, using that word in its widest connotation. Constitution therefore expresses the idea of the totality that has gone to make up the person, and it is the constitution which is acted upon by anything new cropping up in the environment. One of the facets of constitution is physique which is bodily structure and organization, and another facet is that of body composition. Physique and body composition are to some extent related in that fat, for instance, gives a rounded appearance and relative absence of fat gives a thin appearance.

The total weight of the body is made up of the weights of everything belonging to it, but certain parts form by far the greatest amount of the body weight. Old chemical analyses of the body (taken from the Geigy Scientific Tables 5th edition, 1955) give the following for the adult and newborn bodies:

Percentage of total body weight

Tissue	Adult	Newborn
Muscle	28·7	25·1
Skin and subcutaneous tissue	17·4	19·7
Skeleton	11·6	13·7
Blood	4·3	6·5
Fat	18·0	—
	80·0	65·0

These figures show that most of the body weight is found in relatively few tissues.

To some extent weight will be dependent on height; in general the taller people will be heavier. One attempt to relate weight to height is by the ponderal index. This is the height over the cube root of weight.

$$\text{or P.I.} = \frac{\text{Height}}{\sqrt[3]{\text{Weight}}}$$

The cube root of weight is taken since weight is made up of three dimensions whilst height is of only one dimension, so that like can be compared with like. A person of average build will have a P.I. of 12·1 to 13. Heavy people for their height will range from 10·5 to 12 and light people will range from 13 upwards. This is a very crude index and is now not much used and has been superseded by more exact measurements in anthropometric work, and in clinical work where fewer measurements can be taken, weight for height tables have been constructed.

Height will to a large extent govern the weight of the skeleton and provided that the person is not unduly fat it will bear a relation to the weight of the skin and probably muscle. The main variable in people of given heights is therefore usually fat. The measurement of total body fat is difficult. Originally estimates were made by chemical analysis of carcasses, a most time-consuming and elaborate business, and it has seldom been done in Man. Moreover some of the bodies so analysed have been of diseased persons so that the results obtained may not be physiological. Very few analyses of women have been made but they tend to show that the total fat content of a normal woman may be of the order of 20 to 25 per cent, whilst that of men may be of the order of 11 per cent. Naturally there is immense variation round these figures, and in women the fat content may be from 10 to about 40 or even 50 per cent.

Fat tends to be relatively inactive metabolically and so there arises the concept of the lean body mass (L.B.M.) which comprises most of the actively metabolizing tissue. The lean body mass is obviously the total weight minus the weight of fat. It is made up of skeleton, muscle, skin and viscera. It is this group of tissues which is exchanging with the internal environment that is with amino-acids, carbohydrates, fats, vitamins, electrolytes, water, oxygen and all the other materials involved in metabolism. Taking the L.B.M. as a whole it is fairly constant in composition, because individual variations in composition as in the heart and brain are smoothed out by the consideration of the tissues as a whole. It is therefore theoretically possible to have an idea of the metabolism of the L.B.M. by expressing parameters in such a form as (say) the uptake of so much oxygen per gram. To say that the uptake of oxygen is so much per gram of total body weight is not so meaningful since each gram will contain an unknown amount of fat. It is desirable therefore to be able to measure in life either the total amount of fat in the body or its lean body mass. Neither is yet fully possible.

Estimates of total body fat have been made by weighing patients both in air and in water and subsequently making a correction for the amount of air in the lungs. From these measurements it is possible to calculate the specific gravity. The more fat that a patient contains, the less will be the specific gravity. Another method is to calculate the uptake of a fat soluble gas and from this determine the amount of fat in the body. Deducting this value from the total body weight gives the value of the L.B.M.

Alternatively the lean body mass can be calculated by injection of various isotopes such as those of potassium and sodium and calculating their distributions. In general, potassium is an intra-cellular ion and therefore will give some idea of the total functioning cells in the body. In addition isotopes of hydrogen, deuterium and tritium give values for total body water.

These various studies of body composition are still in their infancy and very few have been done in pregnant women, and those have been mainly done for water and electrolytes rather than for fat. Using the few values that are known for the non-pregnant, attempts have been made to correlate the chemical estimates with those derived from body measurements with calipers and rules. There is a rough correlation between measurements of subcutaneous fat in various sites measured by fat calipers and the total body fat, but it is not possible to use the method clinically. If the percentage of body fat could be obtained by some simple method it would be valuable because it might then be possible to predict the values of some other parameters with greater success than is possible at the moment. For instance the water content of the L.B.M. in the non-pregnant is almost exactly 72 per cent and varies very little about this figure in health. Similarly it is found that sodium forms 0·109 gm. per 100 gm. and potassium 0·265 gm. per 100 gm., and calcium 2·01 gm. per 100 gm. and so on with other elements. Predictions from total body weight, because of the variable fat content, have nothing like the accuracy of these figures.

THE WEIGHT GAIN OF PREGNANCY

The weight gain of pregnancy is made up of that due to the fetus, placenta and membranes, liquor amnii, uterus, breasts, blood, tissue

	Weight at term in gm.
Fetus	3300
Placenta	650
Liquor amnii	800
Uterus (without blood)	900
Breasts (without blood)	405
Blood (Plasma and cells)	1250
Extracellular water (tissue fluid)	1195
Fat	4000
	12,500 = 12·5 kg.

fluid and fat. The weights of most of these components have been considered elsewhere in this book with data derived from Hytten and Leitch (1964).

It should be recognized how very variable all these parameters may be. Indeed the total weight gain in completely normal women may vary from 6 to 17 kg. and sometimes more or less than these figures. Taking reasonable extreme ranges the following table can be constructed:

	Weight in gm.	
	Small weight gain	Large weight gain
Fetus	2270	4085
Placenta	340	800
Liquor amnii	250	1200
Uterus	450	1350
Breasts	200	600
Blood	1000	1500
Tissue fluid	500	2000
Fat	500	6000
	5510 = 5·5 kg.	17,535 = 17·5 kg.

It is important to realize that the table is a fictional one in that it does not refer to definite patients but shows the accumulation of weight gain if the ranges of weight of the various components are used. The main message of the table is that the major variants in weight gain are those of fluid and of fat. Other components can be doubled or trebled but the factor of increase in fluid may be four or more and fat might even be increased by a factor of ten or twelve.

If the average weight gain of pregnancy is 12·5 kg. it might be expected that the rate of weight gain per week would be of the order of 500 gm. and if this were so the graph of weight gain against time of pregnancy would be a straight line. However the average weight gain at various times in pregnancy is as follows:

Duration in weeks	Weight gain	Difference
10	450 gm.	
20	4090 gm.	3640 gm.
30	8625 gm.	4535 gm.
40	12,712 gm.	4097 gm.

This shows that there is an increased rate of weight gain between the 20th and 30th weeks as compared with the other ten week periods. The evidence suggests that this increase in weight gain rate at this time is largely due to the deposition of fat. It is thought that this fat is laid down as a store to be drawn upon during lactation. Such deposition of fat may be of extreme importance in areas of the world where the food supply may be poor, but in most civilized countries diet usually tends to be well above subsistence level. Ideally perhaps, the fat laid down in pregnancy ought to be used up during lactation so that the woman returned to her usual non-pregnant weight. However, only about 25 per cent of British women now breast feed their babies and even when they do breast feed they give it up after three or four months. This may be contrasted with Bushwomen in the Kalahari desert who may breast feed their children for up to three years since the food supply from elsewhere

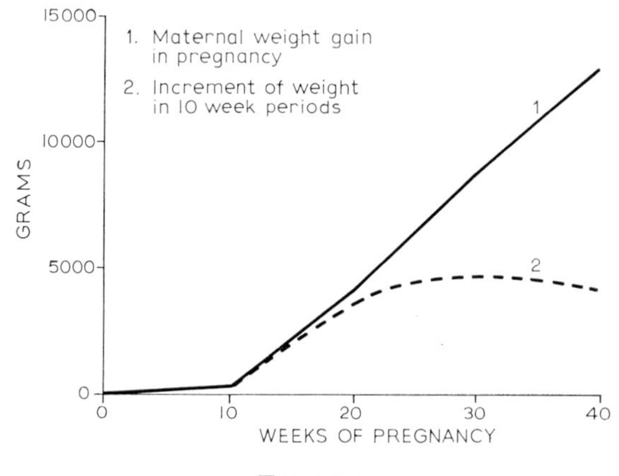

FIG. 25.1.

is so precarious. Moreover diet in British women tends to be above basal needs and so it is usual to find that with each succeeding pregnancy a woman puts on and retains more and more fat. In fact over the course of bearing many children a woman might more than double her original weight. Of course this can be prevented by appropriate dietary adjustments.

Another major variable in weight gain in pregnancy is water. Of course a certain amount of water has to be taken on board to allow for the growth of the fetus, placenta, membranes, uterus, breasts and particularly and obviously for the expansion of the blood volume. The table on p. 284 from Hytten and Leitch (1964) epitomizes the problems in water balance and retention.

Various methods of measurement of the total body water in pregnancy have been made and from the last figure of the table it will be seen that an average at term will be about 7000 ml. The rest of the table in the last column shows that of this 7000 ml. only about 5800 ml. can be accounted for. It is presumed that the remaining 1200 ml. is retained in the extra-cellular fluid. This becomes especially likely when it is remembered that oedema of the ankles and even of the fingers and elsewhere is so very common in pregnancy. Also it will be remembered that the blood is more dilute during pregnancy and it would seem likely that fluid

	Weeks of Pregnancy								
	20			30			40		
	Weight	Water	Water	Weight	Water	Water	Weight	Water	Water
	g.	%	g.	g.	%	g.	g.	%	g.
Fetus	300	88	264	1500	79	1185	3300	71	2343
Placenta	170	90	153	430	85	366	650	83	540
Liquor amnii	250	99	247	600	99	594	800	99	792
Blood free uterus	585	82·5	483	810	82·5	668	900	82·5	743
Blood and fat free mammary gland	180	75	135	360	75	270	405	75	304
Plasma	550	92	506	1150	92	1058	1000	92	920
Red cells	50	65	32	150	65	98	250	65	163
			1820			4239			5805
Measured increment of water			1500			3750			7000

Data from Hytten & Leitch *Physiology of Human Pregnancy*, Blackwell (1964).

under these circumstances would escape from the plasma into the tissue fluid spaces.

The relationship of oedema to fluid retention is of more than academic interest in obstetrics. The syndrome of pre-eclamptic toxaemia consists of hypertension, proteinuria and oedema and the syndrome is associated with an increased rate of fetal death and of maternal death. If it could be fully understood and controlled, childbirth would be safer. It has been assumed that the presence of oedema implies an increased retention of fluid and that the fluid retention may be the cause of pre-eclamptic toxaemia. This has led to attempts therapeutically to control the intake of fluid and of salt or the giving of diuretics to get rid of the excess fluid already present. However, oedema might be due to a redistribution of fluid as between the intra-cellular and extra-cellular fluids and there may not necessarily be an increase in total body water. Also, rather than being a cause of pre-eclampsia water retention might be a result of some other underlying process. Finally oedema may be localized or generalized. In pregnancy it is most commonly found in the ankle region and here it is probably at least in part caused by the increased venous pressure in the legs in pregnancy, which is due to the pressure of the uterus on the inferior vena cava and also the increased venous return from the uterus into the common iliac vein.

Recently the problem of the relation of oedema to fluid retention has come closer to elucidation. And especially the relation of weight gain in pregnancy to oedema has been investigated. About one-third of normal women have some degree of oedema and it is presumed that they do not have abnormal water retention. If there is even a slight degree of hypertension in pregnancy then oedema is found in about two-thirds of such women. If there is proteinuria in addition to hypertension then about 85 per cent of the women will have oedema as well. The rate of weight gain of those who have oedema is in general slightly greater than those who do not develop oedema and this increase is apparent between about 20 and 30 weeks of pregnancy before any clinical oedema is recognizable. Women who are relatively fat early in pregnancy are more likely to develop oedema than those who have a normal weight-for-height ratio.

In the very few patients where total body water estimates have been made the total water increment of pregnancy is found to be about 7 litres. In fatter women the increase is slightly more, perhaps by about one litre. If in addition a woman develops oedema of the ankles and fingers, that is generalized oedema, then she may retain as much as ten or more litres. This is a prodigious performance physiologically and little is known about its mechanisms. In pre-eclamptic toxaemia the water retention tends to be high as with all women who develop oedema but it is still within the normal if upper range.

A further odd feature of oedema is that oedematous women in general tend to have slightly larger babies than those who do not have this sign. The difference is only, however, of the order of 0·2 pounds.

There is obviously a very complex inter-relationship between fat, water, hypertension, proteinuria and fetal nutrition, and the boundary between the physiological and the pathological is very hazy.

Chapter XXVI

THE PUERPERIUM

The puerperium is the time taken after labour for the woman's physiology to return more or less to the non-pregnant state. Obviously if she is lactating the breasts may not return to the non-pregnant condition for some months. The genital organs return to normal in about six weeks and this is conventionally taken as the average length of the puerperium. The time which a woman spends in hospital is called the lying-in period and nowadays is about 7 to 9 days.

The weight gain of pregnancy is apportioned as follows:

	Weight at term in grams
Fetus	3300
Placenta	650
Liquor amnii	800
Uterus without blood	900
Breasts without blood	405
Blood (plasma and cells)	1250
Extracellular water (tissue fluid)	1195
Fat	4000
	12500 = 12·5 kg.

At birth the patient loses fetus, placenta, liquor and blood. If 500 gm. is allowed for the blood loss, a not unusual figure, the weight loss at birth is 4250 gm. During the next ten days one investigation showed a further loss of 2·3 kg. and from the second to the sixth week a further loss of about 0·7 kg. It is probable that the loss in the first ten days is mainly of the excess blood volume and the extracellular fluid, and from the above figures this amounts to 1945 gm. The remainder probably comes from the involution of the uterus and possibly some fat is lost. Further weight loss probably comes from fat, but none of the sums can be very exact when it is realized that the amount of excess tissue fluid taken on board during a normal pregnancy may vary from less than one litre to 3 litres or more.

During the first few days the weight loss is not even. In fact for the first two days the weight may rise by about 1 kg. from the basal weight attained immediately after delivery. This rise in weight is almost certainly due to water retention, similar to that seen after operations in surgery. Such retention is probably due to an output of anti-diuretic hormone from the posterior lobe of the pituitary. If the woman is lactating the weight loss after the second day is greater than if she decides not to breast feed her baby.

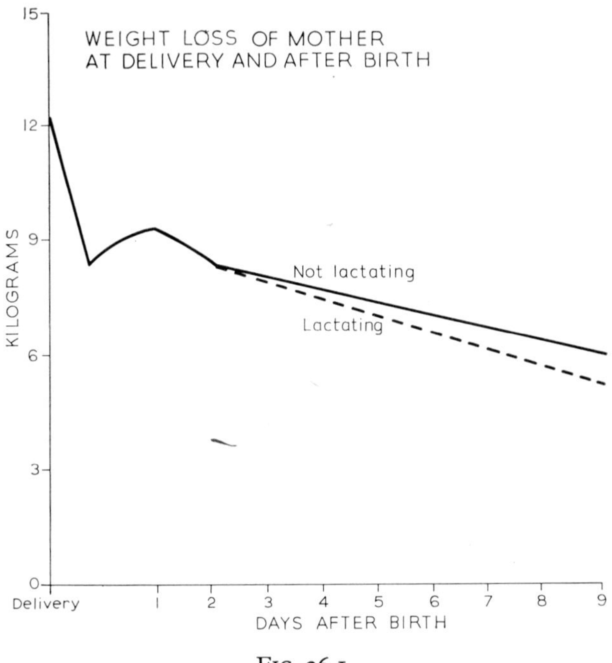

FIG. 26.1.

The major changes of the puerperium are seen in the genital organs and in the breasts.

GENITAL ORGANS

Over the course of six weeks the perineum heals and its tone becomes restored. The vagina which was grossly distended by the passage of the fetus shrinks to its more normal size but usually does not fully return to the virgin state.

The myometrium weighs about 1 kg. at term. It is gradually returned to a weight of about 100 gm. or less by the process of involution. The details of this are quite unknown. As with any wasting disease of muscle there is an increase in the excretion of creatinine in the urine. The rate of involution is quite rapid in the first ten days or so of the puerperium and then slows down so that the uterus is fully involuted at about six weeks post-partum. Myometrial contractions during the puerperium have been considered in Chapter XXII. During the first few days after birth the pressure in the uterine cavity may attain heights of about 60 to 100 mm. Hg. and the peak activity may occur during suckling, due to the circulating oxytocin at this time. Indeed in women who have had more than one baby it is very common for these contractions to be painful and they are called "after-pains".

After delivery of the placenta the uterus contracts down to prevent blood loss by constricting the blood vessels as they course through the uterine wall. But the area of the placental site is roughened to the touch and constitutes a wound which has to be healed. The epithelium in this area is breached and must be repaired. During the process of regeneration and the growth of endometrial epithelium over the site there is some degree of blood loss from the uterus. If the woman remains in bed for several days the blood loss ceases in about ten days. If she gets up and about, which is the usual practice today the loss, which is called lochia, continues for much longer and indeed may persist slightly for up to six weeks. As with any wound the loss of blood is at first bright red but as the loss gets less it first turns dark brown and later becomes serous. Individual patients are very variable in the amount of loss and in its colour characteristics. One estimate of the lochial loss suggests it is about 500 ml. most of which is lost in the first four days.

Histological changes in the endometrium at this time are of interest partly because they show the mechanism of regeneration but mainly because they give an index of the rate of return of ovarian function, since proliferation and secretion, even at this time, must be dependent upon oestrogen alone and oestrogen with progesterone respectively. The endometrium has recently been studied by Sharman (1966). Proliferative endometrium begins to appear as early as the ninth and tenth days of the puerperium, suggesting that ovarian activity returns as early as this. The appearance of secretory endometrium with its evidence that a corpus luteum has formed following ovulation is very irregular, but it is significant in practice that ovulation as judged by this criterion may occur at six weeks after delivery. Moreover lactation does not necessarily suppress ovulation nor menstruation, though in women who are breast feeding the time of return of ovulation tends to be later than in those who do not breast feed. Even in the lactaters though, ovulation may occur by the twelfth week. The importance in practice is the prevalent assumption that women are infertile after the delivery of a baby. But this is only relatively true and therefore those who wish to space their families must use contraceptives from quite an early time after the birth of a baby. It is interesting that in Bushwomen, who because of shortage of food may breast feed their babies for up to three years, the woman is sexually taboo for the whole of this time. If she were not and produced babies regularly every year survival of the race would be prejudiced.

The mechanism of ovulation in those who do not breast feed their babies is presumably exactly as it is in any other non-pregnant woman. But there seems little doubt that in lactation the anterior pituitary gland is very active and is probably producing prolactin which is probably the same as luteotrophin. If this is so it might be expected that the luteotrophin would maintain any corpus luteum that was formed and so suppress any further cycles. But this does not in fact occur and it throws some doubts on the accuracy of present ideas about the control of the ovarian cycle. Apart from the evidence of secretory endometrium that ovulation occurs during lactation, there is also abundant evidence from the fact that women may become pregnant during this time, and this is the only absolutely incontrovertible proof that ovulation has occurred. In some women, however, ovulation and menstruation may be suppressed for as long as ten or twelve months when they are lactating, but why there is such vast individual variation in these functions after delivery is not known.

LACTATION

During pregnancy the gland and duct tissues of the breasts grow under the influence of the rising output of oestrogen and progesterone from the placenta. In addition the breast size increases because of vascular engorgement and probably because of the deposition of fat. Each breast may increase in volume by an average of 200 ml. but the increase may be as much as 800 ml. In women having their first babies the average increase in size declines with age. The nipple and areola also increase in size, and become more mobile. The breasts do not secrete milk during pregnancy because it is thought that the high levels of oestrogen prevent the anterior pituitary from producing the hormone prolactin, which seems to be necessary for the secretion of milk. However, the breasts do produce a certain small amount of watery fluid called colostrum and this can be expressed from the nipple by squeezing with the hands.

With the delivery of the placenta in the third stage of labour the major source of the sex hormones is withdrawn and the blood levels of oestrogen and progesterone fall precipitously. This seems to remove the brake on the anterior pituitary and it can now secrete prolactin (luteotrophin).

For the first three or four days after delivery the breasts secrete colostrum, which is a thin watery fluid which can also be expressed from the breasts during pregnancy, though not in such amount as after the baby is born. Colostrum contains 3 to 7 per cent of protein, 3 to 5 per cent of carbohydrate mainly as lactose, and comparatively little fat. Milk, recognizable as such to the naked eye is expelled from the breasts on about the fourth day after birth and contains 1·25 per cent of protein, 4 per cent of fat and 7 per cent lactose together with small amounts of mineral salts. The protein consists of lactalbumin and casein in the ratio of three to two. Since many babies may have to be fed artificially it is well to know the composition of cow's milk since this is the basis for most artificial feeds even though it may come in dried form. Cow's milk contains 3·75 per cent of fat, 4·75 per cent of lactose and 3·4 per cent of protein of which casein is by far the major proportion. To try to make cow's milk similar to that of human breast milk it is usual to dilute the cow's milk with water and then add sucrose to bring the amount of carbohydrate up to the desired amount. The effect of dilution of the fat is ignored. Proprietary dried milks may make up the carbohydrate with lactose or glucose.

The secretion of milk from the breasts, that is from their glandular tissue, is under the control of prolactin. The delay time of three days before the milk comes in is presumably due to the build up of the action of prolactin on the breasts. However, there may be physiological benefit in colostrum since it contains a high proportion of antibodies. These may be valuable in lower mammals whose digestive systems do not break up the antibodies by enzymatic action, but in the human the enzymes of the stomach destroy antibodies before they can be absorbed.

Although prolactin seems to be responsible for milk production there is a different mechanism for the expulsion of milk into the baby. At first the expulsion is not very effective and the pressure of milk within the breasts causes great engorgement which can be painful. The breasts become hard, tense and tender. It is usual to try to control this condition by the giving of oestrogens which temporarily inhibit the production of prolactin. Recently the giving of oestrogens has been correlated with an increased tendency to thrombosis and a raised level of Christmas factor in the blood. This has called into question the advisability of using oestrogens in the therapy of engorgement and in the suppression of lactation. The cause of the expulsion of milk from the breast is the contraction of myo-epithelial cells which surround the ducts with their contractile processes. These cells contract and raise the pressure in the ducts and force the milk to the nipple under the influence of oxytocin from the posterior lobe of the pituitary. This is an interesting effect for oxytocin's usual action is on the uterus, though after delivery its primary site of action is in the breast. However, whilst the mother is suckling her baby and oxytocin is circulating, she will often feel painful uterine contractions, the so-called "after-pains" and the loss of lochia will increase. The contraction of the myo-epithelial cells causes a rise of pressure in the ducts which may be as great as 50 mm. Hg. and this causes a sensation in the breasts which has been called the "let-down" reflex after the phenomenon known to cattlemen when the teats of a cow are handled and the milk begins to flow easily.

Breast feeding becomes in time a complex reflex arc. The afferent stimulus at first is the contact of the baby with the breast and particularly of its mouthing of the nipple. When the mother becomes conditioned to breast feeding the stimulus of milk secretion is of the Pavlov conditioned reflex type and may be evoked by the passage of time, since most babies are fed fairly regularly, or by the sound of the baby's crying or even by the sight of the baby. On the efferent side of the arc is probably first the secretion of oxytocin from the posterior lobe of the pituitary through the intermediation of the hypothalamus. There is some evidence that the secretion of oxytocin is the stimulus to the anterior lobe for the secretion of prolactin, though it is possible that the hypothalamus directly affects the secretion of prolactin through the portal system of blood vessels. Under the influence of oxytocin the myo-epithelial cells round the breast ducts contract and cause the feeling of the "let-down" whilst the pressure rises in the lumina of the ducts. The prolactin circulating causes the build-up of milk secretion. When the baby sucks at the nipple it forms an area of reduced pressure and the milk can begin to flow. However, it will be realized that the baby does not have to exert much suction, since the milk is actually pumped into his pharynx by the pressure of the milk in the ducts.

The amount of milk produced must satisfy the basic caloric needs of a rapidly growing baby, and the amount needed is about 2½ ounces per pound body weight per day (165 ml. per Kg.). That is an 8 lb. baby will get about 20 ounces of milk per day (600 ml. per day). This amount is not reached all at once but is gradually increased over the first few days until the 7th to 10th day when the needs of the baby are usually fully supplied.

At the present day only about 25 per cent of women breast feed their babies. The reason for this is probably cultural and psychological. There are fashions in breast feeding. In primitive societies breast feeding must be the rule especially when food resources are scarce. It will be recalled that pregnant women probably lay down a store of fat in the 20th to 30th weeks against the coming needs of lactation. Whilst breast feeding the baby the mother has to be relaxed and comfortable and feel free from interruption, because if she is anxious and tense the milk supply fails. It may be that this is caused by the secretion of adrenaline as a response to anxiety.

Chapter XXVII

THE CLIMACTERIC

Somewhere between the ages of 45 and 55 most women pass through the phase of life called the climacteric. Colloquially this is named the "change of life". The essential feature of this phase is the gradual failure of ovarian function. It is one of the intriguing mysteries of physiology why the ovaries should fail at the time that they do. Similarly the delay in the onset of maturity of the ovaries at puberty is of great interest. The ovaries seem to live a life of their own within the general life of the body.

The first function of the ovaries to fail is that of ovulation. The failure is not at all sudden but occurs sporadically so that at first it is probable that ovulation fails in one cycle only. Then it may fail for two or three cycles and finally anovulatory cycles become more frequent than ovulatory ones. It is well known that fertility is very definitely diminished statistically at these ages but from what has been said it will be realized that a woman might become pregnant even at a late age if she happens to have intercourse during an ovulatory cycle.

The pituitary-ovarian cycle during maturity is as follows:

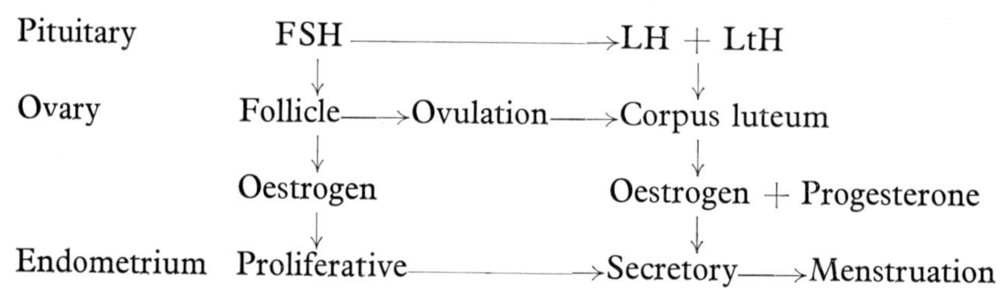

The cycles are kept going by positive and negative feedback between the ovaries and the pituitary. Ovulation holds a key place in this scheme and if it does not take place a corpus luteum cannot form and there can then be no secretion of progesterone. This effectively prevents the functioning of the whole of the right half of the scheme. That is there is no LH and LtH, no corpus luteum, no progesterone and therefore there can be no secretory endometrium. This last is very important since it is by examination of the endometrium at dilatation of the cervix and curettage of the lining of the uterus that inferences can be made about the function of the ovary and hence indirectly of the pituitary. Hormone studies to check the value of these histological studies of the endometrium show them to reflect the hormone status of the patient very accurately. Once ovulation fails to occur in a given cycle or series of cycles the endometrium can only be under the influence of oestrogen. Therefore only proliferative forms of endometrium are seen.

With the failure of ovulation the feedback to the pituitary from the ovary becomes partly deranged. FSH output becomes very irregular. Sometimes it is very high and sometimes it falls to very low levels. The effect of this change is that the follicles of the ovary are also very variable in their output of oestrogen. If the output is very high for some time then the endometrium proliferates greatly and may become hyperplastic. When the oestrogen level in the blood drops then there will be oestrogen withdrawal bleeding from the endometrium. If the circulating oestrogen level is low then the endometrium scarcely proliferates at all and may be atrophic. As the hormone levels rise and fall so there may be episodes of bleeding from the uterus. Nearly all women pass through a phase when their "periods" are hopelessly irregular because of this waxing and waning of the hormone system. It is an axiom of gynaecology, however, that it must never be assumed that such bleeding at the climacteric is due to simple physiological changes since there may be other causes of vaginal bleeding at this time, notably cancer of the cervix or the body of the uterus. Therefore examination under anaesthesia is essential, together with dilatation and curettage. Only then is it possible to be sure that the patient's symptoms are due to physiological variation in hormone outputs from the ovary and pituitary.

Ovulatory cycles give way to anovulatory ones and hormone outputs fluctuate. Gradually over the course of some months the endocrine secretions of the ovary fail too. More specifically the oestrogen output wanes to physiologically negligible levels. Inevitably, therefore, those organs and tissues which are dependent upon oestrogen for their function atrophy and cease their activity. This is most obvious in the endometrium and as the oestrogen level is low this membrane ceases to proliferate and becomes atrophic. There are then no more losses of blood from the uterus and the periods cease. The time of cessation of the periods is called the menopause and the time after this is called the post-menopausal era.

With the fall of oestrogen the pituitary has no negative feedback to reduce its output of FSH so this hormone increases in amount in the urine. There seem to be no physiological effects from this but the fact that it does rise shows that the primary failure at the climacteric is in the ovary and not in the pituitary. It is worth recalling that sexual maturity at puberty is dependent upon the brain whereas failure of sexual function at the menopause resides in the ovary.

The relative withdrawal of oestrogen results in changes in the secondary sex organs. The muscle of the uterus gradually over some years becomes converted to fibrous tissue and in extreme old age the uterus has shrunk so that it can be felt only as a tiny button of tissue at the top of the vagina. The epithelial layers of the vagina disappear and there may be only a layer of basal cells with a little keratin above them. The vaginal skin no longer contains glycogen and the Döderlein's bacilli which in mature life act upon the glycogen to produce lactic acid disappear also. The vagina then has a pH of about 7·2 in distinction from the pH of 4·5 in sexual life. The labia minora shrink to almost nothing over the course of some years, but the labia majora may increase in size due to the deposition of fat which is part of the general tendency to put on fat during this time.

During the climacteric and in the post-menopausal era the fall in oestrogen causes some instability in the vasomotor system and suddenly without warning the woman may have a "hot flush". This is a sensation of sudden heat in the region of the upper chest and it spreads rapidly to involve the neck and head. The woman may feel so hot that she rushes to an open window to cool off. Following this she may perspire a little. The attacks are very variable and may occur once or twice a day and in extreme cases they may occur several times an hour. They are controllable by giving oestrogens. Interestingly not in all cases is there a definite vasomotor flush and although the woman may feel the unpleasant sensations as of blushing there may be no such changes visible to the observer. There is further evidence of the effect of oestrogens on the cardiovascular system in that it is known that women are very much less liable to suffer from coronary thrombosis than men. However, if the ovaries are removed for medical reasons before the climacteric, then women suffer coronary thrombosis as frequently as men. It has also been shown that progesterone has a slight but definite effect upon the action of the heart muscle in a manner similar to that of digitalis.

The glandular tissue and the ducts of the breast atrophy with the withdrawal of oestrogen. The breast size may, however, increase as a result of the deposition of fat which may be generalized. The reason for the increase in body weight after the menopause is not fully known but is obviously due to an excess of intake of calories over the output of energy. It is probable that energy expenditure diminishes with age whilst the former eating habits continue. Also there are often emotional factors in over-eating and many women find the change in their lives from being capable of reproduction to being unable to do so almost more than they can bear. Psychological symptoms are not uncommon at this time of life—and one expression of unhappiness may be over-eating.

Although the ovaries cease to produce hormones the adrenal glands continue. They produce both androgens and oestrogens in small amounts and they can be found by urine assay. The result is a different balance of male and female hormones and so there may be some evidence of relative maleness after the menopause in thinning of the scalp hair and temporal recession as well as the growth of some hair in the beard and moustache areas. The output of these hormones from the adrenal is not enough to prevent osteoporosis which may become widespread in old age. It may result in bone pains and also the bones may easily fracture under stress. Arthritis, however, may not be dependent upon the hormones but rather may be due to other degenerative processes together with the increasing body weight. Later in old age the weight may decrease and the skin then becomes lax and wrinkled.

In the foregoing discussion the terms high and low have been used without giving quantitative values. The following table gives a rough idea of what these terms mean. It is taken from Jeffcoate (1962) and applies to urinary excretion in a 24 hour period.

Gonadotrophins.		
Young children	Negligible.	
Females aged 9–12 years	2–3 HMG units.	
Adult non-pregnant women	5–25 HMG units.	
Women at the climacteric	25–1000 HMG units.	
Total oestrogens.		
Young females	2·0–2·5 μg	
Adult non-pregnant women		
Follicular phase	5–25 μg	Mean 13 μg
Ovulation peak	35–100 μg	Mean 55 μg
Luteal Phase	15–85 μg	Mean 30 μg
Post-menopausal women	3–12 μg	Mean 6 μg.

INDEX

PRINTED IN GREAT BRITAIN BY THE WHITEFRIARS PRESS LTD.
LONDON AND TONBRIDGE